Fish Pheromones and Related Cues

Contributors

Cindy Baker
National Institute of Water and Atmospheric Research Ltd.
Hamilton, New Zealand

Karen M. Cogliati
McMaster University
Hamilton, Ontario, Canada

Lynda D. Corkum
University of Windsor
Windsor, Ontario, Canada

Kjell B. Døving*
University of Oslo
Oslo, Norway

El Hassan Hamdani
University of Oslo
Oslo, Norway

Peter Hubbard
Universidade do Algarve
Faro, Portugal

Stine Lastein
University of Copenhagen
Copenhagen, Denmark

K. Håkan Olsén
Södertörn University
Huddinge, Sweden

Peter W. Sorensen
University of Minnesota
St. Paul, Minnesota, USA

*Deceased

Norm Stacey
University of Alberta
Edmonton, Alberta, Canada

Michael Stewart
National Institute of Water and Atmospheric Research Ltd.
Hamilton, New Zealand

Ashley J.W. Ward
University of Sydney
Sydney, Australia

Brian D. Wisenden
Minnesota State University Moorhead
Moorhead, Minnesota, USA

Preface

It has been known for at least a century that fish, the largest and most diverse group of vertebrates, are strongly influenced by chemicals that they themselves release into the water. Some of these stimuli seem to be discerned by innate mechanisms by all members of their species and are called "pheromones". Other types of related organismal chemical cues appear to have less prescribed actions and may at times be learned. Together, these cues play profound and varied roles in the lives of fish, ranging from avoiding danger to synchronizing reproduction. This book, the first on fish pheromones and related cues, reviews all aspects of conspecific chemical cues in fish. It adopts a broad, systems-level approach to encourage new integrative thinking.

The first chapter was written by the primary editor to define terms and provide common language for the entire book. A second set of chapters reviews how conspecific chemical cues are used by fishes. Both pheromones and related conspecific cues that may be learned are addressed because the distinction can be vague. A third set of chapters addresses mechanisms of pheromone detection and production as well our detection mechanisms. Lastly, applications in fisheries management and culture are addressed. An afterword summarizes some key points and future directions.

A wide range of authors have generously contributed to this book which I hope will invigorate the field and stimulate another book in the near future.

Peter W. Sorensen

Chapter 1
Introduction to Pheromones and Related Chemical Cues in Fishes

Peter W. Sorensen

University of Minnesota, St. Paul, USA

1.1   CHEMICAL INFORMATION TRANSFER IN FISH

Information transfer between fishes, the largest and most diverse group of vertebrates, has long been both of practical importance and a source of wonder given the evolutionary, ecological, and economic importance of this group. Chemicals play an especially significant role in this process, presumably because they function well in vast dark spaces, can encode a great deal of information, are readily soluble, and are inherently "honest." This book specifically addresses how and why chemical information is transferred between fishes of the same species. First I define some basic terminology to promote clarity and then I introduce some other terms, types of chemical cues, and principals along with the chapters which discuss them. Information transfer between different species is not explicitly addressed except how it might occur as part of transfer within a species. This section introduces these terms and issues, and then where more information on them can be found in the book.

1.2   TERMINOLOGY

1.2.1   Overview

Because the terminology used to describe conspecific cues has been used in many ways since the term "pheromone" was first coined 50 years ago (Karlson and Luscher, 1959), I suggest and define some terms in this introductory section to provide clarity. Definitions were chosen for practical reasons and to be consistent with those used by researchers outside the world of fishes. Emphasis is placed on recent work by Tristram Wyatt (2003, 2010). Authors were asked to consider the terminology suggested herein, but not necessarily to use it if their opinions differed. Information transfer between members of different species (kariomones) is not directly addressed in this book.

1.2.2   Pheromones

Following Tristram Wyatt (2010), a "pheromone" may be defined as: "molecules that are evolved cues which elicit a specific reaction, for example a stereotyped behavior and/or development process in a conspecific." This definition is closely based on the original definition by Karlson and Luscher (1959). Key elements are that pheromones are "evolved cues" (stimuli whose production is in some

Fish Pheromones and Related Cues, First Edition. Edited by Peter W. Sorensen and Brian D. Wisenden.
© 2015 John Wiley & Sons, Inc. Published 2015 by John Wiley & Sons, Inc.

1

way adaptive), "species-wide" (used by all members of the species), and that some type of innate recognition is implied. Nevertheless, responses to pheromones can be conditional on context, and/or internal state. Similarly, although pheromone composition (pheromones may be single compounds or mixtures) may vary slightly among individuals, this variation is expected to reflect physiological state (e.g., dominance) and not individual eccentricities or identity. In other words, I believe that the information contained in pheromones should be equally relevant to all individuals of the species (e.g. 'shared' by all members of that species). This type of rigidity appears to be associated with some type of specialized neural mechanisms. In contrast, organismal chemical cues that vary between individuals of the same species (either because of their chemical structures and/or because they are learned) should not considered to be pheromones but 'related conspecific cues.'

Identification of pheromones in fish is difficult and requires isolation of the released chemical(s) and proof that it/they elicit a specific adaptive response via some type of innate (neural) mechanism. This is not to say that learning cannot be involved but it should presumably be highly prescribed so that all members of the species recognize the same cues. Indeed, only a handful of fish pheromones meet these criteria these species include goldfish (*Carassius auratus*) hormonal sex pheromones, sea lamprey (*Petromyzon marinus*) migratory and sex pheromones, the Atlantic salmon (*Salmo salar*) sex pheromone, and perhaps the reproductive cues used by the African catfish (*Clarias gariepinus*). Nevertheless, there is a great deal of circumstantial evidence that pheromones are commonly produced and used by most, if not all, fishes. Their functions are diverse and include conspecific recognition, recognition of reproductive state, and the presence of injured conspecifics. These functions will be reviewed in Section 1.3; but first, a few terms that are commonly used to describe the pheromones in fish and other species are defined.

PRIMER PHEROMONE

A priming pheromone is a conspecific chemical cue that drives an adaptive developmental or otherwise wholly physiological response in an exposed conspecific. All members of a species in the same physiological state should typically be similarly affected. Examples include hormonal cues that drive endocrine changes in exposed conspecifics and alarm cues that change growth characteristics of exposed conspecifics.

RELEASER PHEROMONE

A releasing pheromone is a conspecific chemical cue that drives a rapid, adaptive, and innate behavioral response in a conspecific. All members of a species in the same physiological state should typically be similarly affected. Examples include hormonal metabolites that drive sexual arousal. Pheromones may have both releasing and priming effects.

PHEROMONE MIXTURE

Pheromones may comprise mixtures of chemical cues that may act on their own or synergistically. A "blend" is very specific type of mixture that requires that multiple components be present in very specific ratios for the mixture to have activity. Although commonly described in insects, no examples appear to exist in fish. By contrast, a "complex" is mixture of pheromonal components that can assume different functions depending on mixture composition. Ratios are not necessarily of primary importance but the overall composition is. Recent studies suggest that complexes in goldfish (see chapter 2) may include nonhormonal components that encode species identity and hormonal components that encode sexual condition.

1.2.3 Signature Mixtures

Fish, like all other vertebrates, release and learn to recognize conspecific chemicals for various purposes such as individual and kin recognition. Following Wyatt (2010), the term "signature mixture" is used for non-phermonal but related cues that can be defined as "variable chemical mixtures which are released by organisms and learned by other conspecifics to recognize individual or a member of some type of social group." The term "signature" implies some level of individuality; unlike pheromones, these cues are not anonymous and they are learned. The manner in which they are learned need not be prescribed. Further, their composition is typically complex and variable. In fishes, as in mammals, these cues appear to be commonly used to mediate recognition of individuals within social hierarchies or perhaps other aspect of special value such that they have been subject to recent stress. In mammals, these cues often appear to be genetically based on the major histocompatability complex (MHC) that codes genetic identity, but this possibility has not been fully resolved in fishes. A signature mixture is expected to be highly context dependent and may change with diet and other environmental factors. Recognition appears to require combinatorial responses of broad elements of the olfactory system. Signature mixtures may be found together with pheromones.

Identification of a signature mixture requires isolation of the released chemical(s) and proof that it elicits an adaptive response that is learned. Several examples have been described in fish. For example, Bryant and Atema (1987) show that diet influences production of odors associated with social hierarchy in the bullhead catfish, *Ictalurus nebulosus*, and that amino acids change in urine. Fish have also been show to readily learn to recognize the odors of conspecifics that have been attacked (Chivers and Smith, 1998). These functions will be reviewed in Section 1.3. Many ornamental odors such as those reviewed by Lynda Corkum and Karen Cogliati in Chapter 4 may fit this definition.

1.2.4   Other Definitions Relevant to this Book

A few other definitions associated with production and detection of conspecific chemical cues and signals are defined to promote clarity. Some of these definitions have also been the subject of considerable controversy which is not discussed here as they are defined largely for operational reasons.

Cues refer to any stimulus that elicits a sensory response in an animal's sensory system.

Signals are a prescribed set of cue(s) whose chemical identity has been influenced by evolutionary processes and may thus be considered to be specialized.

Communication may be defined as the exchange of adaptive information (e.g. signals) between two conspecifics.

The olfactory sense is the chemosensory component of the cranial nerve 1 (i.e., taste and common chemical sense are not included). It is also known as the sense of smell.

An odorant is a molecule that binds with olfactory receptor(s) and stimulates the olfactory sense.

An odor is an identifiable suite of odorant(s) that an animal's olfactory system can discriminate.

Fish are chordates with gills and fins that spend most or all of their live in water. (This book will address jawless, cartilaginous, and boney fishes.)

1.3   FUNCTIONS SERVED BY PHEROMONES AND RELATED CUES

1.3.1   Overview

Pheromones and related conspecific cues are known and defined by their biological function(s). Although these functions are diverse, they can be placed into five broad categories as outlined below. Some of these categories are not mutually exclusive, and presumably others may still await discovery.

1.3.2   Alarm Cues

As with other organisms, fish have evolved to recognize and respond to stimuli associated with the risk of predation, of which chemicals released by injured conspecifics are one (Chivers and Smith, 1998; Brown, 2003; Wisenden, Volbrecht, and Brown, 2004; Ferrari *et al.*, 2010). Dozens of examples exist of fish-fleeing areas that contain extracts of damaged conspecific skin odor or reacting in other adaptive manners. In many instances, these responses seem to be species-specific, but this is not always the case. Also, there is evidence that fish can learn to respond to other species if they are damaged. These can be complex multicomponent cues, and there is even evidence that some can serve as primers. For example, the crucian carp, *C. carassius*, becomes more deep-bodied when exposed to damaged conspecific skin (Bronmark and Miner, 1992). Laboratory behavior studies suggest that hypoxanthine-3-N-oxide plays a role in the alarm response (Brown *et al.*, 2000), but this compound is yet to be measured in the water or shown to be detected by the fish nervous system. Quite possibly, multiple cue types are involved. Brian Wisenden reviews alarm responses in a critical manner while evaluating specific evidence of innate versus learned recognition in Chapter 6.

1.3.3   Nonreproductive Recognition and Aggregation

Conspecific recognition is important to fishes, and chemical cues appear to play a significant role in this process, especially amongst fishes that live in dark and/or deeper waters (Hemmings, 1966; Sisler and Sorensen, 2008). One important function is to promote shoaling and aggregation among nonreproductive individuals that seek to find each other to either avoid predation or locate food. Another function is to facilitate migratory orientation by adults or juveniles that seek habitat populated by conspecifics. Freshwater eels (*Anguilla* sp.), chars (*Salvelinus* sp.), galaxids (*Galaxias* sp.), and lampreys (*Petromyzontidae*) use conspecific body odors in this manner (Baker and Montgomery, 2001). It also appears that these conspecific cues may often be mixtures of nonhormonal body metabolites that function together with hormonal pheromones as part of pheromone complexes (Sorensen, Scott, and Kihslinger, 2000; Levesque *et al.*, 2011; Lim and Sorensen, 2011). Both bile acids (Selset and Døving, 1980) and L-amino acids (Saglio and Blanc, 1989) have been implicated in species recognition, but only for the sea lamprey have they been identified and then as sulfated bile sterols (Sorensen *et al.*, 2005). This topic is reviewed by Peter Sorensen and Cindy Baker in Chapter 2.

1.3.4   Individual and Kin Recognition

As is the case with mammals, the complex social systems used by some fish have favored the evolution of chemosensory mechanisms to determine relatedness of conspecifics (Olsén *et al.*, 1998). Functions of these "familial" odors include recognition of young and shoaling/schooling (Ward and Hart, 2003). In North American ictalurid catfish, at least some components of the odor used in individual recognition are L-amino acids (Bryant and Atema, 1987). Studies of salmonids suggest that kin odors are released in the urine and that a gene product associated with the major histocompatability complex (MHC) might be involved (Olsén, Grahn, and Lohm, 2002). The identity of individual, kin-specific odors is unknown although some speculate that peptide may be involved. This type of conspecific cue appears to represent a signature mixture and is reviewed by Ashley Ward in Chapter 5.

1.3.5   Ornamental Odors

Many species of fish are highly territorial and advertize their presence and identity using visual, acoustical, and chemical cues. Some fishes have specialized glands for the production of these cues (Bushmann, Burns, and Weitzman, 2002; Belanger, Corkum, and Zielinski, 2007), but the active components have not yet been identified. Both pheromones and signature mixtures can serve this function. Ornamental odors may assume communicatory roles. This topic is reviewed by Lynda Corkum and Karen Cogliati in Chapter 4.

1.3.6   Reproductive Stimulants

Arguably, the most important event in an organism's life is finding a suitable mate and reproducing. Fish are no exception, and the challenges of life underwater appear to have favored the use of sexual signals including pheromones. A few of these have been identified, and the vast majority appears to be hormonal products and derivatives ("hormonal pheromones") whose production presumably reflects inherent reproductive state and activity. Production, release, and response of select hormonal products have now been demonstrated in a few fishes: the goldfish, common carp (*Cyprinus carpio*), Atlantic salmon, and African catfish (Sorensen and Hoye, 2010). However, hundreds of species of fish from a broad variety of groups have now been shown to detect at least a few hormonal products with high sensitivity and specificity; therefore, the use of hormonal pheromones likely is widespread among fishes. Notable exceptions are a keto bile acid used by male sea lamprey (*P. marinus*; Li *et al.*, 2002) and an unusual amino acid used by ovulated masu salmon (*Oncorhynchus masou*; Yambe *et al.*, 2005). Hormonal pheromones have been especially well described among the minnows and carps where they function as changing mixtures in the contexts of other cues.

Several functions have been elucidated for hormonal sex pheromones. First, there is evidence that at least a few species of fish recognize the gender of maturing conspecifics. For instance, male goldfish release the androgen androstenedione by which females recognize males (Sorensen, Pinillos, and Scott, 2005). In addition, various fishes use priming sex pheromones derived from prespawning hormones to predict spawning and respond with hormonal surges of their own. The best understood of these is 17α,20β-dihydroxy-4-pregnen-3-one that is released along with other conjugates by ovulated female carps detected at picomolar concentrations (Dulka *et al.*, 1987). Hormonal pheromones also mediate mate recognition and sexual encounters between sexually active conspecific fishes. The F prostaglandins that serve to mediate ovulation have an especially prominent role in this process (Sorensen *et al.*, 1988; Stacey and Sorensen, 2009). Interesting questions about hormonal sex pheromones are how they might have come into use, how pheromone identity might relate to reproductive mode, how they might encode species identity, and whether they may influence hormonal function. It is possible that hormonal cues function as part of complexes. This topic is reviewed by Norm Stacey in Chapter 3 in which he addresses some new work exploring evolutionary questions in the African cichlids.

1.4 PHEROMONE IDENTITY, SYNTHESIS, AND RELEASE

The few fish pheromones that have been definitely identified (i.e., isolated and measured in the water and then shown to elicit sensory and biological responses) are relatively simple unspecialized structures. No signature cues, with the possible exception of the relatively simple L-amino acid mixtures employed by Ictalurid catfish (Bryant and Atema, 1987, see above) have been identified. The use of such simple structures in conspecific signaling presumably reflects the origins of these cues as bodily metabolites. To date, F prostaglandins, various C18, C19, and C21 sex steroids, an amino acid, and bile acids have been shown to have pheromonal function in various fishes. Many of these structures are conjugated with sugars or sulfates, perhaps because they increase solubility. These structures were reviewed by Sorensen and Hoye (2010); therefore, they are not reviewed in this book.

1.5 PHEROMONE DETECTION AND PHYSIOLOGICAL RESPONSIVENESS

1.5.1 Overview

Where studied, conspecific chemical cues in fishes have been found to be detected and discriminated by the olfactory system (cranial nerve 1). This also seems to be the case for all vertebrates and, presumably, reflects the inherent ability of this system to encode complex information and rapidly relay it to areas of the forebrain associated with social behaviors. Efforts to understand how social cues are processed in the fish olfactory system have focused on pheromones and the premise that they are discerned by specific components of the olfactory system. Nevertheless, a few studies suggest that this system also encodes signature information as it does in the mammals. Ongoing studies support this possibility, and they are reviewed herein. First, we address olfactory receptors (detection), then discrimination, and last responses ("higher" level function) and how these systems might be evolved.

1.5.2   Pheromone Receptors

The first step in the perception of a chemical cue involves binding of a ligand (odorant) to an olfactory receptor. There is every reason to believe that this is case with pheromones too, but it has yet to be directly demonstrated. Like other vertebrates, fish have many dozens of receptors of several types (Saraiva and Korsching, 2007). Pheromone receptors have unfortunately not been definitely isolated in fishes (although there is speculation (Bazáes, Olivares, and Schmachtenberg, 2013)); therefore, they are not reviewed.

1.5.3   Olfactory Discrimination of Pheromones

Following binding, electrical responses to odorants are transduced via the olfactory nerves whose activity creates neural maps of odor identity in the olfactory bulb. This is how a complex odor is discerned, and presumably pheromones have much simpler and more invariant maps than signature mixtures because only one receptor type is expressed in each olfactory receptor neuron. Various evidences, which include electrophysiological recordings, histological, and neural ablation, suggest that information on sex and alarm pheromones is conveyed by specific subclasses of olfactory neurons that project down the medial portions of the fish olfactory system (Hamdani and Døving, 2007). Although the crucian carp is perhaps the best understood model, there is compelling evidence that the olfactory systems of other fishes function in similar manners. Stine Lastein, El Hasan Hamdani, and Kjell Døving describe in Chapter 8 what we know about the key processes that underlie pheromone discrimination in fishes.

1.5.4   Pheromonal Signaling and Communication

Water-borne pheromones pass readily between conspecifics and present a myriad of opportunities to evolve and change with time. Thus, although many (most) pheromonal cues presumably evolved as unspecialized bodily metabolites whose detection instilled an advantage to the receiver, others with time have assumed secondary roles in which their production comes to impart an advantage to the donor. For example, male tilapia, *Oreochromis mossambicus*, maintain complex hierarchies and nests and have evolved urinary pheromones that convey their status to proximate conspecifics (Barata *et al.*, 2007). This process involves various levels of physiological specialization that may involve specialization of cue production for its own sake and can be considered to be an example of true communication (Wisenden and Stacey, 2005; Stacey and Sorensen, 2009; Wyatt, 2010). Brian Wisenden examines how and why pheromonal cues may have come to be specialized in Chapter 7. New issues about definitions are also raised in the chapter.

1.6   PRACTICAL APPLICATIONS OF FISH PHEROMONES

1.6.1   Overview

Chemical cues, and pheromones in particular, play critical roles in the lives of many fishes. Laboratory and field studies consistently find that fish that experience olfactory damage will often fail to find key habitat or mate. Similarly, other studies show that addition of small quantities of pheromones to the water can exert powerful, adaptive effects. The potency of pheromones and the ease with which they can be added to the water make them excellent candidates for managing fish in aquacultural settings or in the wild (invasive fish in particular). A key component of applying pheromones is to understand their distributions and concentrations. These topics are addressed herein.

1.6.2   Effects of Pollution on the Perception of Conspecific Cues

The olfactory system appears to be exclusively responsible for detection and processing of conspecific chemical cues; yet because olfactory receptors are freely exposed to the water, they are extremely susceptible to environmental damage. In addition, drugs and other water-borne contaminants may specifically disrupt neural function in this sensitive system. Sublethal effects of poor-water quality on chemical information transfer in fishes have been documented (Jaennson *et al.*, 2007). Håkan Olsén reviews this fascinating and important topic along with the effects of pollutants on the olfactory sense in chapter 10.

1.6.3 Application of Pheromones to the Management and Control of Wild Fisheries

Fish pheromones have been shown to exert powerful influences on fish behavior and physiological function at subpicomolar concentrations. They are also easy to apply and, at least in theory, most are environmentally safe because of the specificity of their actions. Management of wild fish is currently challenged by difficulties of censusing fish or in the case of invasive fish, removing them. Fishery agencies are presently examining pheromones for control of the exploding problem of invasive fishes and fishery conservation. The sea lamprey control program has made significant contributions in understanding the biology and application of pheromones to this invasive in the Laurentian Great Lakes. Peter Sorensen examines some of these possibilities in chapter 12.

1.6.4 Measuring and Interpreting Pheromones in the Water

To use pheromones effectively, one needs to know how they are found in natural waters so that levels can be maintained. Two techniques have been developed: radioimmunoassay (Scott and Ellis, 2007) and mass-spectrometry (Fine and Sorensen, 2005). Michael Stewart and Peter Sorensen explore the potential of these techniques and what they have shown in Chapter 9.

1.6.5 Applications of Pheromones in Marine Fish and Their Culture

Pheromones are powerful modulators of fish reproductive behavior and physiology in both fresh and salt water, yet little is known about the latter. Pheromones have also been identified in several species that have commercial importance, some of which will not reproduce without endocrine treatment. Unlike hormones, pheromones can be applied to fish without handling—saving time, money, and stress. Peter Hubbard in Chapter 11 addresses whether and how pheromones are used by marine fishes and how they might be used in aquaculture while focusing mainly on marine species.

1.7 SUMMARY

Chemical information transfer between fishes of the same species can take many forms and exert powerful effects. These effects can be either innate or learned, and all appear to be mediated by the olfactory system that makes them susceptible to damage and manipulation. The complexity of these scenarios requires the use of many terms whose precise meaning should not be overinterpreted because in most conditions they represent continua rather than absolutes.

ACKNOWLEDGMENTS

Peter Sorensen thanks Brain Wisenden and many others including his students and postdocs and his former advisors for their help in two decades of research. Many granting agencies including The National Science Foundation, The National Institutes of Health, Sea Grant, The Minnesota Agricultural Experiment Station, The Great Lakes Fisheries Commission, and Minnesota Environment and Natural Resources Trust Fund have generously supported this research over what seems to be forever.

REFERENCES

Baker, C.F. and Montgomery, J.C. (2001) Species-specific attraction of migratory banded kokopu juveniles to adult pheromones. *Journal of Biology*, **58**, 1221–1229.

Bazáes, A., Olivares, J., and Schmachtenberg, O. (2013) Properties, projections and tuning of teleost olfactory receptor neurons. *Journal of Chemical Ecology*, **39**, 451–464.

Belanger, R.M., Corkum, L.D., and Zielinski, B.S. (2007) Differential behavioral responses by reproductive and non-reproductive male round gobies (*Neogobius melanostomus*) to the putative pheromone estrone. *Comparative Biochemistry and Physiology and Biochemistry: Part A Molecular and Integrative Physiology*, **147**, 77–83.

Bronmark, C. and Miner, J.G. (1992) Predator-induced *phenotypical change in body morphology in crucian carp*. *Science*, **258**, 1348–1350.

Brown, G.E. (2003) Learning about danger: chemical alarm cues and local risk assessment in prey species. *Fish and Fisheries*, **4**, 227–234.

Brown, G.E., Adrian, J.C., Smyth, E. *et al.* (2000) Ostariophysan alarm pheromones: laboratory and field tests of the functional significance of nitric oxides. *Journal of Chemical Ecology*, **26**, 154–163.

Bryant, B. and Atema, J. (1987) Diet manipulation influences social behavior in catfish: importance of body odor. *Journal of Chemical Ecology*, **13**, 1645–1666.

Bushmann, P.J., Burns, J.R., and Weitzman, S.H. (2002) Gill-derived glands in glandulocaudine fishes (teleostei: Characidae: Glandulocaudinae). *Journal of Morphology*, **253**, 187–195.

Chivers, D.P. and Smith, R.J.F. (1998) Predator alarm system in aquatic predator-prey alarm systems. *Ecoscience*, **5**, 338–352.

Dulka, J.G., Stacey, N.E., Sorensen, P.W., and Van Der Kraak, G.J. (1987) A sex steroid pheromone synchronizes male-female spawning readiness in goldfish. *Nature*, **325**, 251–253.

Ferrari, C.O., Elvidge, C.K., Jackson, C.D. *et al.* (2010) The responses of prey fish to temporal variation in predation risk: sensory habituation or risk assessment? *Behavioral Ecology*, **21**, 532–533.

Fine, J.M. and Sorensen, P.W. (2005) Biologically-relevant concentrations of petromyzonol sulfate, a component of the sea lamprey migratory pheromone, measured in stream waters. *Journal of Chemical Ecology*, **31**, 2205–2210.

Hamdani, E.L. and Døving, K.B. (2007) The functional organization of the fish olfactory system. *Progress in Neurobiology*, **82**, 80–86.

Hemmings, C.C. (1966) Olfaction and vision in fish schooling. *Journal of Experimental Biology*, **45**, 449–464.

Jaennson, A., Scott, A.P., Moore, A. *et al.* (2007) Effects of a pyrethroid pesticide on endocrine responses to female odours and reproductive behaviour in male parr of brown trout (*Salmo trutta* L). *Aquatic Toxicology*, **81**, 1–9.

Karlson, P. and Luscher, M. (1959) "Pheromones" a new term for a class of biologically active substances. *Nature*, **183**, 155–156.

Levesque, H., Scaffidi, D., Polkinghorne, C.A., and Sorensen, P.W. (2011) A multi-component species identifying pheromone in the goldfish. *Journal of Chemical Ecology*, **37**, 219–227.

Li, W.M., Scott, A.P., Siefkes, M.J. *et al.* (2002) Bile acid secreted by a male sea lamprey that functions as a sex pheromone. *Science*, **296**, 138–141.

Lim, H.K. and Sorensen, P.W. (2011) Polar metabolites synergize the activity of prostaglandin F2α in a species-specific hormonal sex pheromone released by ovulated common carp. *Journal of Chemical Ecology*, **37**, 695–704.

Olsén, K.H., Grahn, M., Lohm, J., and Langefors, A. (1998) MHC and kin discrimination in juvenile Arctic char (*Salvelinus alpinus* L). *Animal Behavior*, **56**, 319–327.

Olsén, K.H., Grahn, M., and Lohm, J. (2002) Influence of MHC on kin discrimination in Arctic char (*Salvelinus alpinus* L). *Journal of Chemical Ecology*, **28**, 793–795.

Saglio, P.B. and Blanc, J.M. (1989) Intraspecific chemocommunication in immature goldfish, *Carassius auratus* L.: attraction in olfactometer to free amino acid fractions from skin extract. *Biological Behavior*, **14**, 132–147.

Saraiva, L.R. and Korsching, S.I. (2007) A novel olfactory receptor gene family in teleost fish. *Genome Research*, **17**, 1448–1457.

Scott, A.P. and Ellis, T. (2007) Measurement of fish steroids in the water—review. *General and Comparative Endocrinology*, **153**, 392–400.

Selset, R. and Døving, K.B. (1980) Behaviour of mature anadromous char (*Salevenius alpinus* L.) towards odorants of their own population. *Acta Physiologica Scandinavica*, **108**, 113–121.

Sisler, S.P. and Sorensen, P.W. (2008) Common carp and goldfish discern conspecific identity using chemical cues. *Behaviour*, **145**, 1409–1425.

Sorensen, P.W. and Hoye, T.H. (2010) Pheromones in vertebrates, in *Chemical Ecology, Comprehensive Natural Products Chemistry II: Chemistry and Biology*, vol. 4 (ed. K. Mori), Elsevier Press, Amsterdam/ Heidelberg, pp. 225–262.

Sorensen, P.W., Hara, T.J., Stacey, N.E., and Goetz, F.W. (1988) F prostaglandins function as potent olfactory stimulants that comprise the postovulatory female sex pheromone in goldfish. *Biology of Reproduction*, **39**, 1039–1050.

Sorensen, P.W., Scott, A.P., and Kihslinger, R.L. (2000) How common hormonal metabolites function as specific pheromones in the goldfish, in *Proceedings of the Sixth International Symposium on the Reproductive Physiology of Fish* (eds B. Norberg, O.S. Kjesbu, G.L. Taranger, E. Andersson, and S.O. Stefansson), John Grieg AS, Bergen, pp. 125–129.

Sorensen, P.W., Fine, J.M., Dvornikovs, V. *et al.* (2005) Mixture of new sulfated steroids functions as a migratory pheromone in the sea lamprey. *Nature Chemical Biology*, **1**, 324–328.

Sorensen, P.W., Pinillos, M., and Scott, A.P. (2005) Sexually mature male goldfish release large quantities of androstenedione to the water where it functions as a pheromone. *General and Comparative Endocrinology*, **140**, 164–175.

Stacey, N.E. and Sorensen, P.W. (2009) Hormonal pheromones in fish, in *Hormones, Brain and Behavior*, 2nd edn (eds D.W. Pfaff, A.P. Arnold, A. Etgen, S. Fahrbach, and R. Rubin), Elsevier Press, San Diego, pp. 639–681.

Ward, A.J.W. and Hart, P.J.B. (2003) The effects of kin and familiarity on interactions between fish. *Fish and Fisheries*, **4**, 348–358.

Wisenden, B.D. and Stacey, N.E. (2005) Fish semiochemicals and the evolution of communication networks, in *Animal Communication Networks* (ed. P. McGregor), Cambridge University Press, London, pp. 540–567.

Wisenden, B.D., Volbrecht, K.A., and Brown, J.L. (2004) Is there a fish alarm cue? Affirming evidence from a wild study. *Animal Behaviour*, **67**, 59–67.

Wyatt, T.D. (2003) *Pheromones and Animal Behaviour*. Cambridge University Press, Cambridge.

Wyatt, T.D. (2010) Pheromones and signature mixtures: defining species-wide signals and variable cues for individuality in both invertebrates and vertebrates. *Journal of Comparative Physiology A*, **196**, 685–700.

Yambe, H., Kitamura, S., Kamio, M. *et al.* (2005) L-Kynurenine, an amino acid identified as a sex pheromone in the urine of ovulated female masu salmon. *Proceedings of the National Academy of Science*, **103**, 15370–15374.

Chapter 2
Species-Specific Pheromones and Their Roles in Shoaling, Migration, and Reproduction: A Critical Review and Synthesis

Peter W. Sorensen[1] and Cindy Baker[2]

[1]University of Minnesota, St. Paul, USA
[2]National Institute of Water and Atmospheric Research Ltd, Hamilton, New Zealand

2.1 INTRODUCTION

In the vast and dimly lit waters that characterize most aquatic ecosystems, fish need to find and recognize conspecifics for many reasons. Among these are the needs to find each other or shoal to find food efficiently while avoiding predators, to identify nursery habitat (which often contains conspecifics), and to find mates. All of these processes reflect a type of aggregation behavior, and thereby favor the use of conspecific odors. Such odors can travel great distances and convey large amounts of accurate information. Available evidence suggests that recognition of these odors is highly adaptive, common to all members of the species (species-wide), and based on highly prescribed processes that are either innate or involve some type of developmental process: these odors can be considered pheromones (Wyatt, 2010).

This chapter evaluates what we know and what do not know about fish pheromones and how they serve species-wide and often species-specific functions. Species specificity is defining feature of these cues and a focus here yet it presents a paradox because all pheromonal compounds identified to date in fish are relatively common metabolites (Sorensen and Hoye, 2010). We focus on evaluating those experimental scenarios for which there is some type of biochemical as well as behavioral data that explicitly address species specificity - and they are surprisingly rare. First, we define and review the behavioral roles and chemistry of pheromones used in shoaling, then the use of conspecific odors in migration, and then their use in reproduction. Our focus is on describing and deciphering available high-quality data, allowing the data speak for itself. Last, we evaluate common properties of all of these types of aggregation pheromones while considering a unifying theory to explain commonalities and how the same cues may even be used by multiple life history stages and closely related species. The idea of pheromone complexes that are comprised of shifting but distinctive mixtures of compounds compounds is proposed.

Fish Pheromones and Related Cues, First Edition. Edited by Peter W. Sorensen and Brian D. Wisenden.
© 2015 John Wiley & Sons, Inc. Published 2015 by John Wiley & Sons, Inc.

2.2 PHEROMONES AND NONREPRODUCTIVE SHOALING BEHAVIOR

Very few fish live solitary lives; in addition to finding each other to mate, they often aggregate as juveniles to find food and/or habitat and to avoid predators (Pitcher and Parish, 1993). The process of aggregating requires that fish recognize each other; and while many types of sensory cues are involved in this process, odors (distinctive sets of chemical cues that are detected by the olfactory sense) appear to be the most important one. Shoals can be defined as loosely organized aggregations that help fish find food and habitats while avoiding predation (Pitcher and Parish, 1993). Shoals may at times contain other species, especially if these species have similar behaviors and ecological needs (Ward, 2015), and we will touch on that here. In additional, shoals may also contain sexually mature individuals; however, mating is typically not the primary reason for shoal formation. Interpretation of shoals with mature individuals is complicated by reproductive interactions; therefore, we do not discuss this scenario. Highly polarized and synchronized shoals are known as schools and appear to offer further advantages including increased swimming efficiency (Pitcher and Parish, 1993). We do not review schooling *per se* as it is seemingly not fundamentally different from shoaling. Shoaling is especially common among juvenile fish that find food and avoid predation through social interactions.

The nature of shoaling varies with species and situation. While some fishes are very discerning about those they will shoal with (i.e., some select kin or members of the same population), others are less specific and will even shoal with congeners (Ward, Axford, and Krause, 2002; Ward, 2015). Not surprisingly, life history attributes appear to correlate with the specificity and nature of the odors used. Although possible relationships between kin, population, and species odors have not been studied, it seems likely that the former are subsets of the latter. In particular, data exist showing that fish can learn kin odor (Ward, 2015). How kin odor might translate to species-wide (species-specific) odors has seemingly not been studied, although it is interesting because it suggests flexibility in how conspecific identity may be discerned. Nevertheless, for the sake of simplicity, we focus on results of experimental studies that have directly tested the behavioral abilities of immature fishes to discern species identity independent of kinship in preference (head-to-head) tests using natural odors.

While a vast ecological literature suggests that most fishes are attracted to conspecifics and form aggregations at some time in their lives (Liley, 1982), the actual specificity and mechanisms underlying this response has only been illuminated in a few species. While many studies have shown attraction to conspecific odors, only for a handful of species has some type of head-to-head preference test been conducted for assessing the specific ability of immature fishes to discriminate natural odors released by different species. Further, for only one species, the goldfish, *Carassius auratus*, has the chemical characteristics of the conspecific odor that they both respond to and discriminate (i.e. discern from other fish odors including the background) also been described; however, specific compounds have not been identified. In addition, no field tests of natural odors in a natural background have been conducted (Johnson *et al.*, 2009), and our understanding of how immature fishes identify conspecifics using odor is extremely poor. Here, we review what little we know, which all suggests that odor-driven conspecific recognition is a common attribute of fishes and likely involves relatively complex mixtures.

Conspecific recognition and discrimination using odor has been studied in five species of immature fishes. The roach, *Rutilus rutilus*, a European minnow, was the first species to have both the sensory basis of its shoaling behavior and its ability to discriminate examined (Hemmings, 1966). Maze studies demonstrated that it strongly prefers conspecific to heterospecific holding water. Time in the laboratory (i.e., experience) has no effect on this preference, suggesting that this behavior is not situation specific and likely innate (thus species-wide). A similar experiment with the Japanese catfish, *Plotosus anguillaris* (Kinosita, 1972), showed it too can discriminate and select conspecifics using simple body rinses. Similar responses to holding waters have since been noted in juvenile char, *Salvelinus alpinus*, for which amino acids shed via skin mucus were suggested to convey species information but unfortunately have never been tested (Hoglund and Astrand, 1973). Intriguingly, char are also known to recognize kin using, odor and this has been ascribed to learning (Olsen, Grahn, and

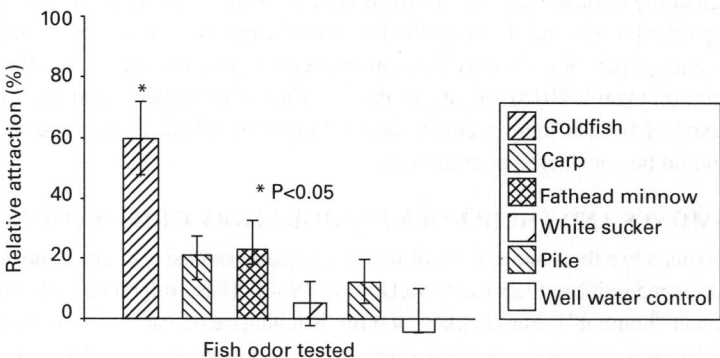

Figure 2.1. Preferences of immature goldfish for rinses of conspecifics and other fish species in a laboratory maze demonstrate the existence of a species-specific pheromone. Sisler, S.P. & Sorensen, P.W. Common carp and goldfish discern conspecific identity using chemical cues. Behaviour 145, 1409-1429. Copyright © Brill (2008).

Lohm, 2002). More recently, Ward, Axford, and Krause (2002) found that a European chub, *Leuciscus cephalus*, prefers to shoal with conspecifics over heterospecifics, and that chemical cues are more important than visual cues. The precise manner with which visual and odor cues complement each other has not yet been addressed nor has the possible role of experience (learning).

A series of recent experiments using the goldfish and its close relative, the common carp (*Cyprinus carpio*), have clearly demonstrated that these species also recognize the odors of their conspecifics and that these odors have many chemical components. These studies used multiple stocks of fishes, suggesting the response is fundamental to these species and species-wide (Sorensen, unpublished data). Initial studies by Sisler and Sorensen (2008) discovered that sexually immature goldfish and common carp strongly prefer rinses of their own species versus those of six other species and that they continue to demonstrate these preferences in direct head-to-head tests (Fig. 2.1). Learning does not appear to have an apparent role in species recognition because cohousing goldfish with other species had no effect on their preferences, while olfactory ablation did block it. The chemical identity of conspecific odor recognized by immature goldfish was later characterized by Levesque *et al.* (2011) who fractionated immature goldfish holding waters and found activity in both the nonpolar and polar fractions, suggesting the pheromone comprises a mixture of compounds. Tests of conspecific bile acids (a class of odors suggested to have pheromonal function; Lastein, Hmadani, and Døving, 2015) suggested these are not important. Cross-tests of sexually mature males and females found all life stages are attracted to and attractive to all others: a common odor is released and recognized throughout the life cycle of this species. Levesque *et al.* (2011) conclude that goldfish employ a "pheromone complex" that contains polar and nonpolar components, the precise composition of which may change (by the addition of hormonal components) with maturational state while key species-specific elements remain constant.

In summary, there is strong evidence that juveniles of many fishes use conspecific odors to mediate species recognition in shoaling and schooling behaviors. These odors seem to resemble kin and population odors; and although none have been identified, it seems that these may frequently be complex mixtures of simple, common metabolites. Further, these studies show that fish may release common suites of compounds throughout their lives that can be discerned by the olfactory sense, and thus lend them different meaning or even allow them to be part of other pheromones (i.e., it could provide the species-specific information that hormonal sex pheromonal cues seem to lack [see Section 2.4; Stacey, 2015]). Such a possibility could explain how in a parsimonious fashion 30 000 species of fish might have each come to have their own pheromones. How these sets of odors are discerned is not yet known; in particular, it is not known whether it might involve some type of highly prescribed learning that occurs as part of early development or even self-referencing such as that suggested for kin odor recognition (Ward, 2015). Whatever the specific circumstances (and there could be many given the diversity of fishes), there seems to be no doubt that shoaling odors are species-wide and adaptive and should be considered pheromones.

2.3   PHEROMONES AND THEIR ROLE IN MIGRATORY ORIENTATION

There are approximately a thousand species of fish that perform extensive directed movements (defined as migrations) at some point in their lives (McDowall, 1988). These migrations take many forms and many involve either "homing" [return to place of birth (hatching); e.g., Pacific salmonids (see below)] or simply locating habitats (and/or locales) that offer special opportunities to feed and/or reproduce (e.g., nurseries). Generally homing involves some type of learning although stock-specific pheromones may also be involved, while habitat recognition can either be innate or learned. In many instances, innately recognized habitats contain conspecifics whose presence, and odor, presumably serve as indicator of habitat quality and might thus have come to assume pheromonal function (Lastein, Hamdani and Doving, 2015). It is very possible that conspecific odors (pheromones) used in long-distance migration are simply a specialized type of long-distance shoaling or aggregation odor, and may not necessarily be different from shoaling cues. Five classic types of migratory life histories (anadromy, catadromy, amphidromy, potadromy, and oceanadromy) have been recognized, as well as partial migration (the trait by which fish populations can show variable tendencies to migrate following the aforementioned schemes). We address the possible roles of species-wide pheromones in all the five strategies in the sections that follow.

2.3.1   The Role of Migratory Pheromones in Anadromy

Anadromy is the most common life history strategy used by migratory fishes. It is best understood in the salmon (Order Salmoniformes, Family Salmonidae) and lamprey (Order Petromyzontiformes, several families). The salmonidae is a large group of fishes, a subfamily of which, the Salmoninae, is especially well known and contains many commercially important species that return to natal streams (i.e., the locations where they hatched) to spawn using odor for orientation. The precise chemical nature of these natal odors appears to vary by species and, in some cases, may involve some type of learning and in some cases could include conspecific odor. The lampreys are an ancient group of jawless, migratory fishes, one of which, the sea lamprey, *Petromyzon marinus*, has been clearly shown to use a larval pheromone comprising biliary sterols to find optimal nursery habitat. This section reviews these systems.

MIGRATORY PHEROMONES IN THE SALMONINAE

Olfactory ablation studies have shown that adult salmoninae use olfactory cues to recognize and relocate their natal spawning grounds to spawn, often after many years at sea or in lakes (Hasler and Scholz, 1983; Stabell, 1984). Olfactory-mediated homing and the possible role of both learned odors and/or conspecific cues which may be pheromones in migratory orientation have been studied in three genera of salmonids: the chars (*Salvelinus* sp.), the Pacific salmon (*Oncorhynchus* sp.), and the trout (*Salmo* sp.). These fishes have also been found to use sex pheromones and kin odors once in streams where possible chemical overlap with migratory odors and pheromones is unclear but interesting to contemplate. We review the roles of conspecific odors in the migrations of all three salmonids here.

Chars

There is strong evidence that the chars use pheromones to return to natal streams as adults. This idea was first espoused by Nordeng (1971, 1977) who proposed the "pheromone hypothesis" (Stabell, 1984). Chars dominate many Northern Hemisphere coldwater streams, where their young spend many years before migrating downstream to the ocean where they grow and return years later. Nordeng (1971) argues that the odor of stream-resident juveniles should be a reliable indicator of spawning (and nursery) habitat and that the homeward migration of adults may be guided by a trail of population-specific odor cues. In 1977, he tested this hypothesis by displacing wild Arctic char (*S. alpinus*) and found that they entered streams that contained their kin. Later, Nordeng and Bratland (2006) found that when both parr (premigratory juveniles) and smolts of Arctic char were transplanted to neighboring rivers and fjords, they homed back to their river of origin; the authors suggested that they were using pheromones. Some other tests of wild fishes have had difficulties repeating this result (Black and Dempson, 1986). Nevertheless, Selset and Døving (1980) have since shown that mature anadromous Arctic char are attracted to the odor of smolts from their own population in the laboratory and that these odors have multiple components (Selset, 1980). They suggest fecal bile acids are the cue and show that bile acids are potent olfactory stimulants (Lastein, Hmadani, and Døving, 2015). Single-unit recording from the char olfactory bulb has also shown that this structure possesses the necessary neural circuitry to discern population-specific odors (Døving, Nordeng, and Oakley, 1974). In summary, although it does appear that conspecific odor guides migratory char to spawning habitat, key questions remain about its precise function. For example, the identity of species-specific cues has not been determined and bile acids release profiles do not appear to be species-specific (Sorensen, unpublished data), suggesting that mixtures may be involved. Possible relationships between kin-specific odors known to be learned (by unknown mechanisms) by shoaling juveniles (Olsén, Grahn, and Lohm, 2002) and those discerned by adults have yet not been addressed, suggesting that learning may also be involved in some fashion. Clearly, more research is needed to determine the precise roles and identities of pheromones in this group of fishes although they do seem to be important.

Pacific Salmon
Pacific salmon evolved in the Pacific Ocean, and they return to natal streams using odors that many suggest they learn as young through imprinting (i.e., as part of highly prescribed—but yet undefined—developmental process that involves learning) (Hasler and Wisby, 1951; Hasler and Scholz, 1983). Transplantation experiments in the field as well as the life history of this group (the young of some species leave rivers before adults return) demonstrate that plants, biofilms, and minerals (and even novel odors) are very likely key contributors to learned stream odor (Scholz *et al.*, 1976; Dittman and Quinn, 1996; Yamamoto, Hino, and Ueda, 2010; Ueda, 2011). However, it seems to us that it is not necessary to rule out conspecific odor as a frequent component of this odor (Hasler and Scholz, 1983; Stabell, 1984) and that it could and should play a role in homestream bouquets. Chemical fractionation of natal-stream water odor suggests great complexity (Idler *et al.*, 1961). Several laboratory studies of coho salmon, *Oncorhynchus kisutch*, lend some support to the possibility that conspecific odors can contribute to natural odors discerned by this group (Quinn, Brannon, and Whitman, 1983; Quinn and Tolson, 1986). Further, while Yamamoto, Hino, and Ueda (2010) suggest that mixtures of amino acids serve as the basis of home stream odor recognition in Pacific salmon, amino acids are also known to be released by fishes including salmon (Hoglund and Astrand, 1973; Hara *et al.*, 1984). There is no known reason that olfactory imprinting would not include conspecific odor if present. Such an odor, would not fit the classical definition of a pheromone. Because it would be adaptive, it could be species-wide (this is not known), and its recognition may occur through a developmental process; accordingly the term "signature mixture" might be appropriate given conspecific odor's seemingly nonuniversal presence. In conclusion, although it is clear that Pacific salmonids learn stream odors for homing and that these odors do not always include include conspecific odors, in other cases they likely do and could be considered a related odor.

Trout
Descriptive studies suggest that Atlantic salmon, *Salmo salar*, imprint and home to natal stream odors like Pacific salmon (Youngson, Jordan, and Hay, 1994). Further, like char, young Atlantic salmon usually spend years in rivers; therefore, it is possible their conspecific odors could be involved in homing. One study of returning adult Atlantic salmon found many to return to hatchery (kin) waters (Sutterlin and Gray, 1973)

while laboratory experiments suggest that young salmon can discern population (kin)-specific odors derived from feces and shoal (Stabell, 1987). Adult Atlantic salmon also appear to discern urinary kin odors (Moore, Ives, and Kell, 1994). The possible role of these odors in migratory orientation is not known. Unfortunately, chemical analysis has not been performed on the odor of natal streams discerned by *Salmo*; therefore, it difficult to reconcile the relationships of kin, species, and natal stream odors; nevertheless, it seems very likely that some type of conspecific odor may help mediate homing in this group too. Clearly, more study is needed and the possibility that conspecific odors play varying, nontraditional roles in this process that could involve highly prescribed learning as in other salmonids warrants investigation.

Migratory Pheromones in the Lampreys

Lampreys are an ancient, 400 million year-old order of jawless fishes whose reliance upon phero-mones for migration and spawning is clearly established. Of the 39 species of lamprey, 35 belong to the Northern Hemispheric family Petromyzontidae, while the two Southern Hemisphere families, Geotriidae and Mordaciidae, contain another three species. Approximately half of all lamprey species in both hemispheres spend their entire lives in freshwater, whereas the other half are migratory and leave freshwaters to parasitize large aquatic animals in lakes and oceans before maturing and returning to fresh waters to spawn and die. Lampreys are not strong swimmers and have evolved life histories that allow them to locate suitable spawning/nursery habitats using the odor (pheromone) of stream-resident larvae and as indicators of quality nursery habitat. The migratory pheromone is best under-stood in the sea lamprey (Sorensen and Hoye, 2007). A multi-component sex pheromone released by males and used by ovulated stream-resident females has also been partially identified (Li *et al.*, 2002).

The Northern Hemisphere lampreys (Petromyzontidae)

Around half of the Petromytontidae family are anadromous; and one species, the sea lamprey, is highly invasive and thus much more studied and understood than the rest. We focus on this species here.

The Sea Lamprey. The sea lamprey evolved in the Atlantic Ocean but became landlocked in the upper Laurentian Great Lakes where it is an invasive species and the subject of a large control program (Sorensen and Hoye, 2007; Moser *et al.*, 2014; Sorensen, 2015). It spawns in freshwater streams that it locates using a multicomponent migratory pheromone released by larvae (Sorensen *et al.*, 2005) before finding nest-building males using another, related multicomponent bile acid-derived sex pheromone (Li *et al.*, 2002). Sea lampreys start life as blind, filter-feeding larvae that spend 3–20 years in streams growing before reaching a critical size and metamorphosing, and then leaving for the lakes (or oceans) where they para-sitize large fishes. After 1–2 years, parasitic lampreys mature, cease feed, and seek streams to spawn. Stream search is guided by a migratory pheromone released by stream-resident larvae that appears to serve as a reliable indicator of spawning and nursery habitat. The discovery and function of this pheromone is reviewed by Sorensen and Hoye (2007, 2010); therefore, we only summarize it here.

A combination of biochemical, behavioral, and physiological studies have clearly demonstrated that the migratory pheromone released by larval sea lampreys contains three sulfated sterols: (1) petromyzonamine disulfate (PADS), (2) petromyzosterol disulfate (PSDS), and (3) petromyzonol sulfate (PS) (Sorensen *et al.*, 2005; Fig. 2.2). These sterols are produced by many lamprey species, whose odor attracts sea lamprey (Fig. 2.3), and probably evolved from relatively common antimi-crobial products (Sorensen *et al.*, 2005). They are specifically detected by the sea lamprey olfactory systems at concentrations below 10^{-13} M—the lowest threshold in a fish (Fine and Sorensen, 2008). Olfactory ablation and trap capture data also clearly indicate that adult sea lamprey rely on a larval pheromone to find spawning rivers and that changing the relative densities of larvae (and thus pher-omone concentration) will change preferences in the laboratory (Fig. 2.3) and lamprey stream choices (Sorensen and Vrieze, 2003; Vrieze, Bjerselius, and Sorensen, 2010, 2011; Sorensen, 2014). Laboratory studies also show that a mixture of these sterols is attractive and that this activity

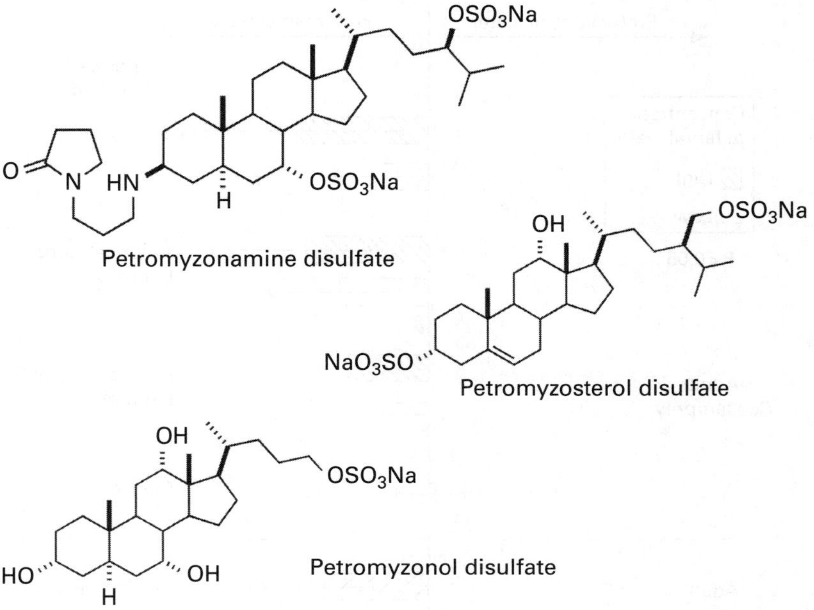

Figure 2.2. Structures that function as principal components of the sea lamprey migratory pheromone.

is synergized by unidentified stream odors (i.e., it is context dependent [Vrieze and Sorensen, 2001]). Recent field tests find that a synthetic mixture of PADS, PSDS, and PS is less attractive than the complete larval odor (Meckley, Wagner, and Luehring, 2012), suggesting that the natural pheromone is actually a rather complex mixture that includes further unknown components. All components of he lamprey migratory pheromone are lamprey-specific sterols that lack species-specific character.

In addition to the migratory pheromone, the sea lamprey is now known to use a sex pheromone at the end of their migratory period; spermiated male sea lamprey release a potent sex pheromone that lures ovulated females to nests (Li *et al.*, 2002). This odor is now known to contain 3-keto petromyzonol sulfate (3kPZS), a derivative of PS and other unknown compounds. While 3kPZS will attract females to nest sites (Johnson *et al.*, 2009), field tests show that it does not hold them there like the natural odor of spermiated males does; this pheromone also is a complex mixture (Johnson *et al.*, 2012; see Sorensen, 2014), which also contains biliary sterols but unknown compounds too.

Other Northern Hemisphere Petromyzontid lampreys. A variety of fisheries, olfactory, and behavioral evidence suggest that the other Northern Hemisphere lampreys use very similar suites of bile acid derivatives as migratory and sex pheromones to the sea lamprey (Fine, Vrieze, and Sorensen, 2004; Gaudron and Lucas, 2006; Robinson *et al.*, 2009; Fig. 2.3). Whether and how these pheromones are species-specific, and the specific roles of complex mixtures in them is not known.

Southern Hemisphere Pouched Lampreys (Geotriidae). In contrast to Northern Hemisphere lampreys, stream selection has not been examined in Southern Hemisphere lampreys, whose origins date to the separation of Gondwana. The pouched lamprey, *Geotria australis*, is the only species in the family Geotriidae. Three species are found in the other family of southern lampreys, the Mordaciidae; but their use of odors has not been studied. The pouched lamprey has a similar life history to the sea

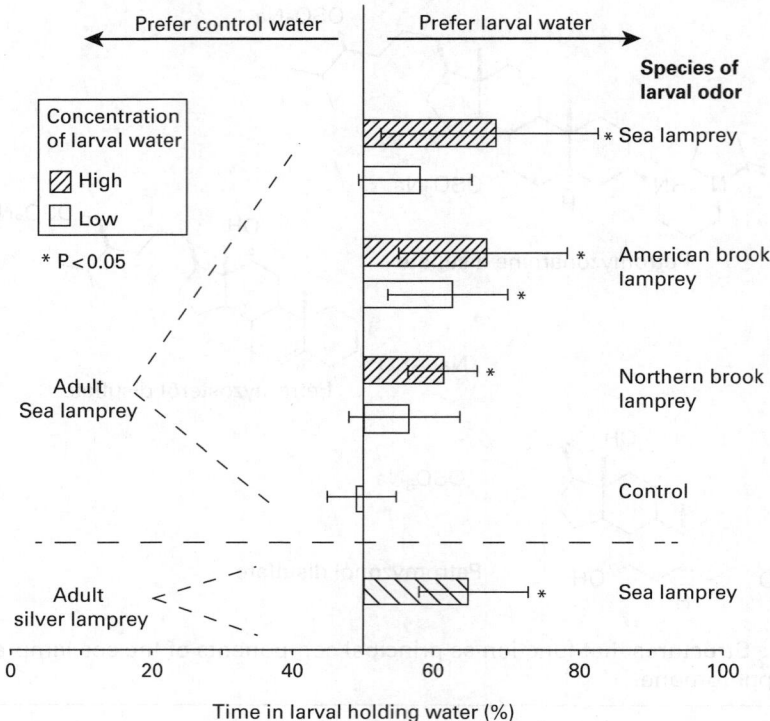

Figure 2.3. Preferences of migratory adult (maturing) sea lamprey to rinses of larvae of their own species and two other species at two concentrations. Reponses of adult silver lamprey are also shown. This pheromone is not species-specific, and it is now known to comprise three sulfate sterols plus at least one other unidentified product. Adopted with permission from Fine, Vrieze and Sorensen (2004).

lamprey with an extended migratory period. Migratory adults of this species also select spawning streams using conspecific cues (Jellyman, Glova, and Sykes, 2002). Following the methods of Sorensen *et al.* (2005), sterol release has been monitored in pouched lamprey ammocetes and large quantities of PS, but no PADS or PSDS, have been detected (Baker *et al.*, 2009). More recent analyses of larval release water using liquid chromatography–mass spectrometry have identified both PS and PADS by pouched lamprey (Stewart and Baker, 2012). This suggests that both Northern and Southern Hemisphere lampreys use similar suites of unspecialized sterols as pheromones. This seems remarkable given the evolutionary time between these species but speaks to the possible universality of metabolite-based pheromones in the fishes.

SUMMARY OF MIGRATORY PHEROMONES USED BY ANADROMOUS FISHES

A wide variety of anadromous fishes use conspecific odors in various ways to find suitable spawning habitats and locations. All appear to be complex mixtures of relatively unspecialized metabolites that often involve biliary products. While the conspecific odors discerned by lampreys and chars are clearly highly adaptive and their recognition seemingly is innate and common across each species, strict species specificity is not always the rule. It is also unclear whether developmentally prescribed imprinting or experience might have a role in pheromone recognition; indeed, this seems likely for the chars and perhaps the Pacific salmonids. Thus, it appears that multi-component conspecific odors that might loosely be considered to be pheromones (although in a nontraditional sense because some type of prescribed learning is likely involved) commonly have roles in migratory anadromous fishes. Only complete identification of these cues can answer these questions.

2.3.2   Migratory Pheromones and Amphridromy

Fish that migrate between fresh and sea water and then return for purposes other than breeding are termed "amphidromous" (McDowall, 1988). Several hundred species fit into this category, and most of these are found in the Southern Hemisphere. For many amphidromous species, larvae develop in the marine environment and then migrate into freshwater as juveniles in search of feeding habitat where they stay through adulthood (McDowall, 1990). During their upriver migrations, conspecific cues including odor appears to be used as an indicator of suitable habitat for colonization, and thus stream selection. These odors are very unlikely to be learnt as the young often leave streams when they hatch so have almost no exposure to conspecific odors. In fact, the eggs of some Southern Hemisphere species develop in a terrestrial environment. We review our understanding of odor usage by the galaxiids and gobies—the only amphidromous fishes that have been studied.

AMPHIDROMOUS GALAXIIDS

Galaxiids belong to the Galaxiidae family, which is found in the temperate zone of the Southern Hemisphere. Most have migratory life histories. Of the five species of migratory galaxiid, four are amphidromous. These are the banded kokopu (*Galaxias fasciatus*), koaro (*G. brevipinnis*), giant kokopu (*G. argenteus*), and shortjaw kokopu (*G. postvectis*). All four species are highly selective in their habitat requirements (McDowall, 1990). Both banded kokopu and koaro that penetrate well inland to high altitudes and conspecific odor have a role in this remarkable process.

Recent studies suggest that during migration, amphidromous galaxiids use conspecific cues as a mechanism for stream and habitat selection. Rowe, Saxton, and Stancliff (1992) discovered that migratory koaro select rivers in a manner that does not correlate with upstream habitat. Instead, laboratory studies show that migratory juveniles of both banded kokopu and koaro discriminate, and are attracted to the odors of adult conspecifics, but not heterospecifics (Baker and Montgomery, 2001; Baker and Hicks, 2003). In the laboratory, conspecific adult odors are strong enough to override avoidance responses to suspended sediment (Baker, 2003).

The galaxiid migratory pheromone has not yet been identified. Interestingly, as with the carps, known bile acids do not seem to be important (Baker et al. 2006) although they are produced and potent olfactory stimuli for this and many other teleost fishes (Hara *et al.*, 1984; Zhang, Brown, and Hara, 2001, Li *et al.*, 2002; Sorensen *et al.*, 2005; Johnson *et al.*, 2009). The bile acids produced by adult banded kokopu are seemingly commonplace (Baker *et al.*, 2006). This does not seem very different than for the salmonids and carps, suggesting that bile acids may nevertheless be components of much more complex pheromonal mixtures.

AMPHIDROMOUS GOBIES

The gobies (fishes from Gobidae family) are the most successful and diverse group of teleost fishes (over two thousand species), and they appear to use pheromones in many aspects of their lives (Corkrum and Coglaiti, 2015). Some of these species are amphidromous, and many tropical islands in both the Northern and Southern hemispheres are inhabited by amphidromous gobies—some of which appear to use pheromone for long-distance orientation. Stream-dwelling adult Hawaiian amphidromous gobies spawn eggs into flowing waters that are then flushed into the ocean where they develop only to return as free-swimming juveniles several months later (Sorensen and Hobson, 2005). Juveniles locate and enter freshwater streams, which in Hawaii terminate at waterfalls, and that the fish scale using suction cups created by fused pelvic fins. Once at the top of waterfalls, juveniles grow and spend the rest of their lives there; a mistake choosing the wrong waterfall (e.g., waterfalls without adult conspecifics at the top) would result in an evolutionary dead end. Studies using artificial waterfalls show that one species, *Awaous guamensis*, selects and scale small rivulets that are scented with conspecific, but not heterospecific, odor (Fig. 2.4).

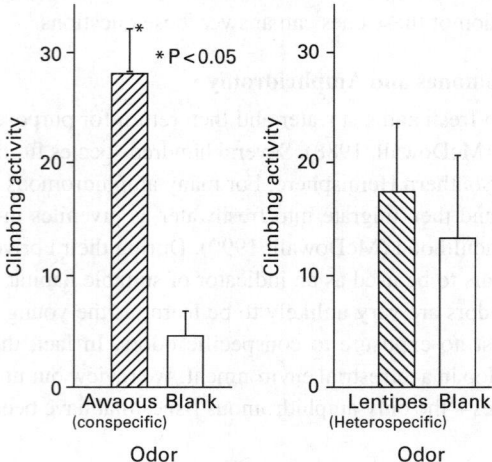

Figure 2.4. Climbing preferences of migratory juvenile *Awaous guamensis*, a Hawaiian goby, to rinses of other conspecifics versus well water, and heterospecific rinses (*Lentipes concolor*) versus well water (Sorensen, unpublished data). This pheromone is species-specific.

Other species of Hawaiian goby also appear to use conspecific odors; and in some cases, the odors of other goby species are also attractive (Sorensen, unpublished data). This is adaptive because gobies share habitats and are not predatory. As with the lamprey migratory pheromone, the goby pheromone is synergized by other unknown stream organics. Together, these results suggest that gobies, like lampreys and kokopu, select freshwater habitat using complex odors that include pheromones. Preliminary chemical characterization of goby holding water suggests that these pheromones are complex because both unidentified polar and nonpolar compounds found in them are behaviorally active (Sorensen, unpublished data).

SUMMARY OF MIGRATORY PHEROMONES USED BY AMPHIDROMOUS FISHES

Several studies demonstrate that both migrating galaxiids and gobies use conspecific odors, which in some cases are species-specific, but others are not. Because responses are highly adaptive and almost certainly instinctual, they are considered pheromones. Less than perfect specificity is likely adaptive because these fishes do not prey on each other and share habitats. Although these pheromones are poorly understood, they, like all pheromones we have discussed, appear to be complex mixtures of many common compounds that may include bile acids and are released by all life history stages.

2.3.3 Migratory Pheromones and Catadromy

Catadromous fishes spawn in the oceans and disperse as larvae to freshwater habitats where they spend most of their lives growing, but they migrate back to the ocean to reproduce. There are a several hundred species that do this in the tropical and temperate oceans of both hemispheres. Finding good-quality freshwater habitat for feeding and later for spawning in the oceans are considerable challenges. Evidence suggests that conspecific odors including pheromones have a role. The best understood example is the freshwater eel, *Anguilla* sp.

ANGUILLID EELS

The family of Anguillidae comprises several dozen temperate–tropical species located in both the Northern and Southern hemispheres. Anguillid eels spawn in mid-ocean current gyres, and their leptocephalus larvae then migrate great distances before metamorphosing and moving into freshwater (Tesch and White, 2008). Once in freshwater streams, larval eels metamorphose into "elvers" that migrate hundreds-to-thousands of kilometers further inland. Inland eels grow and after 3–50 years, mature, and migrate out to the ocean to travel back to current gyres to spawn and die. There is compelling evidence that adult male eels find females in vast oceanic expanses using sex pheromones (Sorensen and Winn, 1984; Hubbard, 2014), but nothing is known about the identity of this sex pheromone; therefore, we focus on the migratory pheromone used by elvers.

It is well established that innately discerned components of conspecific odor (i.e. pheromones) play a role in how American (*Anguilla rostrata*) and European eels (*A. anguilla*) locate freshwater habitat from the oceans. Several studies have shown that natural stream odors (not salinity) guides stream finding and stimulates positive rheotaxis in migrating elvers (Creutzberg, 1961; Sorensen, 1986). As with many amphidromous and anadromous fishes, the odor of freshwater appears complex and originates from both biofilms and conspecifics (Sorensen, 1986). The first evidence for the role of conspecific cues in this bouquet came from Miles (1968) who found that the presence of adult eels increased the attractiveness of stream water to migratory elvers. He suggested that conspecific odor would be an excellent indicator of habitat quality, similar to the scenario suggested for chars, lampreys, and gobies. Subsequent behavioral experiments found that conspecific odor is complemented by nonpheromonal odors derived from biofilms and likely drives short-range attraction within streams (Sorensen, 1986). Other tests have suggested the pheromone has multiple components including bile acids and amino acids (Sola and Tosi, 1993; Sola and Tongiorgi, 1998; Hubbard, 2014). Because Anguillid elvers do not have previous experience with conspecifics before entering streams, this odor must be innately discerned. It is not known if it might be species-specific. Interestingly, the eel pheromone is proving useful to divert upstream swimming elvers away from dams (Sorensen, 2014).

GALAXIIDS

One species of galaxiid, the inanga (*Galaxias maculatus*) is catadromous. The inanga is referred to as "marginally" catadromous because adults spawn within estuarine environments (McDowall, 1990). Laboratory studies suggest that the inanga uses a migratory released by species of galaxiid to find fresh water (adult inanga, banded kokopu and koaro; Baker and Hicks, 2003). This pheromonal odor now appears to only be a component of a complex bouquet as Hale, Swearer, and Downes (2009) find that inanga are attracted to both natural stream odors and conspecific odors. Heterospecific galaxiid attraction may be adaptive because migratory juvenile galaxiids travel in mixed-species shoals when young (McDowall, 1988, 1990). These galaxiid cues do not readily fit the classic definition of a pheromone because of their low specificity; however, their intrinsic biological significance is unquestionable so we consider them pheromones.

SUMMARY OF MIGRATORY PHEROMONES USED IN CATADROMY

Although poorly studied, a diverse variety of catadromous fish use pheromones to guide upstream migration of their larval phases. Most appear to be perceived within the context of other odors, and all are presumably innately discerned and species-wide. The chemical identities of these migratory pheromones are not yet known, but they appear to be complex and may include bile acids. Not all are species-specific, and none appears to be specialized.

2.3.4 Migratory Pheromones in Potadromous Fishes

Potadromous fishes migrate within freshwater systems. There are many such species, especially in large lakes, but they have not been well studied, and some are actually landlocked anadromous species. We only know one freshwater fish that uses pheromones to guide its migration although there likely are others. Foster (1985) describes lab maze experiments which suggest that lake trout,

S. namaycush (a char found in large North American lakes), locate spawning reefs in lakes using the odors of the fecal matter from juveniles. Subsequent studies by Zhang, Brown, and Hara (2001) show that several taurine conjugated bile acids are released and detected by lake trout, but as with Arctic char, confirmation of behavioral activity in the field awaits. It presently seems premature to conclude that this is a pheromone and could not be an entirely learned cue.

2.3.5 Migratory Pheromones in Oceanadromous Fish

Many marine fishes spend their entire life cycles within the sea where they perform extensive migrations; these species are considered oceanadromous. They belong to several taxonomic groups. In some cases, migrations involve long-distant movements between feeding, nursing, spawning, and wintering areas (McKeown, 1984); but in other instances, they involve short-range movement between on- and off-shore habitats. Little is known about the roles of olfactory cues in the first type of process. However, intriguing evidence that various coral reef fishes, whose larvae display a planktonic dispersal phase before recruiting into a reef system, may use pheromones to mediate short-range movements. Because these larvae do not have the opportunity to learn conspecific odors while drifting in the plankton, these cues should be considered a type of migratory pheromone.

PHEROMONES AND LARVAL REEF FISHES

The ability of larval reef fishes to relocate patchily distributed reef environments from off-shore waters suggests habitat selection is an active process (Montgomery, Tolimieri, and Haine, 2001; Gerlach *et al.*, 2007). This is supported by studies which indicate that presettlement larvae are strong swimmers (Stobutzki and Bellwood, 1997; Fisher, Bellwood, and Job, 2000; Fisher *et al.*, 2005) with well-developed sensory systems (Job and Bellwood, 2000; Tolimieri, Jeffs, and Montgomery, 2000; Lecchini *et al.*, 2005; Arvedlund and Takemura, 2006). Although orientation toward reefs likely involves many sensory modalities, olfaction has been implicated in habitat choice for a range of presettlement reef larvae. Atema, Kingsford, and Gerlach (2002) found that plumes of lagoon water extend for kilometers outside of reefs, and that apogonid larval fish can discriminate, and are attracted to, lagoon water over ocean water.

A multitude of chemosensory cues must be present within reef odor plumes, but recent studies highlight the importance of conspecific odor cues in detection and settlement of larvae onto a reef system. Laboratory studies by Lecchini *et al.* (2005) find that *Chromis viridis* larvae prefer conspecific odor over heterospecific odor or coral substrates. Field trials by Lecchini (2004) suggest that 10 of 12 coral reef species detect their settlement location using conspecific odors. Larval damselfish (*Dascyllus* sp.) have also been shown in both aquarium and field trials to prefer reefs that have resident conspecifics over unoccupied or confamilial groups (Sweatman, 1988; Booth, 1992).

Recent studies of subpopulations of coral reef fish species suggest that larvae of at least a few species return to their natal reefs (Almany *et al.*, 2007; Gerlach *et al.*, 2007). Similar to anadromous salmonids, larvae may imprint on reef-specific odors, which likely include conspecific or population-specific pheromones. The use of conspecific odors in homing has been reported in the five-lined cardinalfish (*Cheilodipterus quinquelineatus*), which returned to their home reef sites after being displaced and could differentiate between the chemical cues of conspecifics from two different sites (Døving *et al.*, 2006). Dixson *et al.* (2008) also provides evidence that support the use of reef-specific cues by the clownfish, *Amphiprion percula*, to locate island homes. In this case, recently settled fish exhibited strong preferences for water from reefs with islands, as well as water treated with either anemones or leaves from rainforest vegetation that overhangs island reef waters. For reef fishes with demersal eggs, imprinting on reef-specific cues may be a reasonable mechanism to facilitate homing to natal reefs. However, for reef fish species with pelagic eggs, imprinting on natal reefs may not be plausible, and innately recognized conspecific pheromones might instead play a major role in homing (Montgomery, Tolimieri, and Haine, 2001). It will be interesting to learn more about these seemingly important conspecific odors to determine what they are, how they function, and if they might be considered pheromones.

In summary, intriguing evidence is emerging that coral reef fishes use a variety of conspecific odors that should be considered to be pheromones, but the identities and precise functions of which are a complete mystery. Nevertheless, because many are related to other migratory species and share similar physiologies and ecologies, it seems reasonable to hypothesize that similar types of complex odor cues might be involved.

2.3.6   Summary of Migratory Pheromones

An impressive variety of fishes use conspecific odors released by larvae and young to guide various aspects of their migrations in their chemical complexity and perhaps even components. This makes sense because of the large quantity of specific information that chemosensory stimuli convey, their innate relevance, and their tendency to disperse great distances. Unfortunately, very few migratory fish pheromone systems have been subjected to rigorous biochemical and behavioral analysis. All of the dozen or so that have been studied suggest common use of complex mixtures that may include bile acids and polar components. Some of these appear to be context dependent. While some are clearly species-specific, others are not (e.g., sea lamprey), but all are species-wide. In addition, while some migratory pheromones are clearly innately recognized, it appears that sensitivity to some may be influenced by developmental processes which may be highly prescribed (e.g., salmonids). Some are also critically important to survival (e.g., sea lamprey), whereas others are merely helpful (e.g., galaxiids). How this broad array of pheromones fits into the broader chemical "language" used by these fishes, which includes kin- and sex pheromones, is not clear, but they share biochemical and functional characteristics. The overall picture of migratory pheromones is one of complexity, variety, context, and flexibility as well as importance. To date, evidence suggest these cues vary greatly between groups but are not likely to be novel compounds in any but rather complex mixtures of common products that may at times be learned in highly prescribed manners although innately recognized in others. These cues warrant closer examination from a new perspective. Identifying some will be key to future progress.

2.4   REPRODUCTIVE PHEROMONES

There is strong evidence that most, if not all, reproductively mature fishes rely on sex pheromones to synchronize male–female physiologies and behaviors (Sorensen and Stacey, 1999; Stacey and Sorensen, 2009; Stacey, 2015). So important are these odors that many species experience severe reproductive impairment if their olfactory systems are blocked (Lastein, Hmadani, and Døving, Stacey, 2015). Further, where examined, most natural sex pheromones (the entire odor actually released by mature fish) have generally been found to be species-specific (Liley, 1982; Sorensen and Stacey, 1999; Stacey, Van Der Kraak, and Olsén, 2012). Nevertheless, it is important to appreciate that context and timing/location of reproductive activity can also serve as species-isolation mechanisms (see Sorensen and Stacey, 1999; Stacey, 2014). Real-world tests of sex pheromone function, specificity, and completeness require head-to-head tests in the field, which are rare.

Many compounds with sex pheromonal activity in fishes have now been definitively identified; and with two exceptions, bile acids in sea lamprey (Li *et al.*, 2002) and an amino acid in masu salmon (Yambe *et al.*, 2006) are all relatively common (seemingly unspecialized) metabolic products (Sorensen and Stacey, 1999; Stacey and Sorensen, 2009; Sorensen and Hoye, 2010; Stacey, 2015). In particular, several dozen hormonal products are now known to be detected with extreme sensitivity and specificity by the olfactory systems of dozens of teleost fishes. A notable subset of these products has also been shown to be released and/or to strongly evoke reproductive responses in aquaria (Sorensen and Stacey, 2009; Stacey, 2015). Some of these compounds have priming (endocrinological) effects and others have releasing (behavioral) effects (Stacey, 2015). Yet all are found in numerous species, do not seem to be specialized, and although signal identity may be based on mixture composition (blends) in insects (Sorensen, Christensen, and Stacey, 1998), mixtures only appear to modulate signal strength in fish (Sorensen, Stacey, and Chamberlain, 1988; Sorensen, Pinillos, and Scott, 1995; Stacey, 2015). Many questions about fish sex pheromone function and identity remain. In particular, how can such common products be species-specific? Do they represent entire pheromones or simply parts of them? Herein, we review behavioral data for the few species of mature fishes for which the specificity of natural (complete) sex pheromones has been examined directly using preference tests. A review of our understanding of hormonal pheromones that focuses on synthesized odorants may be found in Stacey (2015); therefore, we focus here on complete (natural) odors.

2.4.1   Overview of Empirical Data on Sex Pheromone Specificity

Species-specificity has not been commonly tested for reproductive pheromones, but where the role of whole fish odors has been examined, specificity is indicated although not always at the species level. This may not be surprising from an evolutionary perspective; but in light of the fact that all identified fish sex pheromones appear to be common metabolites (Sorensen and Hoye, 2010), it is a nevertheless puzzling. However, our knowledge is very fragmentary; although a great deal is known about the olfactory sensitivity of fish to isolated hormonal components in the lab, we know very little about the roles individual compounds serve within natural odors (i.e., there have been few tests to compare component(s) versus the entire natural odor that might have dozens of compounds). Critical review shows that only about a dozen fish odors have been tested and almost none in the field where competition with natural odors illuminates imperfections. No synthesized component(s) has yet been found to have the full behavioral activity of natural pheromones in the field (Johnson and Li, 2010). Herein, we review the data for specificity of the natural pheromonal odors that fish normally release and respond to.

Species-specificity has been examined in half a dozen pheromones naturally released by fish, with the goldfish/carp being one of the best understood and showing specificity. One of the first studies to explicitly test the species specificity of naturally released behaviorally active sex pheromones examined Japan's native bitterlings, species of minnow. Honda (1982) compared behavioral responses of the rose bitterling, *Rhodeus ocellatus*, and the slender bitterling, *Acheliognatus lanceolatus*, and found that males distinguished between the odors of two females in a Y-maze. Intriguingly, these species are now known to detect common F prostaglandins (PGFs), which play a key role in the behavioral activity of the ovulatory releaser but are also common to many fishes (Stacey, 2015). Subsequent studies of two Malaysian gouramis, *Trichogaster trichopterus*, and *T. perctoralis,* showed that, although one species discerned between the odors of ovulated females of both species, the other species, collected from a population that lives in relative isolation, did not (Mckinnon and Liley, 1987). Similar findings of asymmetrical species specificity have also been described for live-bearing swordtails, *Xiphophorus* sp. (Bong, Fisher, and Rosenthal, 2005). Tests of ovulatory European cyprinid odor indicate it to have species-specific priming effects (Stacey, Van Der Kraak, and Olsén, 2012). Unfortunately, although we know many of these fishes such as the European cyprinids detect hormonal products, no biochemical data are available on their entire cue output; therefore, the basis of their preferences and specificity cannot be addressed. Fortunately, this is not true for the carps, which we discuss below.

2.4.2   Releasing Sex Pheromones in the Goldfish and Common Carp

The goldfish and its sister species, the common carp, appear to be the only species pair for which both behavioral and the biochemical bases of species specificity have been characterized. Both species use hormonal pheromones and their hormones and pheromones are nearly identical (Kobayashi, Sorensen, and Stacey, 2002; Stacey, 2015). In both instances, many of their key reproductive hormones are also released directly to the water where some, but not always the most prevalent ones (i.e., production seems unspecialized), serve as sex pheromones (Sorensen and Scott, 1994; Kobayashi, Sorensen, and Stacey, 2002). These species have become models for understanding how relatively common sets of hormonal products might function as potent, species-specific reproductive cues.

Briefly, both carp and goldfish females are seasonal spawners that produce and release a changing suite of dozens of steroid hormone metabolites and PGFs to the water in the weeks surrounding final maturation and spawning (Stacey and Sorensen, 2009). About a dozen of these are specifically detected and discriminated by conspecific (and heterospecific) olfactory systems and used as potent sex pheromones to drive both endocrinological and behavioral synchrony (Kobayashi, Sorensen, and Stacey, 2002; Stacey and Sorensen, 2009; Stacey, 2015). Here, we review two of these hormonal pheromones for which species specificity has been examined: the PGF female spawning pheromone and the androgen-based male sex pheromone (species specificity of the goldfish priming pheromone has not been explicitly addressed)..

At the time of ovulation, female common carp and female goldfish become sexually active because of elevated levels of circulating $PGF_{2\alpha}$ that is also released in their urine along with several PGF metabolites. Males then detect them and use these PGFs as sex pheromones (Sorensen *et al.*, 1988; Kobayashi, Sorensen, and Stacey, 2002; Stacey, 2015). The olfactory sensitivity of male goldfish and carp to PGFs is acute and seemingly identical to each other, and both species will respond with sexual arousal when PGFs are added to tanks of males (Sorensen *et al.*, 1988; Lim and Sorensen, 2011). However, when either male goldfish or common carp (Lim and Sorensen, 2012) are offered the choice of the natural odor of ovulated conspecifics versus that of heterospecifics in mazes, they readily choose the former; their complete, natural pheromones are actually species-specific. Attempts to explain this using chemical fractionation have shown that although females of these species release slightly different profiles of PGFs, PGF ratio is surprisingly unimportant. Rather, unidentified polar compounds are responsible for species-identification (Lim and Sorensen, 2011). Thus, the natural PGF pheromone appears to be a complex species-specific odor comprised of PGFs as well as various, still unknown polar body metabolites. We have tentatively called such multi-component pheromones 'complexes.'

More recent, as yet unpublished, studies of the goldfish male sex pheromone have added to this story. Briefly, mature male goldfish and carp release an androgenic steroid, androstenedione (AD), to the water where males and females encounter it. Although the full function of the male pheromone is not known, we do know that ovulated female goldfish are attracted to the odor of males and then increase urinary PGF release (Appelt and Sorensen, 2007). To test whether AD might be the male cue, and how it could be specific, we have added AD to aquaria containing females in different contexts. While AD is only weakly attractive to sexually receptive females when added on its own, when added within the context of immature conspecific odor, it is highly attractive (Fig. 2.5).

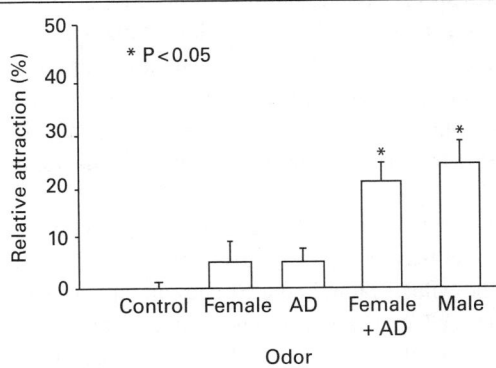

Figure 2.5. Behavioral preferences of sexually receptive female goldfish to the odor of conspecific females, androstenedione (AD) alone (a male hormonal pheromone), AD in combination with female conspecific odor, and the odor of male goldfish; there is synergism between female odors and AD, suggesting that the pheromone normally functions as a "complex" (Lavesque and Sorensen, unpublished data).

Notably, AD is also known to function within the changing mixture of gonadal sex steroids released by preovulatory females (Poling, Fraser, and Sorensen, 2001; Stacey, 2015, Sorensen, unpublished data); once again, odor context is important. We tentatively conclude that the whole (natural) male sex pheromone is a complex that includes the body metabolites earlier identified by Sisler and Sorensen (2008) and which Levesque *et al.* (2011) also suggested to function as part of the species-specific aggregation pheromone (see Section 2.2). This makes sense because it is parsimonious for a fish to develop the ability to discriminate a single (albeit changing) suite of odorants than for it to simultaneously evolve the ability to produce entirely new and highly specialized species-specific products at each life history stage. It also would not appear to be parsimonious for all 30,000 species of fish to have evolved their own completely novel pheromones, and subsets thereof.

2.4.3   Summary for Sex Pheromones

To attract mates, most fishes use species-specific sex pheromones that contain common sex hormone products as well as other metabolites. These sets of cues appear to commonly used by all members of each species, or be species-wide. While mixtures of their components often appear to convey species identity, this is not always true for species that have evolved or function in isolation on their own: species-specificity may be an evolved trait and not always universal. Interestingly, there are no indications that ratios convey species information, but rather the information lies in mixture composition. Recent studies from goldfish and carp also suggest that information on reproductive condition is conveyed by hormonal products, and information on taxon identity may be conveyed by mixtures of bodily metabolites. Bile and amino acids may have roles in these, but it has not yet been shown. Although the role of learning versus innate recognition for fish sex pheromones has not yet been tested, the latter is suggested by the presence of specialized neural pathways for specific hormonal components (Lastein, Hmadani, and Døving, 2015). Clearly, the full complexity and specificity of sex pheromones are as yet poorly understood but appear to best be described as variable and multi-component odors, rather than comprised of a few highly specialized components that always discerned by inflexible neural machinery..

2.5   SUMMARY AND SUGGESTIONS FOR A POSSIBLE UNIFYING THEORY

This chapter has reviewed a wide variety of evidence demonstrating that many species of fish recognize taxon- and often species-specific odors at different stages of their lives for the purposes of shoaling/schooling, migratory orientation, and reproduction. Although the chemical basis of these pheromones has only been examined in a few species, mixtures of common body metabolites that include polar and nonpolar bile acids are implicated in all. Changing mixtures of relatively common hormonal products have also been shown to play key roles in sex pheromones. Bile acids are also implicated in many; but aside from the sea lamprey, definitive behavioral proof of their pheromonal function is still lacking.

Notably, field tests show that no naturally occurring behaviorally active fish pheromone (the complete natural odor) has yet been completely identified (found to function normally in competition with natural cues in the wild). Natural context and mixture composition seem integral to full function. Recent tests using sea lamprey and common carp, the only species for which detailed biochemical profiles have been complied, show that laboratory studies using synthesized cues resulted in overly simplified portrayals of complex pheromonal aromas. This result was especially surprising for the lampreys are ancient and theoretically simple. In the goldfish and common carp, the same complex of cues that attract immature conspecifics also seems to continue to provide context for hormonal sex pheromones. Traditional views that pheromones comprise a few highly specialized products have not been borne out, at least in these studies.

We hypothesize that mixtures of relatively common metabolic products may frequently function as species-specific pheromones in fishes and that the compositions of these mixtures can shift with maturational state and scenario. We tentatively term these odors "pheromonal complexes" and recognize that discrimination of these mixtures may be situation specific and influenced by developmental processes. This appears to describe a previously undescribed type of pheromone, but it appears to make sense for diverse aquatic organisms that evolved to be flexible. By using this type of a pheromonal odor, fish need not have evolved means of producing specialized unique products for each species and each situation but rather the simple ability to recognize small changes in the release patterns of common products. Notably, the neural machinery underlying odor discrimination in fishes is sophisticated enough to permit this solution (Lastein, Hmadani, and Døving, 2015) and the variety of odorants high. We strongly encourage further research into these possibilities. Innovative integrative approaches that include field assays, definitive biochemical analysis, and neural measures of brain function are needed. The future looks very interesting.

REFERENCES

Almany, G.R., Berumen, M.L., Thorrold, S.R. *et al.* (2007) Local replenishment of coral reef fish populations in a marine reserve. *Science*, **316**, 742–744.

Appelt, C.A. and Sorensen, P.W. (2007) Female goldfish signal spawning readiness by altering when and where they release a urinary pheromone. *Animal Behaviour*, **74**, 1329–1338.

Arvedlund, M. and Takemura, A. (2006) The importance of chemical environmental cues for juvenile *Lethrinus nebulosus* Forsskål (Lethrinidae, Teleostei) when settling into their first benthic habitat. *Journal of Experimental Marine Biology and Ecology*, **338**, 112–122.

Atema, J., Kingsford, M.J., and Gerlach, G. (2002) Larval reef fish could use odour for detection, retention and orientation to reefs. *Marine Ecology Progress Series*, **241**, 151–160.

Baker, C.F. (2003) Effect of adult pheromones on the avoidance of suspended sediment by migratory banded kokopu (*Galaxias fasciatus*) juveniles. *Journal of Fish Biology*, **62**, 386–394.

Baker, C.F. and Hicks, B.J. (2003) Response of migratory inanga (*Galaxias maculatus*) and koaro (*Galaxias brevipinnis*) juveniles to adult galaxiid odours. *New Zealand Journal of Marine and Freshwater Research*, **37**, 291–299.

Baker, C.F. and Montgomery, J.C. (2001) Species-specific attraction of migratory banded kokopu (*Galaxias fasciatus*) juveniles to adult pheromones. *Journal of Fish Biology*, **58**, 1221–1229.

Baker, C.F., Carton, F.A.G., Fine, J.M., and Sorensen, P.W. (2006) Can bile acids function as migratory pheromones in banded kokopu *Galaxius fasciatus* (Gray)? *Ecology of Freshwater Fishes*, **15**, 275–283.

Baker, C.F., Stewart, M., Fine, J., and Sorensen, P. (2009) Partial evolutionary divergence of a migratory pheromone between northern and southern hemisphere lampreys. *Challenges for Diadromous Fishes in a Dynamic Global Environment. American Fisheries Society Symposium*, **69**, 845–846.

Black, G.A. and Dempson, J.B. (1986) A test of the hypothesis of pheromone attraction in salmonid migration. *Environmental Biology of Fishes*, **15**, 229–235.

Bong, B.B.M, Fisher, H.S., and Rosenthal, G.G. (2005) Species recognition by male swordtails. *Behavioral Ecology*, **16**, 818–822.

Booth, D.J. (1992) Larval settlement patterns and preferences by domino damselfish *Dascyllus albisella* Gill. *Journal of Experimental Marine Biology and Ecology*, **155**, 85–104.

Corkrum, L.D. and Coglaiti, K.M. (2015) Conspecific odors as sexual ornaments with dual functions in fishes, in *Fish Pheromones and Related Cues* (eds Peter W. Sorensen and Brian D. Wisenden), John Wiley & Sons, Inc., Hoboken.

Creutzberg, F. (1961) On the orientation of migrating elvers (*Anguilla vulgaris* Turt.) in a tidal area. *Netherlands Journal of Sea Research*, **1**, 257–338.

Dittman, A. and Quinn, T. (1996) Homing in Pacific salmon: mechanisms and ecological basis. *Journal of Experimental Biology*, **199**, 83–91.

Dixson, D.L. *et al.* (2008) Coral reef fish smell leaves to find island homes. *Proceedings of the Royal Society B: Biological Sciences*, **275**, 2831–2839.

Døving, K.B., Nordeng, H., and Oakley, B. (1974) Single unit discrimination of fish odours released by char (*Salmo alpinus* L.) populations. *Comparative Biochemistry and Physiology Part A: Physiology*, **47**, 1051–1063.

Døving, K.B., Stabell, O.B., Ostlund-Nilsson, S., and Fisher, R. (2006) Site fidelity and homing in tropical coral reef cardinalfish: are they using olfactory cues? *Chemical Senses*, **31**, 265–272.

Fine, J.M. and Sorensen, P.W. (2008) Isolation and biological activity of the multi-component sea lamprey migratory pheromone and new information in its potency. *Journal of Chemical Ecology*, **34**, 1259–1267.

Fine, J.M., Vrieze L.A., and Sorensen, P.W. (2004) Evidence that petromyzontid lampreys employ a common migratory pheromone that is partially comprised of bile acids. *Journal of Chemical Ecology*, **30**, 2091–2110.

Fisher, R., Bellwood, D.R., and Job, S.D. (2000) The development of swimming abilities in reef fish larvae. *Marine Ecology Progress Series*, **202**, 163–173.

Fisher, R., Leis, J.M., Clark, D.L., and Wilson, S.K. (2005) Critical swimming speeds of late-stage coral reef fish larvae: variation within species, among species and between locations. *Marine Biology*, **147**, 1201–1212.

Foster, N.R. (1985) Lake trout reproductive behavior: influence of chemosensory cues from young-of-the-year by-products. *Transactions of the American Fisheries Society*, **114**, 794–803.

Gaudron, S.M. and Lucas, M.C. (2006) First evidence of attraction of adult river lamprey in the migratory phase to larval odour. *Journal of Fish Biology*, **68**, 640–644.

Gerlach, G., Atema, J., Kingsford, M.J. *et al.* (2007) Smelling home can prevent dispersal of reef fish larvae. *Proceedings of the National Academy of Science USA*, **16**, 858–863.

Hale, R., Swearer, S.E., and Downes, B.J. (2009) Separating natural responses from experimental artefacts: habitat selection by a diadromous fish species using odours from conspecifics and natural stream water. *Oecologia*, **159**, 679–687.

Hara, T.J, Macdonald, S., Evans, R.E. *et al.* (1984) Morpholine, bile acids and skin mucous as possible chemical cues in salmonid homing: electrophysiological evaluation, in *Mechanisms of Migration in Fishes* (eds J.D. McCleave, G.P. Arnold, J.J. Dodson, and W.H. Neill), Plenum, New York, pp. 363–378.

Hasler, A.D. and Scholz, A.T. (1983) *Olfactory Imprinting and Homing in Salmon: Investigations into the Imprinting Process*, Springer, New York.

Hasler, A.D. and Wisby, W.J. (1951) Discrimination of stream odours by fishes and relation to parent stream behaviour. *American Naturalist*, **85**, 223–238.

Hemmings, C.C. (1966) Olfaction and vision in fish schooling. *Journal of Experimental Biology*, **45**, 449–464.

Honda, H. (1982) On the female sex pheromones and courtship behaviour in the bitterlings Rhodeus ocellatus ocellatus and Acheilognathus lanceolatus. *Bulletin of the Japanese Society of Scientific Fisheries (Japan)*.

Hoglund, L.B. and Astrand, M. (1973) Preferences among juvenile char (*Salvelinus alpinus* L.) to intraspecific odours and water currents studied with the fluvarium technique. *Institute Freshwater Research Drottingholm*, **53**, 21–20.

Hubbard, P. (2015) Pheromones in marine fish with comments on their possible use in aquaculture, in *Fish Pheromones and Related Cues*, (eds Peter W. Sorensen and Brian D. Wisenden), John Wiley & Sons, Inc., Hoboken.

Idler, D.R., McBride, J.R., Jonas, R.E., and Tomlison, N. (1961) Olfactory perception in migrating salmon: II studies of laboratory bio-assay for homestream water and mammalian repellant. *Canadian Journal of Biochemistry and Physiology*, **39**, 1575–1584.

Jellyman, D.J., Glova, G.J., and Sykes, J.R.E. (2002) Movements and habitats of adult lamprey (*Geotria australis*) in two New Zealand waterways. *New Zealand Journal of Marine and Freshwater Research*, **36**, 53–65.

Job, S.D. and Bellwood, D.R. (2000) Light sensitivity in larval fishes: implications for vertical zonation in the pelagic zone. *Limnology and Oceanography*, **45**, 362–371.

Johnson, N.S. and Li, W. (2010) Understanding behavioral responses of fish to pheromones in natural freshwater environments. *Journal of Comparative Physiology A*, **96**, 701–711.

Johnson, N.S., Yun, S.-S., Thompson, H.T. *et al.* (2009) A synthesized pheromone induces upstream movement in female sea lamprey and summons them into traps. *Proceedings of the National Academy of Science USA*, **106**, 1021–1026.

Johnson, N.S., Yun, S.-S., Buchinger, T.J. *et al.* (2012) Multiple functions of a multi-component mating pheromone in sea lamprey *Petromyzon marinus*. *Journal of Fish Biology*, **80**, 538–554.

Kinosita, H. (1972) Schooling behavior in marine catfish eel *Plotosus anguillaris*. *Zoological Magazine*, **81**, 241.

Kobayashi, M., Sorensen, P.W, and Stacey, N.E. (2002) Hormonal and pheromonal control of spawning in goldfish. *Fish Physiology and Biochemistry*, **26**, 71–84.

Lastein, S., Hmadani, E.H., and Døving, K.B. (2015) Olfactory discrimination of pheromones, in *Fish Pheromones and Related Cues* (eds Peter W. Sorensen and Brian D. Wisenden), John Wiley & Sons, Inc., Hoboken.

Lecchini, D. (2004) Experimental assessment of sensory abilities of coral reef fish larvae in the detection of their settlement location. *Comptes Rendus Biologies*, **327**, 159–171.

Lecchini, D., Shima, J., Banaigs, B., and Galzin R. (2005) Larval sensory abilities and mechanisms of habitat selection of a coral reef fish during settlement. *Oecologia*, **143**, 326–334.

Levesque, H., Scaffidi, D., Polkinghorne, C.A., and Sorensen, P.W. (2011) A multi-component species identifying pheromone in the goldfish. *Journal of Chemical Ecology*, **37**, 219–227.

Li, W., Scott, A.P., Seifkes, M. *et al.* (2002) Bile acid secreted by male sea lamprey that acts as a sex pheromone. *Science*, **296**, 138–141.

Liley, N.R. (1982) Chemical communication in fish. *Canadian Journal of Fisheries and Aquatic Science*, **39**, 22–35.

Lim, H.K. and Sorensen, P.W. (2011) Polar metabolites synergize the activity of prostaglandin $F_{2\alpha}$ in a species-specific hormonal sex pheromone released by ovulated common carp. *Journal of Chemical Ecology*, **37**, 695–704.

Lim, H.K. and Sorensen, P.W. (2012) Common carp implanted with prostaglandin $F_{2\alpha}$ release a sex pheromone complex that attracts conspecific males in both the laboratory and field. *Journal of Chemical Ecology*, **38**, 127–134.

McDowall, R.M. (1988) *Diadromy in Fishes: Migrations between Freshwater and Marine Environments*, Timber Press, Portland.

McDowall, R.M. (1990) *New Zealand Freshwater Fishes. A Natural History and Guide*, Heinemann Reed, Auckland.

McKeown, B.A. (1984) *Fish Migration*, Timber Press, Portland.

McKinnon, J.S. and Liley, N.R. (1987) Asymmetric species specificity in responses to female sexual pheromone by males of two species of *Trichogaster* (Pisces: *Belontiidae*). *Canadian Journal of Zoology*, **65**, 1129–1134.

Meckley, T.D., Wagner, C.M., and Luehring, M.A. (2012) Field evaluation of larval odor and mixtures of synthetic pheromone components for attracting migrating sea lampreys in rivers. *Journal of Chemical Ecology*, **38**, 1062–1069.

Miles, S.G. (1968) Rheotaxis of elvers of the American eel (*Anguilla rostrata*) in the laboratory to water from different streams in Nova Scotia. *Journal of the Fisheries Board of Canada*, **25**, 1591–1602.

Montgomery, J.C., Tolimieri, N., and Haine, O.S. (2001) Active habitat selection by pre-settlement reef fishes. *Fish and Fisheries*, **2**, 261–277.

Moore, A., Ives, M.J., and Kell, L.T. (1994) The role of urine in sibling recognition in Atlantic salmon *Salmo salar* (L.) parr. *Proceedings of the Royal Society of London. Series B: Biological Sciences*, **255**, 173–180.

Moser, M.L., Almeida, P.R., Kemp, P.S., and Sorensen, P.W. (2014) Lamprey spawning migration, in *Lampreys: Biology, Conservation and Control*, Fish and Fisheries Series (ed. M.F. Docker), Springer.

Nordeng, H. (1971) Is the local orientation of anadromous fishes determined by pheromones? *Nature*, **233**, 411–413.

Nordeng, H. (1977) A pheromone hypothesis for homeward migration in anadromous salmonids. *Oikos*, **28**, 155–159.

Nordeng, H. and Bratland, P. (2006) Homing experiments with parr, smolt and residents of anadromous Arctic char *Salvelinus alpinus* and brown trout *Salmo trutta*: transplantation between neighbouring river systems. *Ecology of Freshwater Fish*, **15**, 488–499.

Olsén, K.H., Grahn, M., and Lohm, J. (2002) Influence of MHC on sibling discrimination in Arctic char, *Salvelinus alpinus* (L.). *Journal of Chemical Ecology*, **28**, 783–795.

Pitcher, T. and Parish, J. (1993) Functions of shoaling behaviors in teleosts, in *Behaviour of Teleost Fishes 2*, (ed. T.J. Pitcher), Springer, London, pp. 369–439.

Poling, K.R., Fraser, E.J., and Sorensen, P.W. (2001) The three steroidal components of the goldfish pre-ovulatory pheromone signal evoke different behaviors in males. *Comparative Biochemistry and Physiology. Part B*, **129**, 645–651.

Quinn, T.P. and Tolson, G.M. (1986) Evidence of chemically mediated population recognition in coho salmon (*Oncorhynchus kisutch*). *Canadian Journal of Zoology*, **64**, 84–87.

Quinn, T.P., Brannon, E.L., and Whitman, R.P. (1983) Pheromones and the water source preferences of adult coho salmon *Oncorhynchus kisutch* Walbaum. *Journal of Fish Biology*, **22**, 667–684.

Robinson, T.C., Sorensen, P.W., Bayer, J.M., and Seelye, J.G. (2009) Olfactory sensitivity of Pacific lampreys to a lamprey bile acid. *Transactions of the American Fisheries Society*, **138**, 144–152.

Rowe, D.K., Saxton, B.A., and Stancliff, A.G. (1992) Species composition of whitebait (Galaxiidae) fisheries in 12 Bay of Plenty rivers, New Zealand: evidence for river mouth selection by juvenile *Galaxias brevipinnis* (Gunther). *New Zealand Journal of Marine and Freshwater Research*, **26**, 219–228.

Scholz, A.T., Horall, R.M., Cooper, J.C., and Hasler, A.T. (1976) Imprinting to chemical cues: the basis for homestream selection in salmon. *Science*, **192**, 1247–1249.

Selset, R. (1980) Chemical methods for fractionation of odorants produced by char smolts and tentative suggestions for pheromone origins. *Acta Physiologica Scandinavica*, **108**, 97–103.

Selset, R. and Døving, K.B. (1980) Behaviour of mature anadromous char (*Salmo alpinus* L.) towards odorants produced by smolts of their own population. *Acta Physiologica Scandinavica*, **108**, 113–122.

Sisler, S.P. and Sorensen, P.W. (2008) Common carp and goldfish discern conspecific identity using chemical cues. *Behaviour*, **145**, 1409–1429.

Sola, C. and Tongiorgi, P. (1998) Behavioural responses of glass eels of *Anguilla anguilla* to non-protein amino acids. *Journal of Fish Biology*, **53**, 1253–1262.

Sola, C. and Tosi, L. (1993) Bile salts and taurine as chemical stimuli for glass eels, *Anguilla anguilla*: a behavioural study. *Environmental Biology of Fishes*, **37**, 197–204.

Sorensen, P.W. (1986) Origins of the freshwater attractant(s) of migrating elvers of the American eel, *Anguilla rostrata* (LeSueur). *Journal of the Environmental Biology of Fishes*, **17**, 185–200.

Sorensen, P.W. (2015) Applications of pheromones in invasive fish control and fisheries conservation, in *Fish Pheromones and Related Cues* (eds Peter W. Sorensen and Brian D. Wisenden), John Wiley & Sons, Inc., Hoboken.

Sorensen, P.W. and Hobson, K.A. (2005) Stable isotope analysis of amphidromous Hawaiian gobies suggest their larvae spend a substantial amount of time in freshwater river plumes. *Environmental Biology of Fishes*, **74**, 31–42.

Sorensen, P.W. and Hoye, T.R. (2007) A critical review of the discovery and application of a migratory pheromone in an invasive fish, the sea lamprey, *Petromyzon marinus* L. *Journal of Fish Biology*, **71** (Supplement D), 100–114.

Sorensen, P.W. and Hoye, T.R. (2010) Pheromones in vertebrates, in *Comprehensive Natural Products Chemistry II*, vol. 4 (eds L. Mander and H.-W. Lui), Elsevier, Oxford, pp. 225–262.

Sorensen, P.W. and Scott, A.P. (1994) The evolution of hormonal sex pheromones in teleost fish: poor correlation between the pattern of steroid release by goldfish and olfactory sensitivity suggests that these cues evolved as a result of chemical spying rather than signal specialization. *Acta Scandinavia Physiologica*, **152**, 191–205.

Sorensen, P.W. and Stacey, N.E. (1999) Evolution and specialization in fish hormonal pheromones, in *Advances in Chemical Signals in Vertebrates* (eds R.E. Johnston, D. Müller-Schwarze, and P.W. Sorensen), Plenum Press, New York, pp. 15–48.

Sorensen, P.W. and Vrieze, L.A. (2003) Chemical ecology and application of the sea lamprey migratory pheromone. *Journal of Great Lakes Research*, **29** (Supplement 1), 66–84.

Sorensen, P.W. and Winn, H.E. (1984) The induction of maturation and ovulation in American eels and the relevance of chemical and visual cues to their spawning behavior. *Journal of Fish Biology*, **25**, 261–268.

Sorensen, P.W., Hara, T.J., Stacey, N.E., and Goetz, F.W. (1988) F prostaglandins function as potent olfactory stimulants that comprise the postovulatory female sex pheromone in goldfish. *Biology of Reproduction*, **39**, 1039–1050.

Sorensen, P.W., Stacey, N.E., and Chamberlain, K.J. (1988) Differing behavioral and endocrinological effects of two female sex pheromones on male goldfish. *Hormones and Behavior*, **23**, 317–332.

Sorensen, P.W., Pinillos, M., and Scott, A.P. (1995) Sexually mature male goldfish release large quantities of androstenedione into the water where it functions as a pheromone. *General and Comparative Endocrinology*, **140**, 164–175.

Sorensen, P.W., Christensen, T.A., and Stacey, N.E. (1998) Discrimination of pheromonal cues in fish: emerging parallels with insects. *Current Opinion in Neurobiology*, **8**, 458–467.

Sorensen, P.W., Fine, J.M., Dvornikovs, V. *et al.* (2005) Mixture of new sulfated steroids functions as a migratory pheromone in the sea lamprey. *Nature Chemical Biology*, **1**, 324–328.

Stabell, O.B. (1984) Homing and olfaction in salmonids: a critical review with special reference to the Atlantic salmon. *Biological Reviews*, **59**, 333–388.

Stabell, O.B. (1987) Intraspecific pheromone discrimination and substrate marking by Atlantic salmon parr. *Journal of Chemical Ecology*, **13**, 1625–1643.

Stacey, N.E. (2015) Hormonal-derived pheromones in teleost fishes, in *Fish Pheromones and Related Cues* (eds Peter W. Sorensen and Brian D. Wisenden), John Wiley & Sons, Inc., Hoboken.

Stacey, N.E. and Sorensen, P.W. (2009) *Hormonal pheromones in fish*, in *Hormones, Brain and Behavior*, 2nd edn, vol. 1 (eds D.W. Pfaff, A.P. Arnold, A.M. Etgen, *et al.*), Elsevier Press, San Diego, pp. 639–681.

Stacey, N.E., Van Der Kraak, G.J., and Olsén, K.H. (2012) Male primer endocrine responses to preovulatory female cyprinids under natural conditions in Sweden. *Journal of Fish Biology*, **80**, 147–165.

Stewart, M. and Baker, C.F. (2012) A sensitive analytical method for quantifying petromyzonol sulfate in water as a potential tool for population monitoring of the southern pouched lamprey, *Geotria australis*, in New Zealand streams. *Journal of Chemical Ecology*, **38**, 135–144.

Stobutzki, I.C. and Bellwood, D.R. (1997) Sustained swimming abilities of the late pelagic stages of coral reef fishes. *Marine Ecology Progress Series*, **149**, 35–41.

Sutterlin, A.M. and Gray, R. (1973) Chemical basis for homing of Atlantic salmon (*Salmo salar*) to a hatchery. *Journal of the Fisheries Research Board of Canada*, **30**, 985–989.

Sweatman, H.P.A. (1988) Field evidence that settling coral reef fish larvae detect resident fishes using dissolved chemical cues. *Journal of Experimental Marine Biology and Ecology*, **124**, 163–174.

Tesch, F.W. and White, R.J. (2008) *The Eel*. Wiley-Blackwell, London.

Tolimieri, N., Jeffs, A., and Montgomery, J. (2000) Ambient sound as a cue for navigation in reef fish larvae. *Marine Ecology Progress Series*, **207**, 219–224.

Ueda, H. (2011) Physiological mechanism of homing migration in Pacific salmon from behavioral to molecular biological approaches. *General and Comparative Endocrinology*, **170**, 222–232.

Vrieze, L.A. and Sorensen, P.W. (2001) Laboratory assessment of the role of a larval pheromone and natural stream odor in spawning stream localization by migratory sea lamprey. *Canadian Journal of Fisheries and Aquatic Science*, **58**, 2374–2385.

Vrieze, L.A., Bjerselius, R.K.A., and Sorensen, P.W. (2010) The importance of the olfactory sense to migratory sea lampreys seeking riverine spawning habitat. *Journal of Fish Biology*, **76**, 949–964.

Vrieze, L.A., Bergstedt, R.A., and Sorensen, P.W. (2011) Olfactory-mediated stream finding behavior of migratory adult sea lamprey (*Petromyzon marinus*). *Canadian Journal of Fisheries and Aquatic Science*, **68**, 523–533.

Ward, A.J. (2015) Intraspecific social recognition in fishes via chemical cues, in *Fish Pheromones and Related Cues* (eds Peter W. Sorensen and Brian D. Wisenden), John Wiley & Sons, Inc., Hoboken.

Ward, A.J., Axford, S., and Krause, J. (2002) Mixed-species shoaling in fish: the sensory mechanisms and costs of shoal choice. *Behavioral Ecology and Sociobiology*, **52**, 182–187.

Wyatt, T.D. (2003) *Pheromones and Animal Behavior*. Cambridge University Press, New York.

Wyatt, T.D. (2010) Pheromones and signature mixtures: defining species-wide signals and variable cues for identity in both invertebrates and vertebrates. *Journal of Comparative Physiology A*, **196**, 685–700.

Yamamoto, Y., Hino, H., and Ueda, H. (2010) Olfactory imprinting of amino acids in lacustrine sockeye salmon. *PLoS One*, **5**, e8633.

Yambe, H., Kitamura, S., Kamiio, M. *et al.* (2006) L-Kynurenine, an amino acid identified as a sex pheromone in the urine of ovulated female masu salmon. *Proceedings of the National Academy of Sciences*, **103**, 15370–15374.

Youngson, A.F., Jordan, W.C., and Hay, W.D. (1994) Homing of Atlantic salmon (*Salmo salar* L.) to a tributary spawning stream in a major river catchment. *Aquaculture*, **121**, 259–267.

Zhang, C., Brown, S.B., and Hara, T.J. (2001) Biochemical and physiological evidence that bile acids produced and released by lake char (*Salvelinus namaycush*) function as chemical signals. *Journal of Comparative Physiology B*, **171**, 161–171.

Chapter 3
Hormonally Derived Pheromones in Teleost Fishes

Norm Stacey

University of Alberta, Edmonton, Canada

3.1 INTRODUCTION TO HORMONAL SEX PHEROMONES IN TELEOST FISH

Living in a medium that typically limits visual information but bathes the olfactory organs in a fortuitous array of water-soluble chemicals, fishes throughout their long evolutionary history have had both obvious cause and ample opportunity to evolve adaptive olfactory-mediated responses to conspecific odors. Although teleost fishes are well known to utilize released conspecific odorants as pheromones for a variety of reproductive functions (Liley, 1982; Burnard, Gozlan, and Griffiths, 2008), the majority of fish pheromone studies have focused on characterizing biological responses rather than chemically characterizing pheromonal components. Nonetheless, there is good evidence that a variety of fishes have evolved to detect and respond to gonadal steroids and prostaglandins (and their precursors and metabolites) as sex pheromones, which hereafter will be termed "hormonal pheromones." Undoubtedly, hormonal pheromone systems have arisen because hormone release to the water is so predictably linked to key reproductive processes in conspecifics (e.g., Scott and Ellis, 2007).

Døving (1976) appears to have been the first to propose that fish might be predisposed to evolve pheromonal responsiveness to released hormones and related products. Døving's hypothesis proved prescient because steroid glucuronides were soon reported to have pheromonal functions in the black goby (*Gobius niger*; Colombo *et al.*, 1980) and zebrafish (*Danio rerio*; Van Den Hurk and Lambert, 1983; reviewed by Van Den Hurk and Resink, 1992). Since then, a number of studies have provided strong evidence that gonadal steroids, prostaglandins, and their precursors and metabolites function as hormonal pheromones with primer (physiological) and releaser (behavioral) effects in fishes from four teleost orders: (1) Siluriformes (African catfish, *Clarias gariepinus*, reviewed by Van Den Hurk and Resink, 1992), (2) Salmoniformes (Atlantic salmon, *Salmo salar*, Moore and Waring, 1996: brown trout, *Salmo trutta*, Moore *et al.*, 2002: lake whitefish, *Coregonus clupeaformis*, Laberge and Hara, 2003: arctic char, *Salvelinus alpinus*, Sveinsson and Hara, 2000), (3) Perciformes (round goby, *Neogobius melanostomus*, Murphy, Stacey, and Corkum, 2001; Belanger *et al.*, 2006), and (4) Cypriniformes (goldfish, *Carassius auratus*, reviewed by Kobayashi, Sorensen, and Stacey, 2002; Stacey and Sorensen, 2005, 2009, and Stacey, 2010): crucian carp, *C. carassius*, Bjerselius, Olsén, and Zheng, 1995 and Olsén, Sawisky, and Stacey, 2006: common carp, *Cyprinus carpio*, Irvine and

Fish Pheromones and Related Cues, First Edition. Edited by Peter W. Sorensen and Brian D. Wisenden.
© 2015 John Wiley & Sons, Inc. Published 2015 by John Wiley & Sons, Inc.

Sorensen, 1993; Stacey *et al.*, 1994; Lim and Sorensen, 2012: oriental weatherfish loach, *Misgurnus anguillicaudatus*, Kitamura, Ogata, and Takashima, 1994a; tinfoil barbs, *Barbonymus* spp., Cardwell *et al.*, 1995). However, only for the goldfish do we have an interrelated series of studies showing (1) how hormonal pheromones are produced and released, (2) the olfactory sensitivities and specificities with which they are detected, and (3) the physiological mechanisms through which they induce their critical biological responses in conspecifics. This chapter therefore devotes considerable attention to the hormonal pheromones of goldfish, after first briefly discussing these three important aspects of hormonal pheromone studies. This chapter does not review either the role of nonsexual steroids, or the sea lamprey (*Petromyzon marinus*) which is both a distant relative of teleost fish and has been found to employ nonhormonal sulfated biliary sterols as sex and migratory pheromones (Li *et al.*, 2002; Sorensen *et al.*, 2005).

3.1.1 Production and Release of Hormonal Pheromones

Key reproductive events in the lives of fish (gamete maturation, ovulation, spermiation, and spawning) are predictably correlated with changes in blood levels of two classes of reproductive hormones that are detected by the olfactory systems of many fishes: F-series prostaglandins (PGFs; Fig. 3.1) and gonadal (i.e., sex) steroids (Fig. 3.2; see the figure legend for full steroid names that are abbreviated in the text). A single report that gonadotropin-releasing hormones (GnRHs) are potent olfactory stimulants in rainbow trout (Andersen and Døving, 1991) has not been confirmed in further studies (Sorensen, unpublished results). Although there is little information about the metabolism and release of PGFs in fish, numerous studies have demonstrated that changes in plasma steroid concentrations are closely correlated with changes in the release of steroids and steroid conjugates to the water; indeed, steroid measurement in holding water is being used as a noninvasive technique to monitor physiological processes in fishes (for review, see Scott and Ellis, 2007).

The range of PGFs known to be detected by fish is presently quite limited (Fig. 3.1): prostaglandin $F_{2\alpha}$ (PGF$_{2\alpha}$) and two of its immediate metabolites, 15-keto-prostaglandin $F_{2\alpha}$ (15k-PGF$_{2\alpha}$; Sorensen

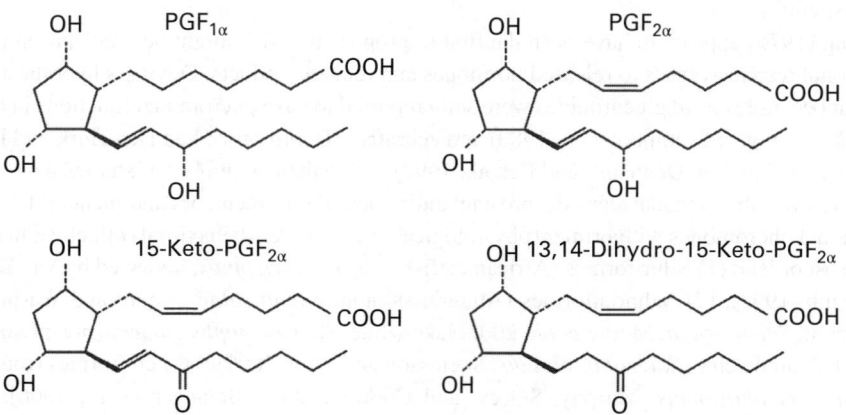

Figure 3.1. F-series prostaglandins that electro-olfactogram (EOG)-recording studies and bioassay studies indicate likely function as hormonal pheromones. prostaglandin $F_{1\alpha}$ (PGF$_{1\alpha}$) [Moore and Waring 1996]; prostaglandin $F_{2\alpha}$ (PGF$_{2\alpha}$), and 15-keto-PGF$_{2\alpha}$ (15-k-PGF$_{2\alpha}$) [goldfish; Sorensen *et al.* 1988]; 13,14-dihydro-15-k-PGF$_{2\alpha}$ [Ogata *et al.* 1994; Kitamura, Ogata, and Takashima, 1994a].

et al., 1988) and 13,14-dihydro-15-keto-PGF$_{2\alpha}$ (13,14-15k-PGF$_{2\alpha}$; Kitamura, Ogata, and Takashima, 1994a), and PGF$_{1\alpha}$ (Moore and Waring, 1996). At least in cypriniforms such as the goldfish, PGF$_{2\alpha}$ is synthesized in large quantity by unknown tissues when ovulated eggs enter the oviduct; is carried in the blood to the brain where it acts rapidly as a hormone, inducing female spawning behavior (Stacey, 1976; Stacey and Peter, 1979); and, during this time, is released with metabolites such as 15k-PGF$_{2\alpha}$ to act as a pheromone that attracts and stimulates males (Sorensen *et al.*, 1988). Because PGFs stimulate female sexual behavior in a variety of fishes (for review, see Stacey and Sorensen, 2005), it seems likely that this hormonal role evolved first, and that pheromonal roles for released PGFs then evolved in mating systems such as that used by goldfish, where chemical cues enhance males' ability to locate ovulated females in dimly lit and chaotic spawning aggregations. A marked departure from this scenario is found in arctic char (*S. alpinus*), where it is reported that females are attracted to PGF released by males (Sveinsson and Hara, 2000).

Fish detect three general categories of sex steroid odorants (reduced biliary sterols are not considered here, but they are potent in many fishes [see Lastein, Hamdani, and Døving, 2014): 18-carbon estrogens (Murphy, Stacey, and Corkum, 2001), 19-carbon steroids that include androgens (Cardwell *et al.*, 1995), and 21-carbon compounds that include the maturation-inducing steroids (MISs) (Sorensen *et al.*, 1987; Nagahama *et al.*, 1994) (Fig. 3.2). Estrogens such as 17β-estradiol (E2) and estrone (E1) increase in female plasma during vitellogenesis (Patiño and Sullivan, 2002). 11-ketotestosterone (11-KT) is generally agreed to be the key fish androgen (Borg, 1994; Pankhurst, 2008), although related precursors such as androstenedione (AD), testosterone (T), and 11β-hydroxytestosterone also are elevated in the plasma during spermatogenesis and testicular maturation. Released androgenic and estrogenic steroids should provide conspecifics with valuable information on both gender and reproductive status, and thus would seem likely candidates for pheromonal function. Indeed, 11-KT is detected by a cyprinid, the tinfoil barb, *Barbonymus schwanenfeldi* (Cardwell *et al.*, 1995), testosterone sulfate (T-s) is detected by the cichlid *Astatotilapia burtoni* (Cole and Stacey, 2006), and E1 and free and conjugated E2 are detected by the round goby (Murphy, Stacey, and Corkum, 2001), *A. burtoni* (Cole and Stacey, 2006), and numerous characiform species (Cardwell and Stacey, 1995; Stacey and Sorensen, 2002). However, only in goldfish (Sorensen, Pinillos, and Scott, 2005; Section 3.2.1) and Atlantic salmon (Moore, 1991; Section 3.3.1) is there compelling evidence for the biological functions of androgenic pheromones.

In female fish the maturation-inducing steroids (MISs) mediate the effect of the preovulatory luteinizing hormone (LH) surge on resumption of meiosis in postvitellogenic oocytes, and therefore typically increase immediately before ovulation and spawning. The key fish MISs appear to be 4-pregnen-17,20β-dihydroxy-4-pregnen-3-one (17,20β-P), and its 21-hydroxylated metabolite 17,20β,21-P (also written as 20β-S although, for clarity, this steroid is abbreviated here as 17,20β,21-P; Fig. 3.2). 17,20β-P appears to be the MIS in cyprinids and salmonids, whereas 17,20β,21-P is likely the MIS for at least some perciforms (Nagahama *et al.*, 1994; Thomas, 2003; Pankhurst, 2008). In male fish, 17,20β-P, 17,20β,21-P, and related free and conjugated 21-carbon steroids also are reported to increase in association with spermiation and spawning in a variety of species (Scott, Sumpter, and Stacey, 2010). 17,20β-P and 17,20β,21-P are often rapidly metabolized, with the result that in the blood plasma they can be at very low concentrations or even undetectable, despite high concentrations of their metabolites (Scott, Sumpter, and Stacey, 2010).

Hormonal compounds such as MISs, androgens, estrogens, and ovulatory prostaglandins are excellent candidates for pheromonal function, primarily because their production and release is reliably linked to reproductive states and key reproductive phenomena of great interest to conspecifics. In addition, however, the steroid hormones and nonhormonal compounds in their hormone synthetic pathways are subject to sulfation and glucuronidation (Fig. 3.2) that greatly increase their water

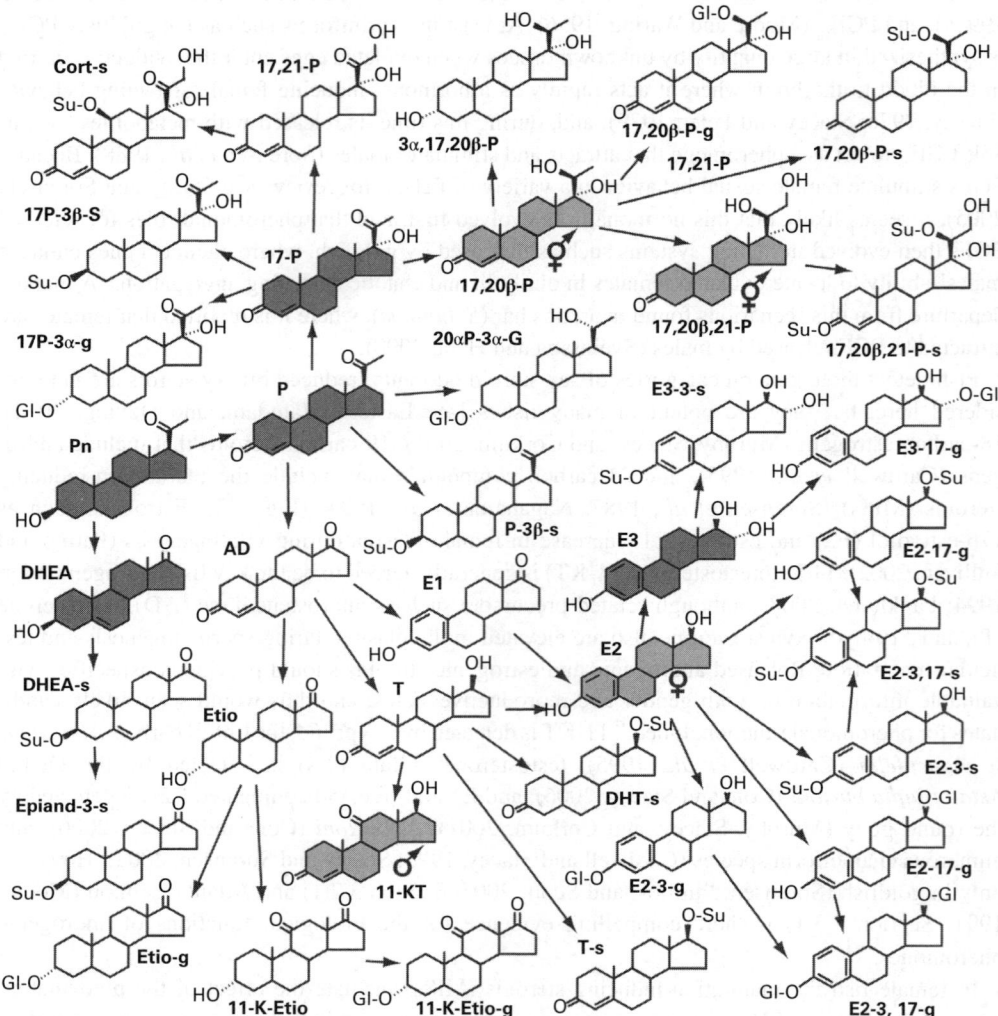

Figure 3.2. Sex steroids and steroid conjugates detected by the fish olfactory system. The figure illustrates some of the steroidal compounds that are potent olfactory odorants in EOG-recording studies, and their possible synthesis pathways. The five compounds with dark grey shading, such as pregnenolone (Pn), progesterone (P), and estriol (E3), have not been reported to be detectable in EOG tests but are shown here as likely intermediates in steroid synthesis pathways. Compounds with light gray shading are common fish sex steroids: 17β-estradiol (E2) is the major fish estrogen, 11-keto-testosterone (11-KT) is the major fish androgen, and 17,20β-P and 17,20β,21-P (chemical names below) appear to be the major maturation-inducing steroids in both males and females. Abbreviations and sources: androstenedione (AD) [goldfish; Sorensen, Pinillos, and Scott, 2005]; cortisol-21-sulfate; 4-pregnen-11β,17,21-triol-3-one-21-sulfate (Cort-s) [cichlid (*Xenotilapia ochrogenys*); Stacey, N.E., unpublished results]; dehydroepiandrosterone-sulfate (DHEA-s); 5-pregnen-3β-ol-17-one-3β-sulfate) [cichlid (*Thoracochromis demeusii*); Stacey, N.E., unpublished results]; 5α-androstan-17β-ol-3-one-17β-sulfate (DHT-s) [cichlid (*Petrochromis ephippium*); Stacey, N.E., unpublished results]; estrone (E1) [round goby; Murphy *et al.* 2001]; 17β-estradiol (E2) [round goby;

solubility, thereby greatly increasing the number of steroidal compounds that could be candidates for pheromonal function and the routes by which they are released (for review, see Scott and Ellis, 2007). Both in rainbow trout (Vermeirssen and Scott, 1996; Ellis, James, and Scott, 2005) and in goldfish (Sorensen *et al.*, 2000), for example, nonconjugated (i.e., free) steroids are released passively across the gills because of the concentration gradient between the blood plasma and the surrounding water. By contrast, the more hydrophilic sulfated and glucuronated steroids are released in the urine and bile, respectively, where at least the urinary steroids are accumulated and released in pulses that can be adjusted in response to social conditions (Appelt and Sorensen, 2007; Barata *et al.*, 2007, 2008). For species that detect both free and conjugated forms of a steroid, the result is that a reproductive event regulated by a single steroid can generate more than one pheromone. In goldfish, for example, the preovulatory surge of 17,20β-P that induces follicular maturation results in release both of free 17,20β-P across the gills and of sulfated 17,20β-P in urine pulses, the free and conjugated forms acting through distinct olfactory receptors and evoking different behavioral responses in males (Poling, Fraser, and Sorensen, 2001; Scott and Sorensen, 1994; Scott and Ellis, 2007; Sorensen *et al.*, 1995, 2000, 2005).

3.1.2 Detection of Hormonal Pheromones

As discussed in detail by Lastein, Hamdani, and Døving, 2014, hormonal pheromones are detected by the olfactory system. The current understanding is that pheromone detection begins when pheromone molecules bind to specific receptor proteins in the cell membrane of olfactory receptor

---

Figure 3.2. (*Continued*) Murphy *et al.* 2001]; 17β-estradiol-3-glucuronide (E2-3-g) [round goby; Murphy *et al.* 2001]; 17β-estradiol-17β-glucuronide (E2-17-Gl) [cichlid (*A. burtoni*); Cole and Stacey 2006]; 17β-estradiol-3,17β-glucuronide (E2-3,17-g) [cichlid (*Oreochromis esculentus*); Stacey, N.E., unpublished results]; 17β-estradiol-3-sulfate (E2-3-s) [round goby; Murphy *et al.* 2001]; 17β-estradiol-17β-sulfate (E2-17-s) [round goby; Murphy *et al.* 2001]; 17β-estradiol-3,17β-disulfate (E2-3,17-s) [round goby; Murphy *et al.* 2001]; estriol-3-sulfate (E3-3-s) [cichlid (*O. stormsi*); Stacey, N.E., unpublished results]; estriol-17β-glucuronide (E3-17-g) [cichlid (*T. demeusii*); Stacey, N.E., unpublished results]; 5α-androstan-3β-ol-17-one-3β-sulfate (Epiand-3-s) [cichlid (*T. demeusii*); Stacey, N.E., unpublished results]; etiocholanolone (Etio) [round goby; Murphy *et al.* 2001]; etiocholanolone glucuronide (Etio-g) [round goby; Murphy *et al.* 2001]; 5β-androstan-3α-ol-11,17-dione (11-K-Etio) [round goby (*Neogobius melanostomus*); Laframboise *et al.* 2011]; 11-keto-testosterone (11-KT) [tinfoil barb (*Barbonymus schwanenfeldii*): Cardwell *et al.* 1995]; 5β-pregnan-3β-ol-20-one-3β-sulfate (P-3β-s) [cichlid (*P. ephippium*); Stacey, N.E., unpublished results]; 5β-pregnan-3α,17,20β-triol (3α,17,20β-P) [catfish (*Brachysynodontis batensoda*); Narayanan and Stacey 2003]; 5β-pregnan-3α,17-diol-20-one-3α-gluc (17P-3α-g) [cichlid (*Astatotilapia burtoni*); Cole and Stacey 2006]; 5β-pregnan-3α,17-diol-20-one-3α-sulfate (17P-3α-s) [cichlid (*P. ephippium*); Stacey, N.E., unpublished results]; 4-pregnen-17,20β-diol-3-one (17,20β-P) [goldfish; Sorensen *et al.* 1987]; 4-pregnen-17,20β-diol-3-one-20β-glucuronide (17,20β-P-g) [goldfish; Sorensen *et al.* 1995b]; 4-pregnen-17,20β-diol-3-one-20β-glucuronide (17,20β-P-s) [goldfish; Sorensen *et al.* 1995b: zebrafish (*Danio rerio*); Belanger , Pachkowski, and Stacey, 2010]; 4-pregnen-17,20β,21-triol-3-one (17,20β,21-P) [goldfish; Scott and Sorensen 1994]; 4-pregnen-17,20β,21-triol-3-one-20β-sulfate (17,20β,21-P-s) [goldfish; Scott and Sorensen 1994]; 4-pregnen-17,21-diol-3-one (17,21-P) [tinfoil barb; Cardwell *et al.* 1995]; 5β-pregnan-3a,20α-diol-3α-glucuronide (20αP-3α-g) [cichlid (*P. ephippium*); Stacey, N.E., unpublished results]; testosterone (T) [Atlantic salmon (*Salmo salar*); Moore and Scott 1991]; testosterone-sulfate (T-s) [cichlid (*A. burtoni*); Cole and Stacey 2006].

neurons (ORNs) located in the sensory epithelia of the olfactory organs. Pheromone-receptor binding then appears to open cation channels that depolarize the ORNs, inducing receptor potentials that are propagated in the ORN axons within the olfactory nerve, and activating ORN synapses with second-order mitral neurons within the olfactory bulbs. Complex synaptic connections among several cell types within the bulbs then generate action potentials that are propagated centrally within the paired olfactory tracts and activate synapses at a number of olfactory terminal fields within the brain (Lastein, Hamdani, and Døving, 2014).

Underwater electro-olfactogram (EOG; Scott and Scott-Johnson, 2002) recording studies have demonstrated that hormonal steroids, prostaglandins, and their precursors and metabolites are detected at low (nanomolar) concentrations by the ORNs of several hundred species ranging from the primitive order Elopiformes (*Megalops cyprinoides*, tarpons; Stacey and Sorensen, 2009; Stacey, 2010), to the highly derived order Perciformes (Gobiidae; Murphy, Stacey, and Corkum, 2001; Cichlidae; Cole, and Stacey, 2006). The major advantage of the simple EOG technique is that it can quickly determine the sensitivity and specificity of a species' olfactory organ to a large number of hormonal odorants (e.g., Murphy, Stacey, and Corkum, 2001; Cole and Stacey, 2006). For example, EOG recordings in goldfish indicate that 17,20β-P, the ovarian MIS in goldfish and many other fishes (Nagahama *et al.*, 1994; Pankhurst, 2008), is detected with remarkable sensitivity and specificity, consistent with the results of bioassays in which known amounts of 17,20β-P are added to aquarium water (Sorensen *et al.*, 1987, 1990). The results of these studies, when combined with others from studies of 17,20β-P release (Stacey *et al.*, 1989; Scott and Sorensen, 1994; Sorensen *et al.*, 2000), enable the calculation of pheromone active space.

EOG recording has also been used for cross-adaptation studies that attempt to determine whether hormonal odorants act through one or more olfactory receptor mechanisms. These studies compare EOG response to a test odorant before and during adaptation to an adapting odorant. If adaptation abolishes response to the test odorant, then it is assumed that both the adapting and test odorants act on a common olfactory receptor and are not discriminated by the fish; alternatively, if EOG responses to test odorants are not reduced by adaptation, it can be concluded that the two odorants likely act on different olfactory receptor mechanisms and, therefore, could be discriminated. The biological significance of EOG cross-adaptation has been clearly demonstrated in the round goby, *N. melanostomus*, where steroidal odorants that are discriminated at the sensory (EOG) level also are discriminated behaviorally (Murphy, Stacey, and Corkum, 2001; Section 3.4.1 and Fig. 3.17).

EOG screening studies have revealed that the kinds of hormonal odorants a species detects can be highly dependent on its phylogenetic position (relationship). Thus, although the range of detected hormonal products often differ greatly between species in different families or orders, the olfactory sensitivities of fish within the same genera or tribes are typically and remarkably similar, suggesting not only that olfactory responsiveness to hormonal compounds could be useful in phylogenetic studies but also that intensive study of a few model species might provide insight into the evolution of hormonal pheromones of closely related fishes. For example, in the cypriniform fishes, which include the carps and minnows, all of more than 75 tested species detect prostaglandins (Stacey and Sorensen, 2009; Stacey, 2010), suggesting that homologues of the female F-prostaglandin pheromone that has been well characterized in goldfish (Sorensen *et al.*, 1988; Kobayashi, Sorensen, and Stacey, 2002) might also be used throughout this large and important freshwater group. And in cichlids, where phylogenetic relationships have been intensively studied, it is possible to use EOG studies to trace the evolutionary history of hormonal pheromones (Sections 3.1.4 and 3.4.2).

Even though EOG studies indicate that hormonal pheromones are widespread amongst teleost fishes, there are two reasons why they likely underestimate both the number of fish that detect hormonal compounds and the range of odorants they detect. First, because the metabolism and release of steroid and prostaglandin hormones is poorly understood in fishes, it is likely that many hormonal odorants are unknown, and therefore not available for testing (e.g., Sorensen *et al.*, 2004; Barata *et al.*, 2008). Second, because the olfactory receptor mechanisms responsive to steroids and prostaglandins can be extraordinarily specific in fishes (e.g., Sorensen *et al.*, 1988, 1990), test compounds that differ even slightly from the natural pheromonal ligand could fail to evoke an EOG response (Fig. 3.3).

3.1.3 Overview of Biological Responses to Hormonal Pheromones

Traditionally, pheromones have been considered to induce two general types of effects (Wilson and Bossert, 1963): (1) releaser effects, which are typically immediate behavioral responses to phero-mone exposure and (2) primer effects, which include a range of slower physiological responses that might take hours or even months to appear. Although this dichotomy can be useful in conceptual-izing how different the effects of pheromones can be, it has had two unfortunate consequences. First, it suggests that the behavioral and physiological responses to pheromones are fundamentally differ-ent, when of course all behavioral responses are the products of physiological systems. Second, it has led to the terms "releaser pheromone" and "primer pheromone" despite the fact that, as noted by Wilson and Bossert (1963) *"it is quite possible for the same pheromone to be both a releaser and a primer."* A clear case in point are the preovulatory steroid pheromones released by female goldfish that induce in males both rapid behavioral effects and more protracted endocrine effects (Sorensen *et al.*, 1987; Poling, Fraser, and Sorensen, 2001; Kobayashi, Sorensen, and Stacey, 2002).

3.1.4 Hormonal Pheromones and Phylogeny

This chapter discusses fish hormonal pheromones from a phylogenetic perspective because of the extremely close relationship between phylogeny and the olfactory detection of hormonal compounds that have been revealed by EOG studies. Hormonal pheromones have been described only in Class Actinopterygii (ray-finned fishes), which contains all but 5% of living fishes (Nelson, 2006) and is dominated by the teleosts—the only fishes in which hormonal pheromones have been reported. Teleosts comprise two small subdivisions of basal fishes (osteoglossomorphs—mooneyes and bony tongues and elopomorphs—ten-pounders, tarpons, and eels) and two major subdivisions of more derived fishes—the ostarioclupeomorphs (or otocephalans) and the euteleosts (Nelson, 2006).

Osteoclupeomorphs include clupeomorphs (herrings; >350 species) and the ostariophysan cypri-niforms (carps and minnows; >3200 species), characiforms (piranha, tetras; >1600 species), siluri-forms (catfishes; >2800 species), gonorhynchiforms (milkfishes; >35 species), and gymnotiforms (New World knifefishes; >170 species). There is no evidence for hormonal pheromones in clupe-iforms, despite attempts to chemically characterize a pheromone from the milt (seminal fluid) of Pacific herring (*Clupea pallasii pallasii*) that has potent releasing effects on spawning behaviors (Carolsfeld, Scott, and Sherwood, 1997). However, there is abundant evidence that hormonal phero-mones are widespread among the ostariophysan cypriniforms, characiforms, and siluriforms (Table 3.1). Of the osteoclupeomorph pheromones, only those of the goldfish and closely related cypriniforms will be discussed in this chapter because previous reviews (Van Den Hurk and Resink, 1992; Sorensen and Stacey, 1999; Stacey and Sorensen, 2002, 2005) have already summarized the earlier studies on zebrafish (Van Den Hurk and Lambert, 1983), catfish (Resink *et al.*, 1989; Narayanan and Stacey, 2003), and characiforms (Cardwell *et al.*, 1995).

The euteleosts, sister group to the osteoclupeomorphs, includes more than 17 500 species in 28 orders, only two of which, the Salmoniformes and Perciformes (gobies and cichlids) are presently known to use hormonal pheromones (Table 3.1).

This chapter focuses on three orders of fishes in which hormonal pheromone studies are ongoing: (1) Cypriniformes, (2) Salmoniformes, and (3) Perciformes. Notably, both osteoclupeomorphs and euteleosts are represented. For summaries of significant hormonal pheromone research that is currently not being pursued, (e.g., zebrafish and African catfish), the reader is directed to previous reviews (Van Den Hurk and Resink, 1992; Stacey and Sorensen, 2002, 2005).

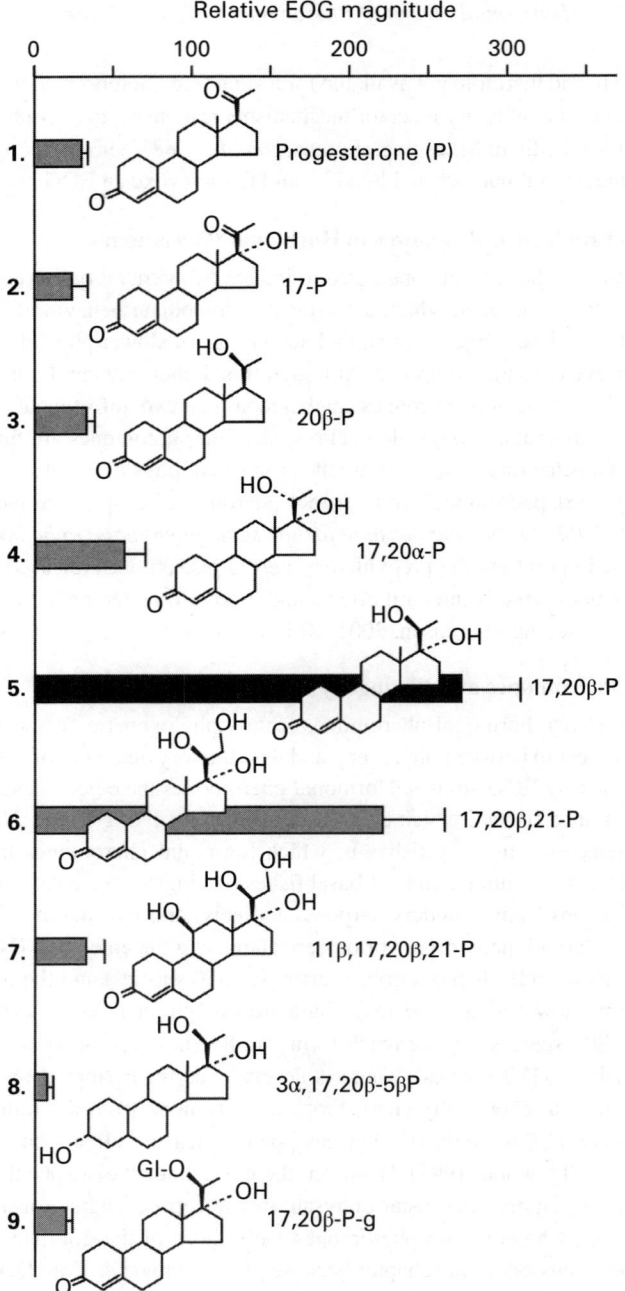

Figure 3.3. Electro-olfactogram responses to various 10^{-8} M steroids show that goldfish's sensitivity to 17,20β-P is highly specific. Progesterone (**1**) is a relatively weak odorant, inducing approximately 25% of the response magnitude of 10^{-5} M L-alanine, and the olfactory potency of the molecule is not significantly affected by addition of α-hydroxyl to carbon 17 (**2**) or β-hydroxyl to carbon 20 (**3**). Potency increases only marginally if α-hydroxyls are added to both carbon 17 and 20 (**4**), but it increases dramatically if the carbon 20 hydroxyl is in the β configuration (**5**). Olfactory potency of the 17,20β-P molecule is diminished by various changes such as addition of hydroxyl to carbon 21 (**6**) and 11 (**7**), reduction of the A ring in the 5β,3α (**8**) or other configurations. Sorensen, P.W., Hara, T.J., Stacey, N.E., *et al.* (1990) Extreme olfactory specificity of the male goldfish to the preovulatory steroidal pheromone 17α, 20β-dihydroxy-4-pregnen-3-one. Journal of Comparative Physiology A, 166, 373–383. With permission from Springer Business and Science Media.

Table 3.1. Electro-olfactogram evidence for detection of prostglandins and steroids in orders of teleost fish[1,2].

Order	Number of extant species	Number of species examined	Common names	Example genera	Evidence for hormonal pheromones		
					PGs	Steroids	
						Unconjugated	Conjugated
Osteoglossomorph Orders							
Elopiformes[3]	8	1	Tarpon	*Megalops*	+[5]	+	0[6]
Osteoclupeomorph Orders							
Cypriniformes	3200	>80	Cyprinids	*Carassius, Danio*	+	+	+
Characiformes	1600	>20	Characins	*Astyanax, Colossoma*	+	+	+
Siluriformes	2800	>20	Catfishes	*Clarias, Synodontis*	+	+	+
Gymnotiformes	170	1	Knifefishes	*Apteronotus*	+	0	0
Euteleost Orders							
Osmeriformes[4]	88	1	Smelts	*Plecoglossus*	+	0	0
Salmoniformes	66	9	Salmonids	*Salmo, Oncohynchus*	+	+	+
Cyprinodontiformes	1000	1	Rivulines	*Aplocheilus*	+	+	
Perciformes	9293	>75	Cichlids, gobies	*Haplochromis, Neogobius*	0	+	+

[1] Systematic terminology from Nelson (2006).
[2] Reprinted with permission from Stacey and Sorensen (2009); Stacey (2009, 2010).
[3] Stacey (unpublished results).
[4] See Kitamura, Ogata, and Takashima (1994b).
[5] Some of the tested species in the taxon responded to at least one compound in this category.
[6] No evidence that species in the taxon detect compounds in this category.

3.2   HORMONAL PHEROMONES IN THE GOLDFISH AND RELATED CARPS

The order Cypriniformes includes two superfamilies: Cyprinoidea (family Cyprinidae: about 2400 species of carps and minnows) and Cobitoidea (about 850 species), including Catostomidae (suckers), Cobitidae (loaches), Gyrinocheilidae (algae eaters), and Balitoridae (river loaches) (Nelson, 2006). EOG screening studies (Stacey, 2010; Stacey and Sorensen, 2005, 2009) show that although PGFs are detected by all tested cypriniforms (>75 species examined), free and conjugated steroids are detected only by cyprinids, where all higher taxa except acheilognathans (five tested species of bitterlings) are responsive. These patterns of olfactory detection suggest that pheromonal PGFs originated in an ancestor common to all cypriniforms, whereas detection of steroids evolved after cyprinoids diverged from cobitoids.

3.2.1   The Goldfish—an Important Model

Most information on the reproductive functions of cypriniform hormonal pheromones comes from the goldfish, the only species for which a complete picture of hormonal pheromone function is beginning to emerge (Fig. 3.4). The variety and nature of hormonal pheromones employed by the goldfish also appears to be used by the crucian carp and common carp; however, because these species are less well understood, they are described here briefly (Sorensen and Stacey, 2004; Stacey and Sorensen, 2005). Goldfish and the closely related crucian and common carps all display little sexual dimorphism; live in mixed-sex, apparently unstructured aggregations; and employ a promiscuous or polygynandrous mating system (Taborsky, 2001; Taylor and Knight, 2008), involving intense sperm competition. All three species undergo vitellogenesis in winter and early spring, and several times during spring and summer spawn large numbers of adhesive, undefended eggs in aquatic vegetation. In goldfish, field and laboratory studies indicate that females are triggered to ovulate by two key environmental cues: (1) rising water temperature and (2) the presence of the aquatic vegetation that serves as spawning substrate (Stacey *et al.*, 1979a; Kobayashi *et al.*, 2008). Although the synchronous ovulations observed in goldfish (Kobayashi, Aida, and Hanyu, 1988), common carp (Sorensen, unpublished observations), and crucian carp (Olsen and Stacey, unpublished observations) might simply result from multiple females responding to a common environmental cue, there is evidence that preovulatory females releasing 17,20β-P (which also acts as a female pheromone that acts on males) stimulate ovulation in additional females (Kobayashi, Sorensen, and Stacey, 2002). These mechanisms to promote ovulatory synchrony presumably evolved because releasing large numbers of eggs swamps the ability of predators to eat the undefended eggs and young and because, in mating systems involving intense sperm competition, female fertility is unlikely to be constrained by sperm availability.

At ovulation, which occurs late in the night, males compete for spawning access to females as they repeatedly enter aquatic vegetation over several hours to oviposit. It seems likely that the evolution of hormonal pheromones in these three carps has been strongly influenced by the fact that they typically spawn in dim and turbid water and have mating systems in which a male's reproductive success depends solely on the number of eggs he fertilizes. Consistent with the strong correlation between phylogeny and hormonal pheromone detection in many fishes, EOG, behavioral, and physiological studies indicate that male crucian and common carp detect and respond to the same steroidal and prostaglandin odorants that constitute the periovulatory pheromones released by female goldfish (Bjerselius and Olsén, 1993; Irvine and Sorensen, 1993; Stacey *et al.*, 1994; Bjerselius, Olsén, and Zheng, 1995; Olsén, Sawisky, and Stacey, 2006). Despite the fact that goldfish and common carp clearly discriminate each other's reproductive (Lim and Sorensen, 2011) and nonreproductive (Sisler and Sorensen, 2008) body odors, it seems likely that their use of common hormonal odorants might provide an explanation for the natural hybridizations reported among these species (Taylor and Mahon, 1977; Hänfling *et al.*, 2005).

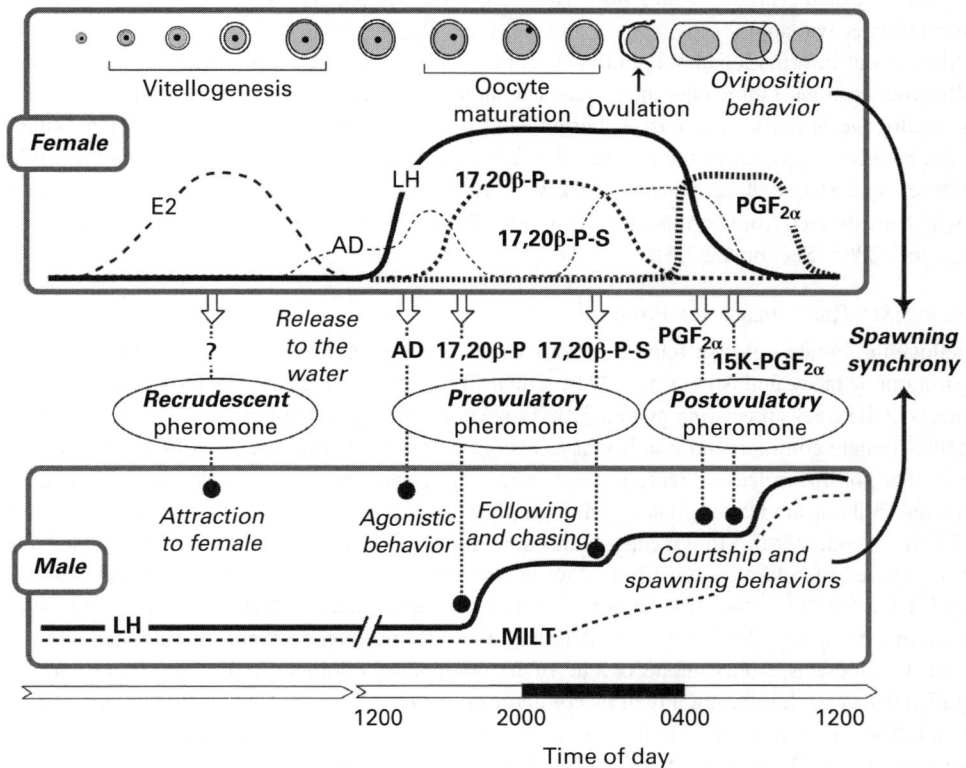

Figure 3.4. Schematic model of the hormonal pheromones released by periovulatory female goldfish and their primer and releaser effects on males. Estradiol-17β (E2) in vitellogenic females stimulates urinary release of an unidentified recrudescence pheromone that attracts males. In females that have completed vitellogenesis, exogenous cues induce a luteinizing hormone (LH) surge that in turn stimulates the release of a dynamic preovulatory pheromone containing androstenedione (AD), the maturation-inducing steroid 4-pregnen-17,20β-diol-3-one (17,20β-P), and its sulfated metabolite, 17,20β-P-s. Early in the female's LH surge, her released AD induces agonistic behaviors among males. Later, as the 17,20β-P:AD ratio of the female's pheromone increases, males respond by following and chasing conspecifics and by increasing their plasma LH concentrations, which in turn increases movement of sperm from the testes to the sperm ducts, and increases sperm quality. Late in the female's LH surge, 17,20β-P-s dominates her preovulatory pheromone mixture, enhancing its behavioral and endocrine effectiveness. At ovulation, eggs in the oviduct stimulate synthesis of PGF$_{2\alpha}$ which acts in the brain to trigger female sex behavior, and is released with its major metabolite (15k-PGF$_{2\alpha}$) as a postovulatory pheromone stimulating both male spawning behaviors and additional LH increase. Kobayashi, M., Sorensen, P.W. & Stacey, N.E. (2002). Hormonal and pheromonal control of spawning behavior in the goldfish. Fish Physiology and Biochemistry, vol. 26, issue 1. Copyright © 2002, Springer.

The goldfish currently provides the best understood model of hormonal pheromone function—studies having revealed the partial chemical characterization of the pheromones, the timing of their synthesis, their routes of release, the physiological mechanisms mediating their effects, and the reproductive benefits these provide. The following discussion of cypriniform hormonal pheromones will therefore deal primarily with goldfish, but refer to additional cypriniform species where relevant information is available.

Most of our information about goldfish hormonal pheromones pertains to females, which sequentially produce at least three pheromones: an uncharacterized recrudescence pheromone released during vitellogenesis, and two hormonal pheromones that are released only during the brief periovulatory period between the onset of the preovulatory LH surge and completion of oviposition (Kobayashi, Sorensen, and Stacey, 2002; Stacey, 2010). Male goldfish also release sex pheromones, at least some of which are derived from hormones (Stacey *et al.*, 2001; Fraser and Stacey, 2002; Sorensen, Pinillos, and Scott, 2005; Section 6.2.1.5).

THE FEMALE RECRUDESCENCE PHEROMONE

Vitellogenic female goldfish release a recrudescence pheromone, previously termed "maturation" pheromone (Stacey and Sorensen, 2002), which likely enables males to identify and remain near females in the weeks preceding ovulation and spawning (Stacey and Sorensen, 2009). It is not known whether female common and crucian carps also release a recrudescence pheromone. The first evidence that goldfish release a recrudescence pheromone came from studies showing that males are attracted to the urine from females with high plasma E2 concentrations (Yamazaki and Watanabe, 1979; Yamazaki, 1990). More recent studies have shown that male goldfish exhibit social behaviors when exposed to holding water from either vitellogenic females or ovariectomized females treated with E2, but fail to respond to the odors of ovariectomized females or spermiated males (Kobayashi, Sorensen, and Stacey, 2002; Fig. 3.5). Although these results suggest that E2 or a metabolite might be the recrudescence pheromone or one of its component odorants, EOG recordings show that goldfish detect neither E2 nor any of its common sulfates or glucuronides (Sorensen *et al.*, 1987 and unpublished). It is also possible that the recrudescence pheromone is an unknown E2 metabolite. However, because E2 treatment changes goldfish kidney structure and the pheromone is released in urine (Yamazaki and Watanabe, 1979), it seems most likely that rather than being derived from E2, the recrudescence pheromone is instead synthesized in the kidney under E2 stimulation.

THE FEMALE PREOVULATORY STEROID PHEROMONE

In female goldfish that have completed vitellogenesis, a remarkable cascade of endocrine and physiological changes that culminate in spawning begins when two key environmental cues—increasing water temperature and aquatic vegetation (spawning substrate)—trigger a neuroendocrine reflex, generating a photoperiodically synchronized LH surge that rapidly increases synthesis of the MIS 17,20β-P, the key component of the preovulatory steroid pheromone (Stacey *et al.*, 1979a,b; Kobayashi, Sorensen, and Stacey, 2002; Canosa, Stacey, and Peter, 2008); periovulatory LH and steroid changes in common carp are very similar in timing to those of goldfish (Santos *et al.*, 1986), but this has not been studied in crucian carp.

Ovulation occurs approximately 12 h after the start of the LH surge, at which point movement of eggs to the oviduct stimulates synthesis of $PGF_{2\alpha}$, which then acts in the brain within minutes to stimulate female oviposition behavior (Stacey, 1976; Stacey and Peter, 1979; Sorensen *et al.*, 1995a). Males, which have been stimulated by the preovulatory pheromone to attend females prior to ovulation, now begin to compete vigorously for access to ovulated females during the several hours of brief and repeated spawning acts. Although casual observation might suggest that these spawning behaviors are unspecialized and chaotic, they belie a surprisingly complex suite of pheromone-driven physiological and behavioral processes that begin with the preovulatory surge in female LH and end when spawning ceases approximately 15 h later (Fig. 3.4). During this brief time, hormonal pheromones function in three social contexts: (1) from female to male, (2) between females, and (3) between males. In the former situation, which is best understood, females first release steroid

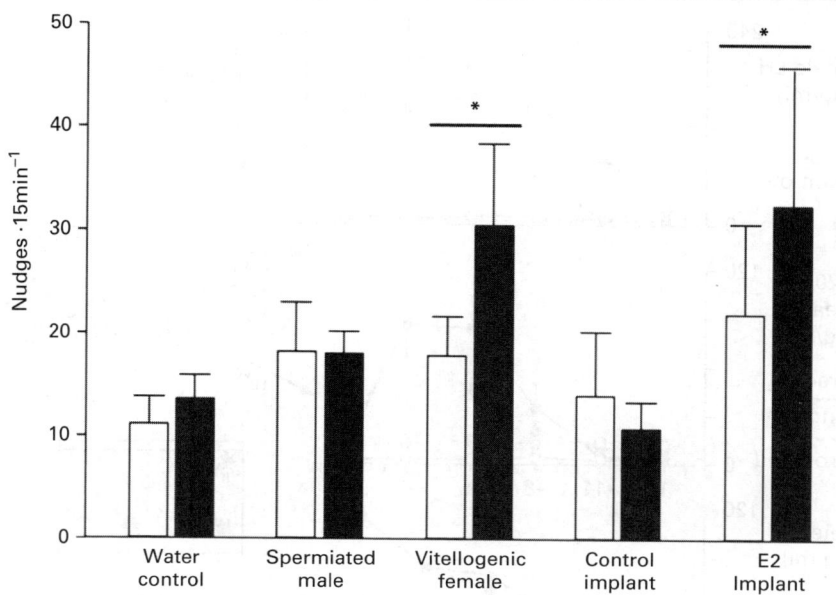

Figure 3.5. Behavioral evidence for a recrudescence pheromone in goldfish. Groups of five spermiating male goldfish (*N*=10 groups per treatment) were held in aerated 70 l test aquaria overnight and observed continuously for male–male nudging (physical contact/inspection, a behavior typical of courtship; Poling, Fraser, and Sorensen, 2001) during a 15-min prestimulus period (clear bars). Immediately following this first observation period, male groups were observed for an additional 15-min observation period (black bars) during which they were exposed to one of five odor treatments added to the test aquaria (100 ml/min) by peristaltic pump: (1) well water control, (2) spermiated male odor, (3) vitellogenic female odor, (4) odor of ovariectomized females that had been implanted (3 months' post surgery) for 30 days with silicone capsules containing 50 μl castor oil, and (5) odor of ovariectomized females that had been implanted with silicone capsules containing 200 ng of E2 dissolved in castor oil. Test odors were prepared by holding pairs of fish overnight in 2 l of well water. *, *P* <0.05: Wilcoxon matched pairs test. Kobayashi, M., Sorensen, P.W. & Stacey, N.E. (2002). Hormonal and pheromonal control of spawning behavior in the goldfish. Fish Physiology and Biochemistry, vol. 26, issue 1. Copyright © 2002, Springer.

odorants in the preovulatory period when the LH surge is stimulating final oocyte maturation, and after ovulation release prostaglandin odorants during spawning.

Preovulatory steroid pheromone release begins in the afternoon, soon after the preovulatory LH surge begins, and continues until ovulation occurs the following morning (Stacey *et al.*, 1979a,b, 1989; Canosa, Stacey, and Peter, 2008; Fig. 3.4). If males are exposed to females undergoing a preovulatory LH surge, or their odor, their plasma LH increases within minutes, thereby stimulating steroid changes that increase milt volume within several hours (Kobayashi, Aida, and Hanyu, 1986a,b; Dulka *et al.*, 1987a; Stacey *et al.*, 1989) (Fig. 3.6); these male responses are abolished if anosmia is induced by cutting the olfactory tracts (Kobayashi, Aida, and Hanyu, 1986b). Although not nearly so well studied as male goldfish, male common and crucian carp appear to have similar responses to preovulatory females (Bjerselius and Olsén, 1993; Irvine and Sorensen, 1993; Bjerselius, Olsén, and Zheng, 1995; Stacey *et al.*, 1994; Olsén, Sawisky, and Stacey, 2006).

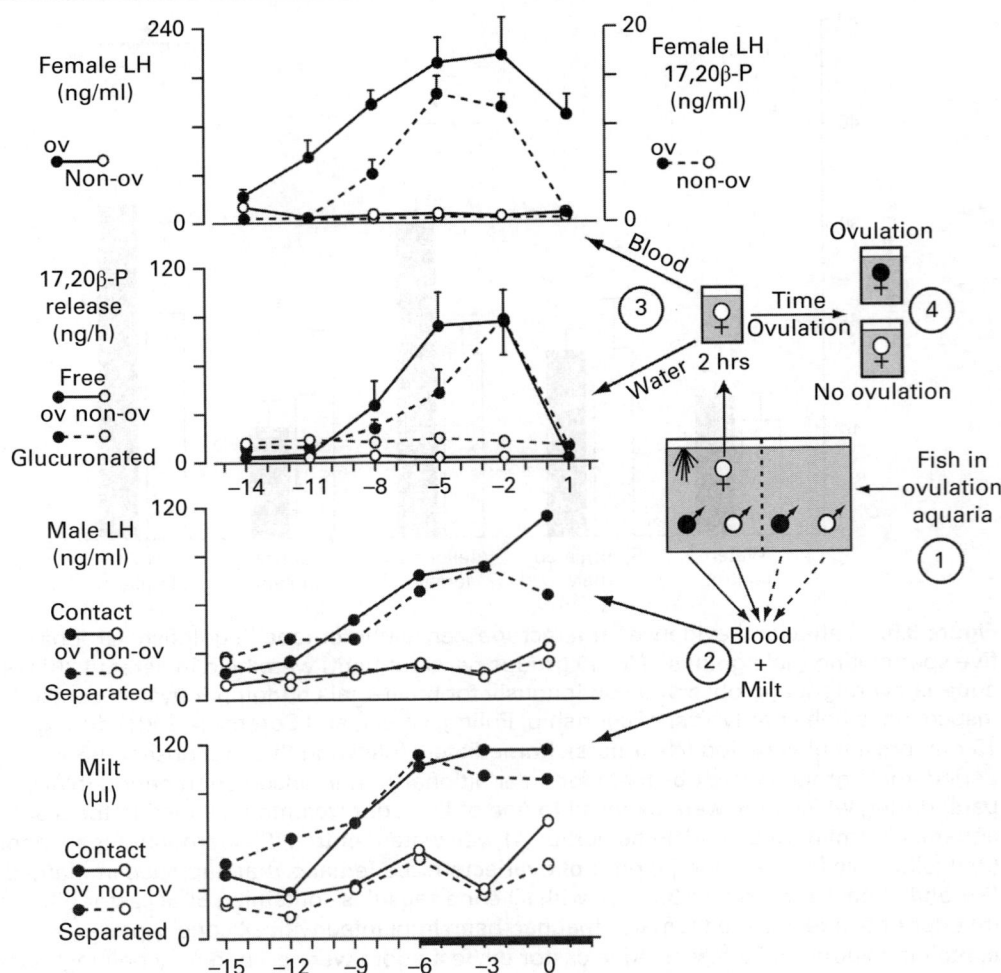

Figure 3.6. Male goldfish increase their blood LH concentrations and milt volumes if they are held with ovulating females, or separated from them by a perforated barrier. The right-hand side illustrates the experimental procedure. (1) Male and female goldfish were held overnight in warm water with aquatic vegetation, conditions that induced ovulation in half of the females. Half of the males were held behind a perforated barrier (separated), whereas the other half were in direct contact with females. (2) Males were removed from the experimental aquaria and immediately sampled for LH and milt. (3) Females were removed from the experimental aquaria, held for 2 h to obtain holding water for radioimmunoassay measurement of free and glucuronated 17,20β-P, and then bled for radioimmunoassay measurement of LH and 17,20β-P. (4) Females were checked frequently to determine if and when they ovulated. Graphs on the left-hand side of the figure illustrate the parameters measured. Adapted with permission from Stacey and Sorensen (2005, 2009) and Stacey (2009, 2010).

Chemical analyses of holding water show that females during their preovulatory LH surge release dozens of gonadal steroids (Scott and Sorensen, 1994; Sorensen and Scott, 1994). Further, approximately half a dozen of these steroids stimulate the olfactory systems of goldfish and are therefore likely to induce biological responses (Sorensen and Scott, 1994). However, it seems likely that three of these released steroids—17,20β-P, 17,20β-P-s, and AD—are key components of the preovulatory steroid pheromone, and induce reproductive responses in males by acting via separate, specific, and highly sensitive (pM threshold) olfactory receptors (Sorensen *et al.*, 1990, 1995b; Sorensen, Pinillos, and Scott, 2005). Although it has been convenient to refer to these three major odorants as a single preovulatory steroid pheromone, the female goldfish in fact releases multiple preovulatory odors by controlling both spatial and temporal aspects of pheromone release. 17,20β-P and AD, for example, are released continuously across the gills, whereas 17,20β-P-s is released in urine pulses (Sorensen *et al.*, 2000; Scott and Ellis, 2007). Further, these three steroid odorants have different release patterns: AD, 17,20β-P, and 17,20β-P-s reach peak release rates during the early, middle, and late portions of the preovulatory period, respectively (Stacey *et al.*, 1989; Scott and Sorensen, 1994; Fig. 3.4), thereby changing the function of the preovulatory pheromone by altering the ratios of its steroid components.

Early in the preovulatory period, for example, the low 17,20β-P:AD ratio of the female's odor dampens the male's LH response (Stacey, 1991), but increases agonistic behavior among males (Poling, Fraser, and Sorensen, 2001). Later in the female's LH surge, an increasing 17,20β-P:AD ratio stimulates prolonged following and inspection behaviors among males (Poling, Fraser, and Sorensen, 2001) and triggers an LH increase that increases milt volume and enhances paternity (Dulka *et al.*, 1987a; DeFraipont and Sorensen, 1993; Zheng *et al.*, 1997). Toward the end of the female's LH surge, increasing urinary release of 17,20β-P-s maintains elevated LH levels in males while inducing intense but brief bouts of male chasing (Sorensen *et al.*, 1995b; Poling, Fraser, and Sorensen, 2001).

In addition to being spatially separated by release in urine and across the gill, the preovulatory sex steroid odorants also are likely perceived in different social contexts. For example, olfactory detection thresholds from EOG studies (Sorensen *et al.*, 1987, 1990; Sorensen, Pinillos, and Scott, 2005) indicate that male goldfish should be able to detect AD and 17,20β-P only in the immediate vicinity of the female, whereas they should detect 17,20β-P-s in urine patches at a distance from the female (Sorensen *et al.*, 2000; Scott and Ellis, 2007; Table 3.2).

Changes in LH and milt are convenient measures of pheromonal 17,20β-P activity, but are only the early stages of a complex male response that ultimately and dramatically enhances the male's reproductive success (paternity) during sperm competition (Zheng *et al.*, 1997). This pheromonal effect on paternity appears to result from the combined effects of increased competitive behaviors, sperm motility, and sperm release, because all these measures of male reproductive performance are increased by 17,20β-P exposure (DeFraipont and Sorensen, 1993; Zheng *et al.*, 1997; Hoysak and Stacey, 2008; Fig. 3.7). Importantly, similar paternity effects occur if sperm from 17,20β-P-exposed and control males compete for *in vitro* fertilizations (Zheng *et al.*, 1997; Fig. 3.7), providing strong evidence that a major effect of the preovulatory pheromone is increased sperm quality. It is therefore significant that 17,20β-P, which increases rapidly in male goldfish plasma in response to pheromonal 17,20β-P-induced LH increase (Kobayashi, Aida, and Hanyu, 1986a; Dulka *et al.*, 1987a), also enhances milt hydration, spermiation, and sperm motility in a variety of other fishes (Thomas, 2003; Pankhurst, 2008; Scott, Sumpter, and Stacey, 2010). Effects on sperm motility are likely of key importance to mating success and have been shown to be mediated in salmon and eel by classical genomic mechanisms acting on the sperm duct to increase seminal pH (Miura *et al.*, 1992), and in sciaenids and flatfish by direct nongenomic action on sperm membrane progesterone receptors (Thomas, 2012).

Pheromonal effects of 17,20β-P in goldfish are not only restricted to the male but also include the induction of ovulation in females (Kobayashi, Sorensen, and Stacey, 2002). Because male goldfish release extremely small quantities of 17,20β-P and related C21 steroids (Sorensen, Pinillos, and Scott, 2005), it is very likely that the preovulatory female is the source of this pheromonal 17,20β-P effect on other females. Indeed, 17,20β-P released by females during final oocyte maturation seems to synchronize ovulations reported in laboratory goldfish (Kobayashi, Sorensen, and Stacey, 2002; Sorensen, unpublished results) and cultured common carp (see Stacey *et al.*, 1994). The function of

Table 3.2. Theoretical active spaces of pheromonal compounds released across the gills (G) and in the urine (U) of periovulatory female goldfish (F) and mature male goldfish that are either sexually inactive (M) or spawning (Msp)[1].

Sex	Compound	Release route	Release duration (h)[2]	Peak release (ng·h⁻¹)	EOG threshold (log M)	Active space (l·h⁻¹)[3]	Reference
M	AD	G	Continuous	~50	~10^{-11}	~17	5
Msp	AD	G	Variable[4]	~750	~10^{-11}	~250	5
F	AD	G	6	~90	~10^{-11}	~30	6
F	17,20β-P	G	12	~60	~5×10^{-12}	~35	6
F	17,20β,21-P	G	3	~60	~5×10^{-11}	~3	6
F	17,20β-P-s	U	9	~65	~10^{-11}	~8 × 3	6–8
F	17,20β,21-P-s	U	9	~150	~5×10^{-11}	~8 × 1	6–8
F	15 K-PGF$_{2a}$	U	Variable[4]	~750	~5×10^{-12}	~40 × 11	8–10

[1] Reprinted with permission from Stacey (2009, 2010).
[2] The time that the compound should be detectable, when neither the source fish nor the receiver are moving, and the odor is being dispersed by diffusion only.
[3] Defined as the calculated volume of water in which a pheromone of known release rate will be detectable, and calculated as the quantity of pheromone (moles) released per hour divided by the EOG detection threshold of the receiver. Calculations based on data from Scott and Sorensen (1994), Sorensen, Pinillos, and Scott (2005), and Sorensen *et al.* (1988, 2000).
[4] Deteremined by the time (typically several hours) required to complete oviposition.
[5] Sorensen *et al.* (2005).
[6] Scott and Sorensen (1994).
[7] Sorensen *et al.* (1995b).
[8] Appelt and Sorensen (2007).
[9] Sorensen *et al.* (1988).
[10] Sorensen (unpublished results)

such ovulatory synchrony is unknown, but may serve either as a predator swamping strategy or to reduce the potentially disruptive effects of high male: female sex ratios on female spawning success.

THE FEMALE POSTOVULATORY PROSTAGLANDIN PHEROMONE

Release of the preovulatory steroid pheromone ceases within minutes of ovulation, at which point the female goldfish becomes sexually active and begins to release the postovulatory prostaglandin pheromone. The cause of this dramatic state change is the movement of ovulated oocytes into the ovisac, seemingly stimulating a synthesis of PGF$_{2\alpha}$ (Stacey and Liley, 1974; Sorensen *et al.*, 1988, 1995a) that in turn rapidly exerts two critical and almost simultaneous effects. First, PGF$_{2\alpha}$ that enters the circulation acts hormonally in the brain to stimulate female spawning behavior; this is easily demonstrated by giving either systemic (Stacey, 1976) or intracerebroventricular PGF$_{2\alpha}$ injections (Stacey and Peter, 1979) to nonovulated females, which soon perform spawning behaviors that are normal in all aspects except that no eggs are released. Second, recently ovulated females begin to release from their gills and in their urine PGF$_{2\alpha}$, 15k-PGF$_{2\alpha}$, and likely other unidentified PGF metabolite odorants as a postovulatory pheromone that exerts not only releaser effects on male courtship and spawning behaviors but also primer effects on male LH concentration and milt volume that

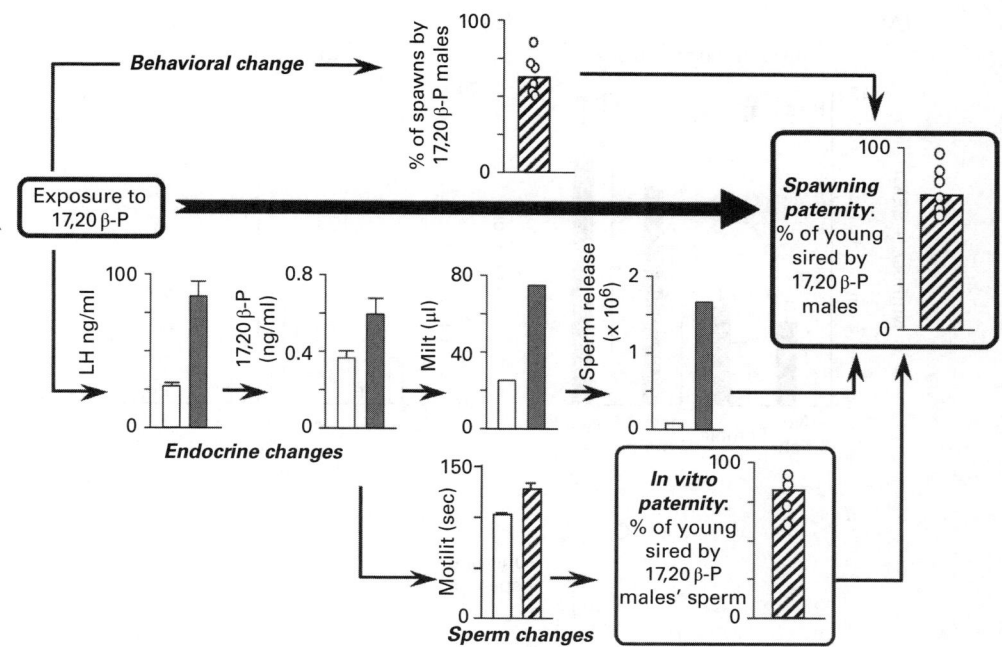

Figure 3.7. Pheromonal 17,20β-P induces both primer and releaser effects that enhance paternity in male goldfish. In each of six, paired spawning competitions, DNA paternity studies showed that a male exposed overnight to pheromonal 17,20β-P sired a significantly greater percentage of the offspring than did his control male competitor (box at right). Each circle represents the percentage of paternity acquired by each 17,20β-P-exposed individual and the hatched bar represents the mean percentage of the 17,20β-P-exposed group. This striking effect on paternity likely is mediated by both behavioral and physiological changes. For example, in each of the six, paired competitions (top), 17,20β-P-exposed males performed a greater proportion of the total male spawning acts than did control males (again, each circle represents one 17,20β-P-exposed individual). Moreover, these pheromone-exposed males (hatched and shaded bars) underwent rapid endocrine changes that enabled them to release dramatically more sperm than control males (open bars; middle and bottom graphs). Finally, it is likely that pheromone-induced 17,20β-P synthesis by the testis increases indices of sperm quality such as sperm motility. Such pheromone-induced sperm changes may account for the finding that in four, paired *in vitro* competitions, sperm from 17,20β-P-exposed males fertilized a greater percentage of the eggs than did sperm from control males. Reprinted from Stacey N.E. & Sorensen P.W. (2011) Hormonal pheromones. In: Encyclopedia of Fish Physiology: From Genome to Environment, Vol. 1, (ed. A.P. Farrell). Pp. 1553–1562. Elsevier, San Diego.

likely serve to maintain milt stores during spawning (Sorensen *et al.*, 1988, 1989; Zheng and Stacey, 1996, 1997; Appelt and Sorensen, 2007; Figs. 3.4 and 3.6).

Unlike the situation with the preovulatory steroid pheromone, where release appears to be passive and unspecialized, there is evidence for evolutionary specialization in release of the postovulatory prostaglandin pheromone. Although PGFs appear to be released across the gills in an unspecialized manner, they are also released in urine pulses that increase in frequency as the female is selecting a spawning site in floating vegetation (Fig. 3.8), suggesting that the release mechanism is specialized to attract males to the site (Sorensen *et al.*, 1995a, 2000; Appelt and Sorensen, 2007).

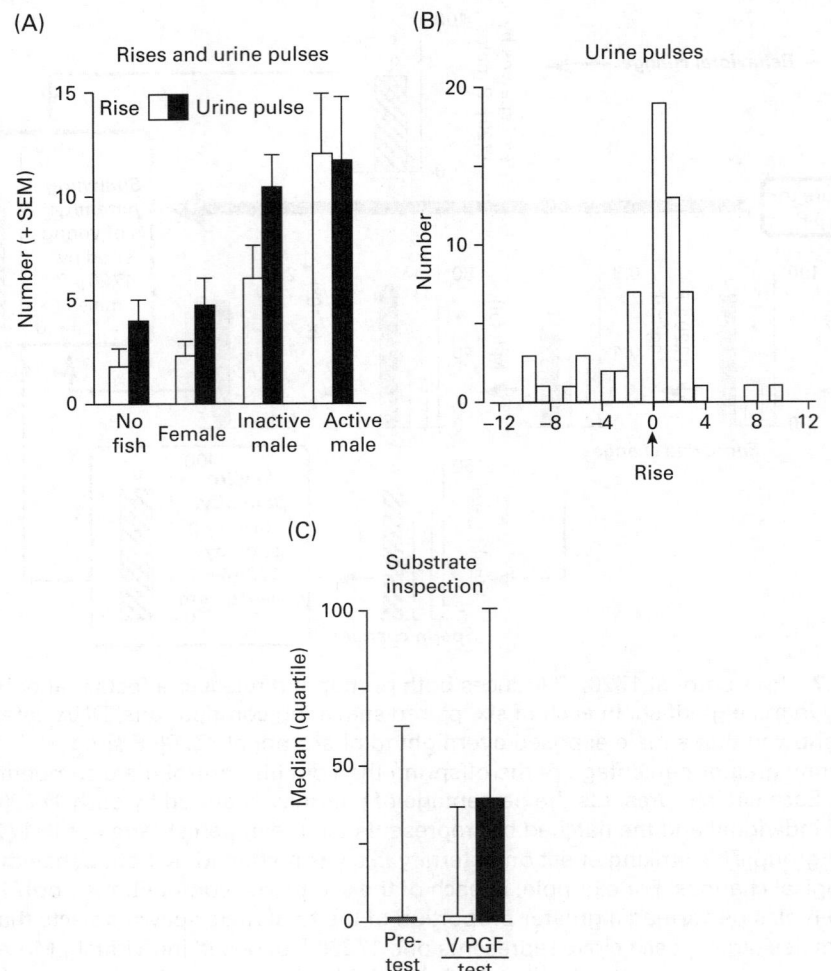

Figure 3.8. Female goldfish are specialized for signaling with the postovulatory prostaglandin pheromone by altering the frequency and place that they urinate. (A) Immediately before each spawning act, females *rise* into floating spawning substrate. During a 60-min test, both rising and urine pulses increased in the presence of an inactive or active male, but not in the presence of a female. (B) Females urinate more frequently in the few seconds just preceding and following rising behavior. Bars indicate the total number of urine pulses released by eight females tested individually in the presence of sexually active males during a 60-min test. (C) When the urinary pheromone 15-keto-PGF$_{2\alpha}$ (PGF) was not present during a 15-minute pretest, male goldfish displayed little inspection of a spawning substrate. However, during a 15-min test, males dramatically increased inspection of the substrate if it was associated with PGF, but not if it was associated with a control vehicle (V). Adapted with permission from Appelt, C.W. & Sorensen, P.W. (2007) Female goldfish signal spawning readiness by altering when and where they release a urinary pheromone. Animal Behavior, 74, 1329-1338. Elsevier, San Diego. Redrawn with permission from Stacey, N.E. & Sorensen, P.W. (2009) Fish hormonal pheromones. In: Hormones, Brain, and Behavior, Vol. 2, (eds. D. W. Pfaff, A. P. Arnold, A. M. Etgen, et al.), 2nd edn. pp. 639-681. Elsevier, San Diego; Stacey N.E. & Sorensen P.W. (2011) Hormonal pheromones. In: Encyclopedia of Fish Physiology: From Genome to Environment, Vol. 1, (ed. A.P. Farrell). pp. 1553–1562. Elsevier, San Diego.

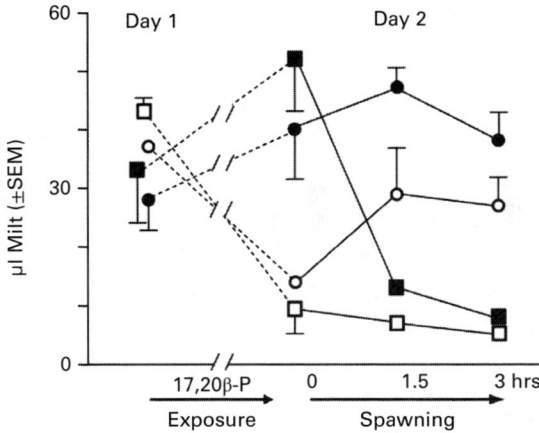

Figure 3.9. Synergistic effects of the female preovulatory and postovulatory pheromones on milt production in male goldfish. Following the stripping of milt on Day 1, males that are not exposed to 17,20β-P on Day 1 (clear symbols) have consistently reduced milt volumes on subsequent strippings on Day 2 (clear squares) unless they are allowed to spawn with PGF$_{2\alpha}$-injected females (clear circles). By contrast, males that are exposed to 17,20β-P following milt stripping on Day 1 (filled symbols) increase their milt volumes on Day 2, but this increase is not maintained on subsequent strippings (black squares) unless they are allowed to spawn with PGF$_{2\alpha}$-injected females (black circles). Redrawn with permission from Stacey *et al.* (1987).

It seems likely that release of prostaglandins both across the gill and in urine effectively creates a complex pheromonal signal with multiple actions. PGF$_{2\alpha}$ (nM detection threshold) is released both across the gills and in urine, whereas virtually all 15k-PGF$_{2\alpha}$ (pM detection threshold) is released in urine. Therefore, even though there is no evidence that these two PGFs have different effects on males, they act through separate and relatively independent olfactory receptor mechanisms (Sorensen *et al.*, 1988) and are expected to be encountered in different social contexts. PGF$_{2\alpha}$ may assist the male to maintain contact with an ovulated female in a spawning group, because its release across the gills should create an active space that is detectable only in the immediate vicinity of the female (Sorensen *et al.*, 2000). On the other hand, 15k-PGF$_{2\alpha}$ might function primarily to inform males of an ovulated female in the general vicinity, or to locate a spawning site in vegetation (Appelt and Sorensen, 2007).

Although the preovulatory and postovulatory pheromones are chemically and temporally distinct, they clearly have complementary effects on the spawning male goldfish. In particular, it is clear that being stimulated by the preovulatory pheromone not only renders the male capable of releasing more sperm at the onset of spawning (Zheng *et al.*, 1997; Hoysak and Stacey, 2008), but it also increases his ability to replenish releasable sperm stores during spawning, in response to the postovulatory PGF pheromone and other cues provided by the ovulated female (Fig. 3.9).

NEURAL AND NEUROENDOCRINE MECHANISMS MEDIATING RESPONSES TO FEMALE PHEROMONES

Although it had been proposed (Demski and Northcutt, 1983) that the terminal nerve (cranial nerve 0), which is anatomically associated with the olfactory nerve (cranial nerve 1), is a chemosensory system specialized for pheromone detection, terminal nerve ablations in fish do not induce noticeable reproductive deficits (Kobayashi *et al.*, 1994, 1997a; Yamamoto *et al.*, 1997), and single unit, extracellular recordings in goldfish showed that pheromones and other odorants that affect the activity of neural units in the olfactory bulbs do not affect the activity of terminal nerve neurons

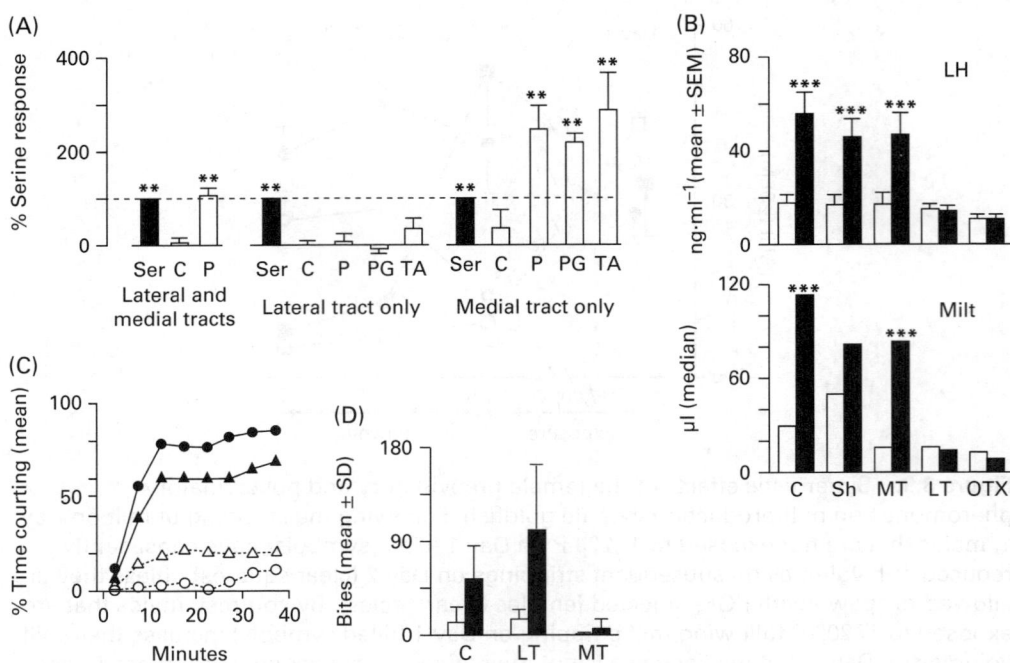

Figure 3.10. Evidence that the medial (MT) and lateral (LT) olfactory tracts have distinct functions in the goldfish. (A) Integrated electrical activity recorded in male goldfish by hook electrodes from either one complete olfactory tract (LT and MT), or a single lateral or medial tract, and expressed as a percentage of the response to a 10^{-5} M L-serine standard (Ser). Ser induced activity in all three preparations although a control methanol solution (Con) was ineffective. 17,20β-P (P), prostaglandin F$_{2α}$ (PG), and taurocholic acid (TA) induced activity only in the MT preparation. **, significantly different from Con; $P<0.01$. Redrawn with permission from Sorensen *et al.* (1991). (B) Effect of exposure to 17,20β-P (5×10^{-10} M; filled bars) or ethanol control (empty bars) on luteinizing hormone (LH; after 1 h post exposure) and milt volume (after 24 h exposure) in intact control (C) and sham-operated (Sh) fish, and in fish in which the LTs have been cut (MT), but not in fish in which the medial tracts have been cut (LT) or anosmic fish (OTX) in which all olfactory tracts have been cut. *, *** — significantly different from ethanol control; $P<0.05$ and 0.0001. Redrawn with permission from Dulka and Stacey (1990). (C) Courtship behavior of male goldfish 2 d after receiving one of four olfactory tract (OT) surgeries: Unilateral OT cut (filled circles); unilateral OT cut plus contralateral LT cut (filled triangles); unilateral OT cut plus contralateral MT cut (empty triangles); bilateral OT cut (empty circles). Redrawn with permission from Stacey and Kyle (1983). (D) Feeding (biting) responses before (open bars) and immediately after (filled bars) addition of a food extract in crucian carp that received a sham operation (C) or had both medial olfactory tracts (LT) or both lateral olfactory tracts (MT) cut. El Hassan Hamdani, Alexander Kasumyan, Kjell B. Døving. Is Feeding Behaviour in Crucian Carp Mediated by the Lateral Olfactory Tract? Chemical Senses, vol 26. Copyright © 2001, Oxford University Press.

(Fujita *et al.*, 1991). Indeed, as discussed in detail by Lastein, Hamdani, and Døving, 2014, it is clear that the olfactory system mediates responses to pheromones; pheromonal inputs from olfactory bulb to the brain are carried in the medial olfactory tracts, whereas food odor inputs are carried in the lateral tracts (Fig. 3.10).

Although the female's preovulatory and postovulatory pheromones induce qualitatively similar increases in male LH concentration and milt volume, they exert these effects through very different mechanisms. For example, pheromonal 17,20β-P increases LH and milt by activating a simple neuroendocrine reflex, because it is effective not only in males held in groups but also in single males isolated from all social stimuli (Sorensen *et al.*, 1989; Fraser and Stacey, 2002). By contrast, the postovulatory prostaglandin odorants do not increase LH and milt in isolated males, but are effective only when males are either tested in groups or are spawning with a $PGF_{2\alpha}$-injected female (Sorensen *et al.*, 1989; Zheng and Stacey, 1996). Together, these results suggest that the postovulatory prostaglandin odorants induce their endocrine effects indirectly through the sociosexual behaviors they trigger among conspecifics.

Because the female's preovulatory steroid and postovulatory prostaglandin pheromones affect LH and milt through different mechanisms, they have been studied with different experimental protocols. Studies of the preovulatory pheromone have simply added sex steroids to aquaria containing single or grouped males because this pheromone acts directly on the individual male to trigger a neuroendocrine reflex. However, in experiments of the postovulatory pheromone, which evidently acts indirectly through changes it induces in conspecifics, males have been tested either in the presence of other males with which they interact during pheromone exposure, or with $PGF_{2\alpha}$-injected female conspecifics with which they interact sexually as a result of the PGF odors the females release. For this reason, males in postovulatory pheromone studies are described as being exposed to "social stimuli" or "spawning stimuli," rather than to "pheromonal stimuli" (because it is not known which aspects of the complex of sensory stimuli are responsible for the observed treatment effects).

The mechanisms mediating the effects of the preovulatory and postovulatory pheromones differ not only in their requirements for social context but also in how they stimulate sperm movement to the ducts (Fig. 3.11). The key difference is illustrated by the fact that if the pituitary gland is removed to eliminate the endogenous source of LH, male goldfish are able to increase milt volume in response to spawning stimuli but not in response to 17,20β-P exposure (Dulka *et al.*, 1987a; Zheng and Stacey, 1996). These experiments demonstrate that even though the postovulatory pheromone normally increases both LH and milt in intact males (Sorensen *et al.*, 1989), the milt increase can occur through an LH-independent, extra-pituitary pathway—a finding consistent with the results of experiments that examined the effects of water temperature on pheromone-induced milt responses. For example, at the normal spawning temperature of 20° C, the latency to the milt increase induced by pheromonal 17,20β-P is at least 3–4 h, whereas the latency to the milt increase induced by spawning stimuli is much less than 1 hr and, unlike the 17,20β-P-induced milt increase, is not increased at lower water temperatures (Kyle *et al.*, 1985; Zheng and Stacey, 1996). The short latency of the PGF pheromone-induced milt increase suggests not only that it would be capable of replenishing sperm stores during spawning (as seen in Fig. 3.9) but also that it might be mediated by a neuromuscular mechanism that moves milt from the testes to the sperm ducts (see Dulka and Demski, 1986), a possibility that has not been investigated.

The preovulatory and postovulatory pheromones also differ in the mechanisms by which they increase male LH (Fig. 3.11). As in many teleost fishes (Zohar *et al.*, 2010), LH secretion in goldfish is regulated through direct innervation of the pituitary gonadotropes by preoptic-hypothalamic neurons that stimulate and inhibit LH release through secretion of GnRH and dopamine (DA), respectively. Pheromonal 17,20β-P evidently acts through both pathways to stimulate LH release, whereas pheromonal PGFs appear to act only through the GnRH pathway.

For example, injecting an inhibitory GnRH analogue to antagonize the action of endogenous GnRH blocks the male's LH response to both 17,20β-P and spawning stimuli, indicating that both the preovulatory and postovulatory pheromones increase LH by increasing endogenous GnRH release (Murthy *et al.*, 1994; Zheng and Stacey, 1997). The inhibitory GnRH analogue also blocks

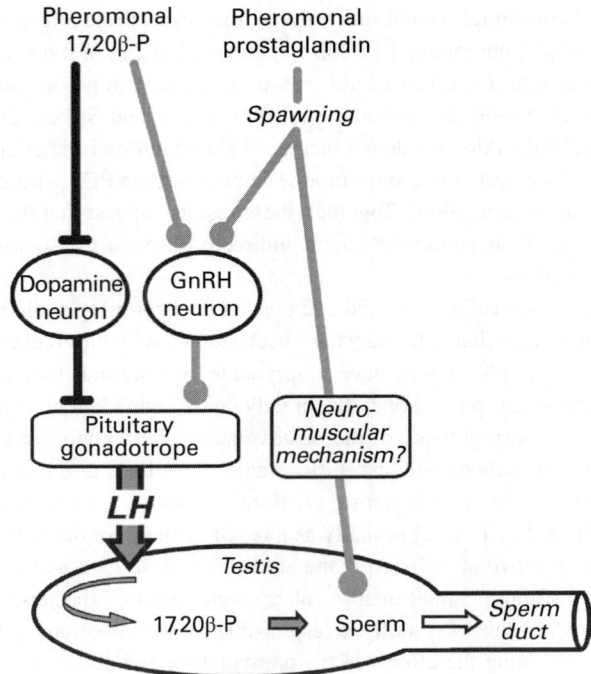

Figure 3.11. The preovulatory steroid pheromone and postovulatory prostaglandin pheromone of goldfish increase releasable sperm through different mechanisms. Pheromonal 17,20β-P can act directly on individuals, whereas pheromonal prostaglandin acts through the spawning stimuli that it induces. Although both pheromones increase LH concentration in males by actions on gonadotropin-releasing hormone (GnRH) neurons, only pheromonal 17,20β-P increases LH by inhibition of hypothalamic dopaminergic neurons. In addition, prostaglandin-induced spawning stimuli rapidly increase releasable sperm through an unknown extra-pituitary pathway that could involve testicular smooth muscle. Stacey N.E. & Sorensen P.W. (2011) Hormonal pheromones. In: Encyclopedia of Fish Physiology: From Genome to Environment, Vol. 1, (ed. A.P. Farrell). pp. 1553–1562. Elsevier, San Diego.

17,20β-P-induced milt increase but does not block spawning-induced milt increase, providing further evidence that the postovulatory pheromone can increase milt through an LH-independent mechanism (Zheng and Stacey, 1997). Chung-Davidson *et al.* (2008) also report that GnRH mRNA is differentially affected by 17,20β-P and $PGF_{2\alpha}$ exposure, although the relevance of these findings to specific reproductive function remains to be explored.

Several lines of evidence indicate that in addition to acting through GnRH, pheromonal 17,20β-P increases LH by disrupting dopaminergic inhibition of pituitary gonadotropes. Brief 17,20β-P exposure both elevates LH for up to 2 h and reduces pituitary DA turnover (Dulka *et al.*, 1992). However, in fish pretreated with dopamine type-2 receptor agonists to maintain inhibition of gonadotropes, 17,20β-P-induced LH and milt responses are blocked, whereas spawning-induced responses are unaffected (Zheng and Stacey, 1997). Finally, even though exposure to suprathreshold concentrations of 17,20β-P induces no further increase in LH (Dulka *et al.*, 1987a), combined exposure to 17,20β-P and $PGF_{2\alpha}$ has a synergistic effect on LH secretion (Sorensen *et al.*, 1989), providing further evidence that the preovulatory and postovulatory pheromones increase LH through distinct mechanisms.

Unfortunately, nothing is known about how 17,20β-P-s, which acts through different olfactory receptors than 17,20β-P (Sorensen *et al.*, 1995b), increases LH and milt, or how AD inhibits the endocrine-gonadal response to 17,20β-P (Stacey, 1991). Nonetheless, our currently rudimentary understanding of the goldfish preovulatory and postovulatory pheromones (Figs. 3.4 and 3.11) reveals that these female hormonal odors trigger an unexpectedly complex suite of physiological and behavioral responses in males that evidently have evolved as tactics for sperm competition. Throughout the periovulatory period, males respond to rapidly changing female hormonal odors that exert releaser effects on intra- and intersexual behavioral interactions, while also inducing multiple primer effects that first increase both the quantity and quality of releasable sperm in anticipation of imminent spawning opportunities, and then activate both endocrine and apparently nonendocrine mechanisms that likely function to replenish sperm stores during spawning. In addition, males regulate releasable sperm not only in response to the relatively well-characterized preovulatory steroid and postovulatory prostaglandin odors but also in response to poorly understood cues from males, as briefly discussed in Section 3.2.1.

THE MALE GOLDFISH PHEROMONE(S)

Unlike the situation with female goldfish, where all three known pheromones have stimulatory effects, chemical cues released by male goldfish induce both stimulatory and inhibitory effects. However, only the inhibitory cue (likely AD) is known to be a hormonal pheromone and, in comparison to the preovulatory and postovulatory pheromones of female goldfish, is poorly understood.

AD is a potent goldfish odorant (10 pM EOG detection threshold) and the most abundant of more than a dozen steroids released by male goldfish (Sorensen, Pinillos, and Scott, 2005). Although mature but sexually inactive males release very large quantities of AD ($\sim$50 ng h^{-1}), this release increases more than 10-fold shortly after they begin to spawn (Sorensen, Pinillos, and Scott, 2005). Importantly, aquarium studies show that water-borne AD induces both stimulatory releaser and inhibitory primer effects on males: AD quickly increases intermale agonistic behavior (Poling, Fraser, and Sorensen, 2001), but also inhibits milt production induced by pheromonal 17,20β-P (Stacey, 1991). Recent evidence (Levesque and Sorensen, unpublished) strongly suggests that AD released from males attracts ovulated, sexually receptive females.

Various experiments using grouped and isolated male goldfish show that uncharacterized male cues change milt volume in ways that appear adaptive in sperm allocation and sperm competition. Although there is no direct evidence that pheromonal AD is responsible for these milt changes, a pheromone operating in the males' immediate vicinity is suspected. For example, grouped males that have not been interacting sexually with females, and therefore have basal plasma LH concentrations, evidently inhibit milt production in their tank mates because, if isolated from the group, their milt volume increases dramatically for several days and returns to basal levels following regrouping (Fraser and Stacey, 2002; Fig. 3.12A). This milt increase in the absence of male cues can be blocked by water-borne AD (Stacey, 1991), and is induced by mechanisms distinct from those inducing milt responses to the female preovulatory and postovulatory pheromones because it has a longer latency (12–24 h) and is not associated with LH increase (Stacey *et al.*, 2001; Fraser and Stacey, 2002).

In addition to suppressing their milt production in the presence of other mature males with basal LH, males increase their milt production in the presence of "stimulatory" males that have either been exposed to 17,20β-P or injected with gonadotropin (Fig. 3.12B and C)—a response that is not associated with LH increase (Stacey *et al.*, 2001; Fraser and Stacey, 2002). The cue(s) produced by stimulatory males is unlikely to be 17,20β-P or other free and conjugated C21 steroid odorants (Sorensen and Scott, 1994; Sorensen *et al.*, 1995) because male goldfish release negligible quantities of these steroids (Sorensen, Pinillos, and Scott, 2005). Moreover, males with elevated LH greatly increase

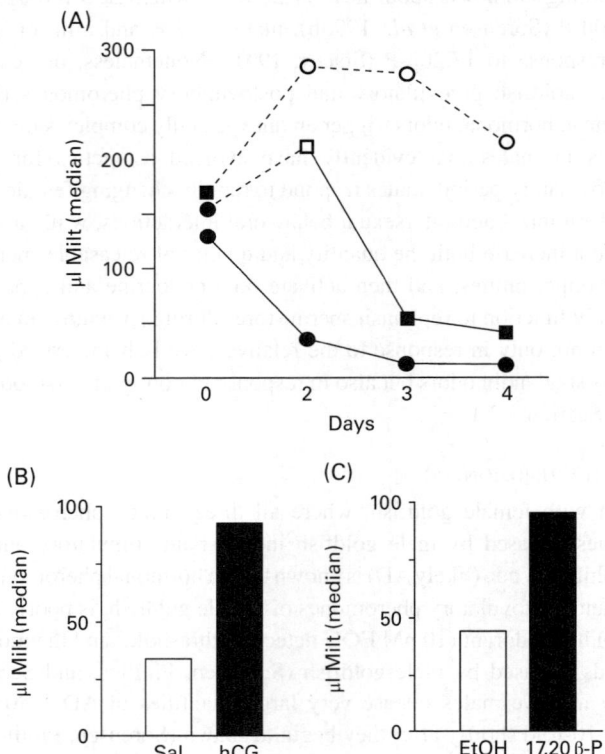

Figure 3.12. In male goldfish, milt production can be both stimulated and inhibited by cues from other males. (A) If groups of males (filled symbols) are stripped of milt on Day 1, they replace smaller volumes on Day 2 if returned to their groups (filled circles) and larger volumes if held in isolation (open circles). Males that are stimulated by isolation on Day 1 (open square) are inhibited if returned to their groups on Day 2 (filled squares). Redrawn with permission from Fraser and Stacey (2002). (B) Milt volumes of males held for 12 h with a human chorionic gonadotropin (hCG)-injected male are larger than those of males held for 12 h with a saline-injected male (Sal). Redrawn with permission from Stacey *et al.* (2001). (C) Milt volumes of males held for 12 h with a male previously exposed to 17,20β-P are larger than those of males held for 12 h with an ethanol-exposed male (EtOH). Fraser, E.J. & Stacey, N.E. (2002) Isolation increases milt production in goldfish. *Journal of Experimental Zoology*, 293, 511–524.

their release of AD (Sorensen, Pinillos, and Scott, 2005), which should have the effect of suppressing, rather than increasing, milt in other males (Stacey, 1991).

Despite a lack of information on the stimulatory and inhibitory cues released by mature males, it is clear that these cues induce conspecific milt changes that are as consistent and dramatic as those induced by the female periovulatory pheromones. Moreover, these studies of male–male interactions suggest that the preovulatory pheromone of female goldfish increases milt production in two fundamentally different ways: (1) directly, by stimulating males that enter its active space and (2) indirectly, by inducing these stimulated males to release unknown factors that in turn stimulate additional males. The selective pressure of sperm competition thus evidently has resulted in mechanisms that the male goldfish not only to increase his absolute fertility by responding to ovulatory females cues but also to increase his relative fertility by responding to cues from male competitors;

the latter effect is somewhat analogous to the eavesdropping behaviors that have evolved in visually and acoustically mediated communication networks (McGregor and Peake, 2000; Wisenden and Stacey, 2005).

A GOLDFISH CHEMICAL NETWORK

Having no sensory experience of another species' pheromones, we have a natural tendency to conceptualize them simply as acting between an individual sender and receiver. However, this is at odds with the goldfish, where the proximity of individuals in social groups and the size of pheromone active spaces enable hormonal pheromones to simultaneously synchronize various reproductive processes among the numerous individuals forming local aggregations. Although our current understanding of these pheromonal functions is rudimentary, it is clear that what might at first appear to be a rather simple and unspecialized mating system (nonterritorial, promiscuous, and nonparental) is in fact likely regulated by a chemically coordinated social network with two surprising features (Fig. 3.13). First, it is complex, enabling females to achieve ovulatory synchrony and males to adaptively adjust milt production to the perceived level of sperm competition through cues provided by both males and females. Second, it is very dynamic, because individuals can rapidly change their pheromonal status from receivers to senders.

Prior to spawning, vitellogenic females release an uncharacterized recrudescence pheromone that attracts males, which in turn are both sources and receivers of inhibitory cues that suppress milt production in their local group through tonic reciprocal inhibition. Provided that environmental cues such as increasing temperature and spawning substrate trigger an ovulatory LH surge in even a single female, this stable, prespawning social situation can be perturbed in two ways. First, release of the

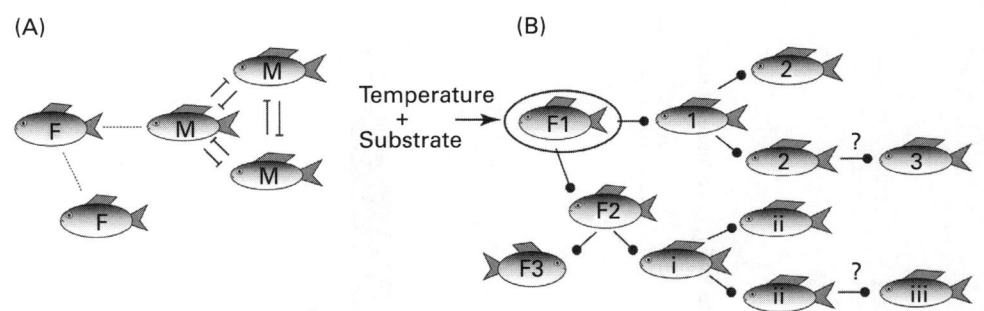

Figure 3.13. Proposed mechanism for pheromone-mediated synchronous ovulation and mass spawning in goldfish. In the nonspawning condition of early spring (A), when water temperature is low and aquatic vegetation (spawning substrate) has not begun to grow, males (M) are in a basal endocrine state and release unknown cues that suppress milt volume in other males. Laboratory experiments suggest that the transition to the spawning condition (B) can occur when increasing water temperature, and appearance of aquatic vegetation trigger an ovulatory LH surge in a single female (F1), which then releases a preovulatory steroid pheromone that in turn triggers ovulatory LH surges in additional females (F2 and F3). The preovulatory pheromone also directly increases releasable sperm in males (1 and i) by increasing LH, and indirectly increases sperm in additional males (2 and ii) through an unknown mechanism; it is not known whether such indirectly stimulated males can increase releasable sperm in additional males (3 and iii). Stacey, N.E., Chojnacki, A., Narayanan, A., et al. (2003) Hormonally-derived sex pheromones in fish: exogenous cues and signals from gonad to brain. Canadian Journal of Physiology and Pharmacology, 81, 329–341. NRC Press.

ovulatory female's steroidal preovulatory pheromone likely has the potential to trigger additional LH surges in females she encounters, thereby greatly amplifying the active space of the preovulatory pheromone. Second, males exposed to the preovulatory pheromone also exhibit a rapid LH increase that will increase the quality and quantity of their releasable milt within hours and in the process also transforms the male into a source of an (unknown) stimulatory cue that increases releasable sperm in additional males. Although it is possible that some elements of goldfish reproduction have been influenced by domestication, EOG and other studies suggest that the closely related crucian and common carps have remarkably similar hormonal pheromone systems.

3.2.2 Hormonal Pheromones in Other Cypriniforms

Although it is not known whether the carp hormonal pheromone system is found in more distant-related cypriniforms, it is expected to be the case. It seems likely that the postovulatory goldfish PGF pheromone is ubiquitous among cypriniforms, given that all of more than 75 species subjected to EOG testing detect PGFs (Stacey and Sorensen, 2005, 2009; Stacey, 2010), and that males spawn with $PGF_{2\alpha}$-injected females in both cyprinoid (Java barb, *B. gonionotus*; Liley and Tan, 1985 and Cardwell *et al.*, 1995: redtail sharkminnow, *Epalzeorhynchos bicolor*; Belanger, Pachkowski, and Stacey, 2010) and cobitoid species (pond loach, *M. anguillicaudatus*, Kitamura, Ogata, and Takashima, 1994a). In the loach, as in the goldfish (Kobayashi, Sorensen, and Stacey, 2002), intact but not anosmic males court and spawn with nonovulated females that have been injected with PGF (Kitamura, Ogata, and Takashima, 1994a). *Misgurnus* release negligible amounts of PGFs prior to ovulation; when ovulated, they release large amounts of 13,14-dihydro-15-keto-$PGF_{2\alpha}$ ($>2\,\mu g \cdot h^{-1}$; Fig. 3.14), less $PGF_{2\alpha}$, and virtually no 15k-$PGF_{2\alpha}$ (Ogata, Kitamura, and Takashima, 1994), consistent with EOG studies

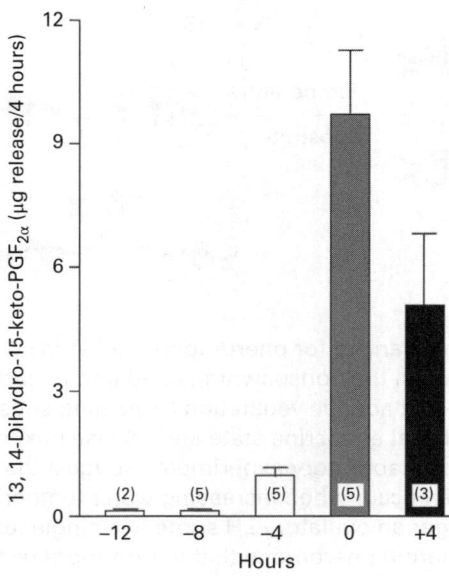

Figure 3.14. Release of immunoreactive 13,14-dihydro-15-keto-prostaglandin $F_{2\alpha}$ by female oriental weatherfish loach (*Misgurnus anguillicaudatus*) induced to ovulate by human-chorionic gonadotropin (hCG) injection. At 4 h intervals following hCG injection, fish were checked for ovulation, which was first detected in the 0 h sample. Ogata, H., Kitamura, S. & Takashima, F. (1994) Release of 13,14-dihydro-15-keto-prostaglandin $F_{2\alpha}$, a sex pheromone, to water by cobitid loach following ovulatory stimulation. Fisheries Science, 60, 143–148.

(Kitamura, Ogata, and Takashima, 1994a), indicating that males are much less sensitive to $PGF_{2\alpha}$ or 15k-$PGF_{2\alpha}$ (1 nM detection threshold) than they are to 13,14-dihydro-15-keto-$PGF_{2\alpha}$ (1 pM detection threshold).

Unlike the situation with PGFs, which are expected to be detected by all cypriniforms, EOG screening studies indicate that steroids are not detected by cobitoids (11 species representing all 4 cobitoid families have been tested (Stacey and Sorensen, 2005, 2009; Stacey, 2010)). By contrast, steroids are detected by all major lineages within superfamily Cyprinoidea, with the exception of Acheilognathinae (five bitterling species tested) (Cardwell, Dulka, and Stacey, 1992; Cardwell *et al.*, 1995; Freidrich and Körsching, 1998; Pinillos *et al.*, 2002; Lower, Scott, and Moore, 2004; Stacey and Sorensen, 2005, 2009 and unpublished; Belanger, Pachkowski, and Stacey, 2010; Stacey, 2009, 2010). EOG screening studies likely provide a distorted picture of cypriniform steroid detection patterns, because they have tested only a small fraction of extant cypriniforms, and have not included all important steroid odorants detected by cypriniforms.

3.3 HORMONAL PHEROMONES IN SALMONIFORMES

Despite their economic importance as sport and cultured species, the relatively small-order Salmoniformes (containing only the family Salmonidae) comprises only about 65 species in three subfamilies: Thymallinae (graylings), Coregoninae (whitefishes), and Salmoninae (char, trout and salmon) (Nelson, 2006). Hormonal pheromones have been studied only in whitefishes and salmonins. The sole coregonin study clearly shows that lake whitefish (*C. clupeaformis*) detect two $PGF_{2\alpha}$ metabolites (15k-$PGF_{2\alpha}$ and 13,14-dihydro-$PGF_{2\alpha}$) that act through one olfactory receptor mechanism, increase male whitefish locomotor behavior, and induce larger magnitude EOG responses in males than in females (Hara and Zhang, 1997; Laberge and Hara, 2003) (Fig. 3.15A). In contrast to coregonins, salmonins are relatively well understood, as discussed below.

3.3.1 Genus *Salmo*: Atlantic Salmon and Brown Trout

Hormonal pheromone research in subfamily Salmoninae has focused on Atlantic salmon and brown trout, in which similar pheromone systems might have facilitated natural hybridizations (Olsén *et al.*, 2000; Garcia-Vazquez *et al.*, 2004). Studies in both species have exploited a convenient laboratory model, the precociously mature male parr.

PHEROMONAL PROSTAGLANDINS IN GENUS *SALMO*

In EOG recordings, mature Atlantic parr are very sensitive to PGFs, and increase plasma hormones in response to water-borne $PGF_{2\alpha}$ (Moore and Waring, 1995, 1996; Olsén *et al.*, 2001); however, the source of $PGF_{2\alpha}$ is not clear. For example, Moore and Waring (1996) suggest that the PGF priming activity is in ovulated female urine, which they found to have significant amounts of immunoreactive $PGF_{2\alpha}$. By contrast, Olsén *et al.* (2001, 2002) report that ovarian fluid has much more $PGF_{2\alpha}$ than ovulated urine, which in their studies had weak priming activity and strong releasing activity. Despite such inconsistencies, these studies all indicate that a PGF pheromone in ovulatory fluid and/or urine has pheromonal priming effects in precocious male Atlantic salmon.

Similarly, ovulated female odor induces in male brown trout endocrine priming effects that appear to be due to PGFs. EOG studies are in general agreement both on the relative olfactory potencies of PGFs (e.g., $PGF_{2\alpha} \geq PGF_{1\alpha} > 15k\text{-}PGF_{2\alpha} > 13,14\text{-dihydro-}15k\text{-}PGF_{2\alpha}$), and the presence of a single olfactory receptor mechanism for the detected PGFs (Essington and Sorensen, 1996; Moore and Waring, 1996; Hara and Zhang, 1997; Moore *et al.*, 2002; Laberge and Hara, 2003). As with Atlantic salmon (Olsén *et al.*, 2001, 2002), brown trout ovulatory fluid contains far more immunoreactive $PGF_{2\alpha}$ ($>175\,ng\,ml^{-1}$) than does ovulated urine ($<5\,ng\,ml^{-1}$), and both increase plasma hormones and/or milt (Olsén *et al.*, 2000; Moore *et al.*, 2002). By contrast, there is conflicting evidence as to

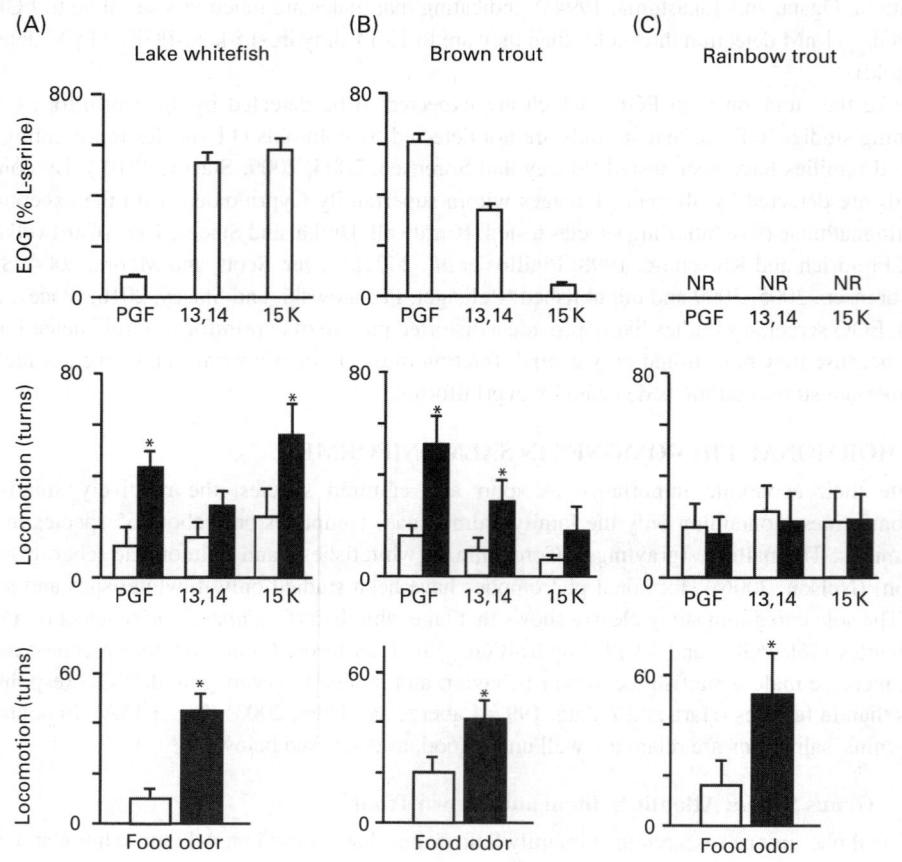

Figure 3.15. Relationship between olfactory and behavioral responsiveness to prostaglandin (PG) in three salmonids. Top panels: At 10 nM concentration, rainbow trout exhibit no EOG response (NR) to $PGF_{2\alpha}$ (PGF), 13,14-dihydro-15-keto-$PGF_{2\alpha}$ (13,14), or 15-keto-$PGF_{2\alpha}$ (15 K), whereas brown trout and lake whitefish respond with markedly different EOG magnitudes (mean + SEM; as a percentage of 10 mM L-serine standard). Middle panels: With the exception of 13,14-dihydro-15-keto-$PGF_{2\alpha}$ in lake whitefish (which induced large EOG response but no behavioral response), swimming turns following exposure to detectable PGFs (filled bars) were significantly greater than prior to PGF exposure (open bars). Bottom panels: In the same apparatus used to test PGF responses, all species exhibited more swimming turns following food-odor exposure (filled bars) than prior to food-odor exposure (open bars). Laberge, F. & Hara, T.J. (2003) Behavioral and electrophysiological responses to F-prostaglandins, putative spawning pheromones, in three salmonid fishes. Journal of Fish Biology, 62, 206–221.

whether EOG responsiveness changes with maturation. Moore *et al.* (2002), for instance, report that $PGF_{2\alpha}$ induces EOG response only in mature brown trout parr (immature parr failed to respond even at 10 µM), whereas other studies in brown trout (Essington and Sorensen, 1996; Laberge and Hara, 2003) and Arctic char (Sveinsson and Hara, 2000) report equivalent PGF-induced EOG responses in undifferentiated juveniles and adults of both sexes. Regardless of these inconsistencies, PGFs induce both releaser and primer effects in brown trout. Exposure to PGFs increases locomotor behavior in mature bourgeois male trout (Fig. 3.15B) and induces digging and nest probing (prespawning behaviors) in females (Laberge and Hara, 2003), whereas $PGF_{2\alpha}$ and $PGF_{1\alpha}$ increase plasma hormones and milt volume of mature male trout parr (Moore *et al.*, 2002), as they do in Atlantic salmon parr (Moore and Waring, 1996).

PHEROMONAL STEROIDS IN GENUS *SALMO*

In contrast to the situation with pheromonal PGFs, it is much more difficult to interpret the evidence for steroidal pheromones in Atlantic salmon and brown trout. Moore and Scott (1991), for example, report that precocious Atlantic salmon parr have extreme EOG sensitivity to T, but only for a brief period prior to spawning, a change in olfactory sensitivity that has not been observed in brown trout or any other fish. T stimulates positive rheotaxis (upstream movement in a current) in male Atlantic parr (Moore, 1991), but does not affect plasma hormones or milt volume (Waring *et al.*, 1996). Of particular concern is the report that ovulated female urine dramatically increases olfactory response to 17,20β-P-s within minutes (Moore and Scott, 1992). It is important to confirm this seemingly anomalous finding, because it has not been reported in any other vertebrate, and it is unknown what biological effect water-borne 17,20β-P-s might have on the precocious Atlantic parr (Waring and Moore, 1995; Waring *et al.*, 1996).

It will also be important in future studies of brown trout and Atlantic salmon not only to determine the specific PGFs detected, their origins and routes of release, and their specific pheromonal functions, but especially to confirm findings (Essington and Sorensen, 1996; Hara and Zhang, 1997; Laberge and Hara, 2003) that brown trout do not detect steroids such as T and 17,20β-P-s that have been proposed to be pheromones in Atlantic salmon (Moore, 1991; Moore and Scott, 1991, 1992). Such apparently striking differences in steroid detection between congeners are markedly different from what has been reported in cypriniforms (Sorensen *et al.*, 1990; Irvine and Sorensen, 1993; Sorensen and Scott, 1994; Stacey and Sorensen, 2005, 2009) and other groups (Narayanan and Stacey, 2003).

3.3.2 Genus *Salvelinus* (Chars)

Although EOG screening studies of chars have not tested the range of steroids and prostaglandins used for cypriniforms, it appears that this group commonly detects PGFs and etiocholanolone glucuronide (Etio-G) (Essington and Sorensen, 1996; Hara and Zhang, 1997; Laberge and Hara, 2003); brook char (*S. fontinalis*) also are reported to detect testosterone glucuronide (T-g; Essington and Sorensen, 1996). EOG responsiveness to $PGF_{2\alpha}$ is unaffected by gender or maturity in brook char (Essington and Sorensen, 1996) and Arctic char (Sveinsson and Hara, 2000), consistent with one study of brown trout (Laberge and Hara, 2003), but in marked contrast to others in brown trout (Moore *et al.*, 2002) and Atlantic salmon parr (Moore and Waring, 1995). Although there is no information on the possible pheromonal functions of Etio-g and T-g in *Salvelinus*, Sveinsson and Hara (2000) have proposed, based on their findings that ovulated char are attracted to $PGF_{2\alpha}$ and that males release immunoreactive PGFs, that PGFs function in Arctic char as a male pheromone affecting females, the reverse function of that proposed for brown trout (Moore *et al.*, 2002), where postovulatory PGFs are thought to affect males. This apparently large difference in spawning pheromones is surprising considering that brook char and brown trout will hybridize (Sorensen *et al.*, 1995c).

3.3.3 Genus *Oncorhynchus* (Pacific Salmon)

With respect to putative steroidal pheromones, *Oncorhynchus* species appear broadly similar to other salmonids in being relatively insensitive to steroids. EOG studies, for example, indicate that immature rainbow trout (*Oncorhynchus mykiss*) detect T (Pottinger and Moore, 1997), whereas mature rainbow trout and chinook (*O. tshawytscha*) and amago salmon (*O. rhodurus*) detect only Etio-G (Kitamura, Ogata, and Takashima, 1994b; Hara and Zhang, 1997; Laberge and Hara, 2003; Stacey and Sorensen, 2005, 2009), which is also detected by lake whitefish, brown trout, and brook trout

(Essington and Sorensen, 1996; Hara and Zhang, 1997; Laberge and Hara, 2003). Finally, as reported in mature male crucian carp (Bjerselius, Olsén, and Zheng, 1995); precociously mature male chinook salmon parr avoided 17,20β-P in a choice maze, whereas age-matched immature males did not (Dittman and Quinn, 1994); importantly, however, it has not yet been shown that chinook salmon detect 17,20β-P.

With respect to their use of PGF pheromones, *Oncorhynchus* sp., are clearly different than other salmonids. For example, just as EOG studies consistently report that PGFs are detected by coregonins (lake whitefish), *Salmo* (brown trout, Atlantic salmon), and *Salvelinus* (brook trout, Arctic char) (Essington and Sorensen, 1996; Moore and Waring, 1996; Sveinsson and Hara, 2000; Laberge and Hara, 2003), they are consistent in failing to detect PGF-induced responses in rainbow trout, chinook salmon, and amago salmon (Kitamura, Ogata, and Takashima, 1994b; Hara and Zhang, 1997; Laberge and Hara, 2003; Stacey and Sorensen, 2005, 2009) (Fig. 3.15C).

Despite their insensitivity to PGFs, male *Oncorhynchus* sp. respond to chemical cues from ovulated conspecifics. Ovulatory urine contains immunoreactive $PGF_{2\alpha}$ in masu salmon (*Oncorhynchus masou*) and, in rainbow trout and masu salmon induces releaser (attraction) and endocrine primer effects (Scott, Liley, and Vermeirssen, 1994; Vermeirssen *et al.*, 1997; Yambe *et al.*, 1999; Vermeirssen and Scott, 2001; Yambe and Yamazaki, 2000, 2001a, 2001b; Yambe *et al.*, 2003). In masu salmon, both the releaser (Yambe *et al.*, 2006) and primer effects (Yambe *et al.*, 2008) of urinary pheromone can be replicated by the tryptophan metabolite, L-kynurenine. This appears to be the first instance in which a nonhormonal teleost fish sex pheromone has been characterized, although there are reports that amino acids can induce sexual behaviors in bitterlings (Kawabata, 1993).

From current understanding of salmonid phylogeny (e.g., Crespi and Fulton, 2004), it might be inferred that olfactory sensitivity to PGF odorants was present in a common ancestor to all salmonids, and that the insensitivity to PGFs in Pacific trout and salmon (*Oncorhynchus* species) is either due to loss of receptors, or the retuning of the receptor to PG compounds that have not yet been tested (Fig. 3.16A). Corollaries of this line of reasoning are that olfactory sensitivity to PGFs is also likely to be present in the graylings (subfamily Thymallinae) (Fig. 3.16A) and that the olfactory PGF receptor in Salmoniformes might be homologous with that in the ostariophysans (cypriniforms, characiforms, etc.) (Fig. 3.16B).

3.4   HORMONAL PHEROMONES IN PERCIFORMES

Despite the fact that perciforms provided the first example of a hormonal pheromone in fish (Etio-g in the black goby; Colombo *et al.*, 1980), they have until recently been the focus of relatively little hormonal pheromone research. This is especially surprising given that the more than 10 000 species of perciform fishes comprise approximately 40% of living teleosts, are the dominant vertebrates in freshwater and marine ecosystems, and are of immense ecological and economic importance (Nelson, 2006).

Virtually all information on hormonal pheromones in perciforms comes from studies of the round goby (Murphy, Stacey, and Corkum, 2001; Corkum *et al.*, 2008) and cichlids such as *A. burtoni* (Cole and Stacey, 2006). In addition, studies of other perciforms (Mozambique tilapia, *Oreochromis mossambicus*; Frade *et al.*, 2002: Eurasian ruffe, *Gymnocephalus cernuus*; Sorensen *et al.*, 2004) are a cautionary reminder of how little is known about pheromones in this group.

In EOG tests, for example, Mozambique tilapia did not detect any of a relatively small number of steroids, despite evidence that males release a urinary pheromone that could be a sulfated aminosterol (Frade *et al.*, 2002; Barata *et al.*, 2008). In ruffe, laboratory studies showed that females during their preovulatory LH surge produce a pheromone that is very likely to be steroidal, despite the fact that EOG studies show that this species does not detect any of a large number of free and conjugated steroids (Sorensen *et al.*, 2004). Such findings provide strong evidence that EOG screening studies

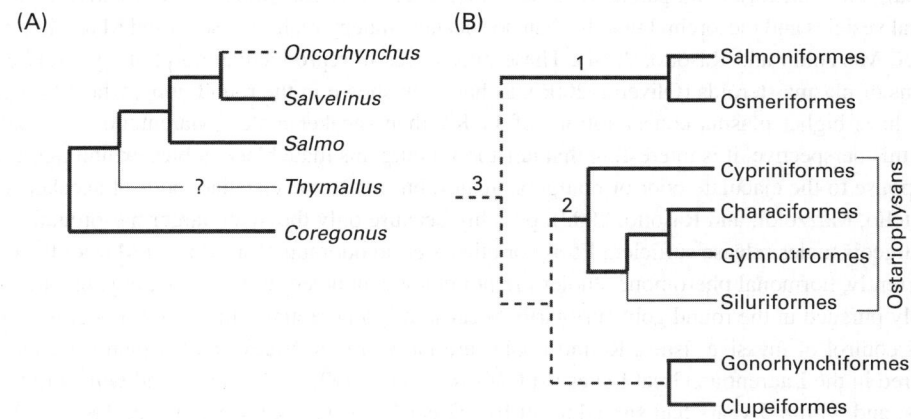

Figure 3.16. Relationship between phylogeny and ability to detect F-series prostaglandins (PGFs) in electro-olfactogram (EOG) studies; salmoniform phylogeny (A) from Crespi and Fulton (2004) and higher order phylogeny (B) from Saitoh *et al.* (2003). (A) Sensitivity to PGFs (thick line) may have been present in the common ancestor of salmoniforms, among which thymallins (graylings) are predicted to detect PGFs (thin line); *Oncorhynchus* spp. appear not to detect PGFs (dotted line). See text for sources of EOG studies. (B) Olfactory detection of PGFs by salmoniforms, cypriniforms, and related orders might have a common origin. (1) PGF detection is seen in Osmeriformes (Kitamura, Ogata, and Takashima, 1994b), suggesting salmoniform and osmeriform PGF detection is homologous. (2) In EOG studies, all cypriniforms, characiforms, and gymnotiforms that have been tested, and some siluriforms (Stacey and Sorensen 2002, 2009) detect PGFs, suggesting a common origin for this trait among all ostariophysans. (3) If PGF detection arose in a common ancestor to ostariophysans and salmoniforms–osmeriforms, then gonorhynchiforms (milkfish) and clupeiforms (herring) would be predicted to detect PGFs. See text for sources of EOG studies. Stacey, N.E. (2010) Hormonally derived sex pheromones in fishes. In: Hormones and Reproduction of Vertebrates, Vol. 1, (eds. D.O. Norris & K.H. Lopez). pp. 169–192. Elsevier, San Diego.

alone, in the absence of whole-odor bioassays, are likely to significantly underestimate the prevalence of hormonal pheromones in fishes.

3.4.1 Gobies: Family Gobiidae

In the Gobiidae, a primarily marine family with as many as two thousand species, mating systems typically involve male territoriality, attraction and courtship of ovulated females, and paternal nest defense, social interactions that can be mediated by intraspecific visual, acoustic, and chemical cues (Tavolga, 1956; Rollo *et al.*, 2007; Meunier *et al.*, 2009; Yavno and Corkum, 2010). The first evidence that hormonal pheromones are involved in coordinating goby reproductive behaviors comes from the pioneering black goby studies of Colombo, Belvedere, and Pilati (1977) and Colombo *et al.* (1980) who reported that (1) males synthesize large amounts of 5β-reduced androgen glucuronides and sulfates in the mesorchial gland, a nonspermatogenic, Leydig cell-rich area of the testis, and (2) adding one of the major mesorchial gland products (Etio-g) to artificial nests induces ovulated females to approach and oviposit in the nests (see also Chapter 11).

More recent studies of age-dependent alternate reproductive strategies suggest that Etio-g also might mediate male–male interactions in black gobies. For example, male black gobies that exhibit the bourgeois (territorial and parental) strategy invest relatively less in testes, but relatively more in seminal vesicles and mesorchial glands, than do sneaker strategy males (Rasotto and Mazzoldi, 2002; Immler, Mazzoldi, and Rasotto, 2004). These differences in reproductive morphology likely affect patterns of plasma steroids (Oliveira, 2006), as has been shown in the round goby, where bourgeois males have higher plasma concentrations of 11-KT than sneaker males (Marentette *et al.*, 2009). From this perspective, it is interesting that territorial bourgeois male black gobies exhibit aggression in response to the ejaculate odor of bourgeois males, but not to the ejaculate odor of sneaker males (Locatello, Mazzoldi, and Rasotto, 2002), possibly because only the much larger mesorchial glands of bourgeois males release sufficient Etio-g or other steroid odorants (Rasotto and Mazzoldi, 2002).

Currently, hormonal pheromone studies are not being conducted in the black goby, but are being actively pursued in the round goby, primarily because it is a potentially important model for pheromonal control of invasive fishes. Round gobies are native to the Black and Caspian seas, but first appeared in the Laurentian Great Lakes (St Clair River) in 1990, likely introduced with ship ballast waters, and within 5 years had spread to all five Great Lakes (Corkum, Sapota, and Skora, 2004). Round gobies in the lakes are of concern for a variety of reasons, particularly because they prey on eggs and young of indigenous fishes and, because they feed on benthic organisms in contaminated sediments, can accelerate movement of contaminants through food webs. Round gobies spawn multiple times over an extended summer breeding season, and their mating system is typical of most gobies, parental males defending cryptic nests to which females are attracted for oviposition (Corkum, Sapota, and Skora, 2004).

Initial EOG studies (Murphy, Stacey, and Corkum, 2001) found that male and female round gobies do not detect any known PGFs, but exhibit equivalent olfactory responses to more than a dozen free and conjugated 18-, 19-, and 21-carbon sex steroids, including Etio-g (Fig. 3.17A). EOG cross-adaptation studies (Murphy, Stacey, and Corkum, 2001) then showed that these detected steroids were discriminated by at least four olfactory receptor mechanisms, for which the most potent of the tested odorants were etiocholanolone (Etio), E2-3g, estrone (E1), and dehydroepiandrosterone-3-sulfate (DHEA-s) (Fig. 3.17B).

Murphy, Stacey, and Corkum, (2001) were not able to observe reproductive responses when steroid odorants were added to aquaria; however, they fortuitously discovered that most steroid odorants increased the frequency of ventilation (opercular and buccal pumping), which likely functions to increase water flow through the olfactory organ (Nevitt, 1991; Belanger *et al.*, 2006). Moreover, despite the fact that steroid odorants induced equivalent EOG responses in males and females, the ventilation responses to steroid odorants were dramatically sexually dimorphic. For example, females increased ventilation only when exposed to Etio (and other steroids shown by cross-adaptation tests to act *via* the Etio receptor), whereas males increased ventilation in response to Etio, E1, and E2-3g (Murphy, Stacey, and Corkum, 2001) (Fig. 3.17C).

The steroid-induced ventilation increase in the round goby is a transient response that adapts within minutes, even in the continued presence of odorant, and therefore provides a simple behavioral cross-adaptation assay that has demonstrated two fundamental and important aspects of the ventilation response. First, the behavioral cross-adaptation tests indicate that males discriminate at a central level the same steroid odors that EOG cross-adaptation tests indicate they discriminate at the peripheral sensory level (Murphy, Stacey, and Corkum, 2001; Fig. 3.17D). Second, female gobies that receive androgen implants not only exhibit the male-typical pattern of hyperventilation in response to Etio, E1, or E2-3g, but also discriminate behaviorally among these steroids in cross-adaptation tests (Murphy and Stacey, 2002; Fig. 3.17E). Given that steroid-induced EOG responses are sexually isomorphic (Murphy, Stacey, and Corkum, 2001; Fig. 3.17A), these results in androgenized females indicate that androgens act centrally to induce the sexually dimorphic hyperventilation response. More recent studies (Belanger, Corkum, and Zielinski, 2007) confirm that E1 stimulates

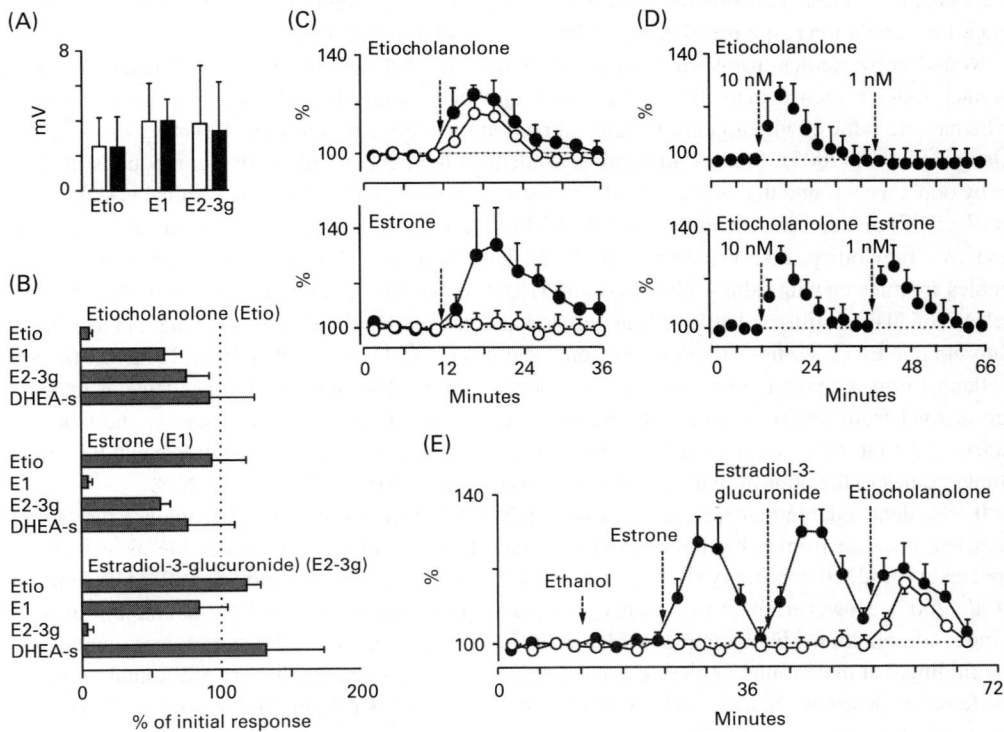

Figure 3.17. Electro-olfactogram (EOG) and behavioral responses (mean + standard error of the mean or SEM) of round gobies to water-borne steroids. (A) EOG responses to etiocholanolone (Etio), estrone (E1), and 17β-estradiol-3-glucuronide (E2-3g) are equivalent in males (filled bars) and females (open bars). (B) EOG cross-adaptation studies indicate that Etio, E1, and E2-3g are each detected by separate olfactory receptor mechanisms. Top: During adaptation to 100 nM Etio, EOG responses to a 10 nM Etio pulse are drastically reduced, whereas responses to pulses of E1, E2-3g, and dehydroepiandrosterone-s (DHEA-s) are unaffected. Middle and bottom: similar results during adaptation to E1 and E2-3g. (C) Both males (filled circles) and females (empty circles) exhibit a transient increase in ventilation frequency (presented as a percentage of mean pre-exposure frequency) when exposed to 10 nM water-borne Etio (arrow), whereas only males respond to 10 nM E1. (D) Males that are behaviorally adapted by prolonged exposure to 10 nM Etio do not increase their ventilation rate in response to a 10% increase in Etio concentration, but do increase their ventilation rate in response to an equivalent amount of E1. (E) Females implanted with methyltestosterone capsules for 18 days (filled circles) do not increase ventilation when exposed to ethanol vehicle (EtOH), but they do increase ventilation when exposed to 1 nM E1, E2-3g, and Etio. By contrast, control females implanted with empty capsules (open circles) respond only to Etio. Murphy, C.A., Stacey, N., & Corkum, L.D. (2010) Putative steroidal pheromones in the round goby, *Neogobius melanostomus*: olfactory and behavioral responses. Journal of Chemical Ecology, 27, 443–470. Springer.

ventilation in male gobies and provide further indirect evidence that sensitivity of the ventilatory response to E1 increases during the reproductive season. For example, males with large testes (1.5% of body mass), that were caught and tested in the spawning season, increased ventilation at a threshold E1 concentration of 10^{-11} M; whereas males with small testes (0.24% of body mass), that were caught and tested outside the spawning season, required a higher E1 concentration (10^{-9}M) to elicit the ventilation response (Belanger, Corkum, and Zielinski, 2007).

Round goby studies using whole body odors indicate that the odor of mature males induces a greater EOG response than the odor of immature males and also attracts vitellogenic females, whereas the odor of vitellogenic females attracts nonvitellogenic females (Bélanger *et al.*, 2004; Gammon *et al.*, 2005). Studies attempting to identify the behaviorally active compounds in these goby odors have found that, *in vitro*, both the testis (Arbuckle *et al.*, 2005) and seminal vesicle (Jasra *et al.*, 2007) metabolize AD into a number of 5β-reduced and 11-oxygenated steroids including Etio and two previously unknown fish steroids: 11-keto-Etio and 11-keto-Etio-s. Moreover, if male gobies are injected with salmon GnRH (sGnRH) to stimulate the pituitary–gonadal axis, they increase release of 11-keto-Etio and at least four sulfated and glucuronated metabolites (Katare *et al.*, 2011). Subsequent EOG studies (Laframboise and Zielinski, 2011) report that both 11-keto-Etio and 11-keto-Etio-s are potent odorants that act through separate olfactory receptor mechanisms that also are distinct from that mediating EOG response to Etio and Etio-g (Murphy, Stacey, and Corkum, 2001). Thus far, however, it remains unknown what steroid or steroid mixture attracts mature female round gobies to the odor of mature males (Gammon *et al.*, 2005; Corkum *et al.*, 2008).

It is understandable that research on round gobies has focused on male pheromones that attract females, because from a biological control perspective this likely is a vulnerable aspect of this species' reproductive biology (Bélanger *et al.*, 2004; Corkum, Sapota, and Skora, 2004; Gammon *et al.*, 2005). However, there is evidence that round goby pheromones mediate interaction in other gender combinations. For example, evidence for male responses to female pheromones comes from the finding that male round gobies detect and respond to E1 and E2-3g that are presumably released by females (Murphy, Stacey, and Corkum, 2001; Belanger, Corkum, and Zielinski, 2007), and Tavolga's classic studies of frillfin gobies (Tavolga, 1956).

3.4.2   Cichlids: Family Cichlidae

The family Cichlidae, with more than 1300 species distributed in Africa, South and Central America, and the Indian subcontinent, is the most diverse non-Ostariophysan teleost family in freshwater, and has generated much interest among evolutionary biologists because of its multiple and rapid adaptive radiations, particularly in the African rift valley lakes of Malawi, Tanganyika and Victoria (Salzburger *et al.*, 2005; Nelson, 2006), the site of the "East African radiations" (Schwarzer *et al.*, 2009).

Cichlids typically exhibit complex social and parental behaviors, a situation where it is anticipated that selection would have favored sensory specializations that facilitate social interactions among conspecifics. Although in this context cichlid visual communication likely has received the most attention (e.g., Oliveira *et al.*, 2001; Seehausen *et al.*, 2008), cichlids also utilize both audition (Amorim *et al.*, 2008; Van Staaden and Smith, 2011) and olfaction (Frade *et al.*, 2002; Miranda *et al.*, 2005; Barata *et al.*, 2007, 2008) to mediate interactions among conspecifics.

SEX PHEROMONE STUDIES IN CICHLIDS

Although there is no definitive behavioral evidence that cichlids use hormonal pheromones, there is strong circumstantial evidence for a number of species. In two well-studied African cichlids, the Mozambique tilapia and *A. burtoni*, there is strong evidence that sex pheromones are released in urine. Mozambique tilapia are native to lakes and rivers of southeast Africa, and *A. burtoni* is endemic to Lake Tanganyika; both species are maternal mouthbrooders in which males defend territories in leks to which females come to spawn disproportionately with dominant males (Miranda *et al.*, 2005; Maruska and Fernald, 2012).

In the Mozambique tilapia, urine of preovulatory females induces larger EOG responses in males than does urine from postovulatory females, and males increase urination rates in response to preovulatory, but not postovulatory females (Miranda *et al.*, 2005). Further, in male tilapia, concentrations of several free and conjugated steroids are positively correlated with male dominance status (Oliveira *et al.*, 1996), and males respond to intruder males by increasing uation rate with a pattern that is temporally distinct from the urination response to females (Almeida *et al.*, 2003; Barata *et al.*, 2007). In *A. burtoni*, Maruska and Fernald (2012) compared the effects of conspecific visual stimuli and social interaction on male urination rates, and found results very similar to those reported in Mozambique tilapia. Male *A. burtoni* increase urination both in response to cues from dominant males and preovulatory females, but not in response to cues from postovulatory (brooding) females (Maruska and Fernald, 2012).

Although no cichlid pheromone has yet been chemically characterized, Barata *et al.* (2008) have proposed that male Mozambique tilapia release a urinary sulfated amino-sterol that functions as a pheromone to signal dominance status to females (a sea lamprey pheromone has recently been shown to be a sulfated amino-sterol (Sorensen and Hoye, 2007)). In addition, EOG studies of *A. burtoni* (Cole and Stacey, 2006) clearly show that, as in the round goby (Murphy, Stacey, and Corkum, 2001), the olfactory organ detects a diverse variety of 18-, 19-, and 21-carbon steroids but is insensitive to a variety of F-series prostaglandins. However, although the round goby detects both conjugated and unconjugated steroids, *A. burtoni* appears to detect only conjugated (glucuronidated and sulfated) steroids, which EOG cross-adaptation and binary mixture tests indicate act through five olfactory receptor mechanisms capable of discriminating both the position and type of conjugate: for example, 3-glucuronide; 17-glucuronide; 3-sulfate; 17-sulfate; 3,17-disulfate (also see Section 3.4.2). Presently, there is no evidence that the steroid conjugates detected by *A. burtoni* induce a biological response, or even that they are synthesized and released. However, the number and variety of steroids that are discriminated (Cole and Stacey, 2006) and the fact that they are conjugated (and thus likely to be released in controlled urine pulses: Barata *et al.*, 2007; Scott and Ellis, 2007; Katare *et al.*, 2011; Maruska and Fernald, 2012) suggest that *A. burtoni* has evolved a complex hormonal pheromone system.

Male Nile tilapia (*Oreochromis niloticus*) have been reported to both increase their milt volume and sperm motility in response to 17,20β-P exposure (Pinheiro, Souza, and Barcellos, 2003) and to increase courtship if females are injected with 17,20β-P (Souza *et al.*, 1998), but only if nares are intact (untreated). These findings appear to conflict with the results of EOG recording studies in *A. burtoni* (Cole and Stacey, 2006) and a number of *Oreochromis* species (*O. aureus, esculentus, mossambicus*, and *tanganicae*; see Frade *et al.*, 2002; Stacey, 2009, 2010 and unpublished results), where 17,20β-P did not induce an EOG response and only conjugated steroids were detected. However, it is possible that the reported effect of 17,20β-P on milt in Nile tilapia (Pinheiro, Souza, and Barcellos, 2003) is due not to olfaction, but rather to uptake of 17,20β-P from the water and subsequent endocrine action (Vermeirssen and Scott, 1996; Scott, Pinillos, and Huertas, 2005; Maunder *et al.*, 2007). In addition, it is possible that male Nile tilapia responded to 17,20β-P-injected females (Souza *et al.*, 1998) because the females metabolized the injected 17,20β-P and released detectable conjugates.

EVOLUTION OF OLFACTORY SENSITIVITY TO SEX STEROIDS IN AFRICAN CICHLIDS

A consistent theme throughout this chapter is that a fish's phylogenetic position is a strong predictor of the kinds of prostaglandins and sex steroids it can detect. Indeed, olfactory responses to hormonal compounds appear to be sufficiently stable over evolutionary time that, given a group of fishes whose phylogenetic history is well understood, it should be feasible to observe the origin and evolution of hormonal pheromone systems. Cichlid fishes provide excellent models for this kind of study. Not only are the phylogenetic relationships of many major cichlid taxa well studied (Salzburger *et al.*, 2005; Koblmüller, Sefc, and Sturmbauer, 2008; Schwarzer *et al.*, 2009), but cichlids are also speciose, widely distributed, and diverse in their mating systems, and thus are rich sources of evolutionary material. Finally, cichlids typically reproduce readily in captivity, greatly facilitating the study of hormonal pheromone function.

Currently, four major clades of cichlids are recognized (Sparks and Smith, 2004; Genner *et al.*, 2007), the basal Etroplinae and Ptychochrominae found in India and Madagascar, and two derived sister groups, the African Pseudocrenilabrinae and the New World Cichlinae. Both Mozambique tilapia and *A. burtoni* are pseudocrenilabrins, and *A. burtoni* is a member of the "modern haplochromines" (Salzburger *et al.*, 2005), the remarkably diverse (>1000 spp.) crown group of pseudocrenilabrins that has undergone an explosive speciation unmatched by other cichlid lineages. Cole and Stacey (2006) examined olfactory detection of hormonal compounds in *A. burtoni* to test the hypothesis that complex pheromone systems might have played a role in the East African cichlid radiations by facilitating chemically mediated conspecific recognition and reproductive isolation. Having found that *A. burtoni* detects and discriminates a number of conjugated steroids, the focus of my ongoing work has been to determine when and where these olfactory sensitivities arose in the evolutionary history of cichlids.

Initial EOG screening studies tested a variety of cichlids from Africa, Madagascar, India, and the New World with a large number of free and conjugated sex steroids (Cole and Stacey, 2006) at the nominally high concentration of 10 nM. Only African species (the pseudocrenilabrin clade) exhibited EOG responses and, as with *A. burtoni*, they responded only to conjugated steroids (Stacey, 2009, 2010, and unpublished). Since then, these EOG studies have been conducted on more than 75 species of African cichlid, including at least one species from all but three of the 22 currently recognized pseudocrenilabrin clades. Although the overall results are too complex for this general review and still incomplete, several key findings are worth noting because they appear to provide our only insights into how sex pheromone systems can evolve in fish.

First, *Heterochromis multidens*, the basal African cichlid (Schwarzer *et al.*, 2009), is the only African species tested that does not detect any steroid conjugates; thus, the ability to detect steroid conjugates likely arose more than 50 million years ago (mya), between the time that the Heterochromini and Tylochromini clades split from the remainder of the pseudocrenilabrin clade (Fig. 3.18). Although there is no evidence that this, or any steroid detection in cichlids, is involved with pheromonal function, the very great likelihood that all extant pseudocrenilabrins except for *Heterochromis* detect steroid conjugates suggest that this olfactory innovation has had adaptive value throughout the evolutionary history of the pseudocrenilabrins.

The first olfactory receptor (3G-1) for sex steroid conjugates to evolve in African cichlids appears to be most sensitive to $5\beta,3\alpha$-steroids with the glucuronide attached to carbon 3. However, it is crucial to note that this conclusion is based on the activities of the relatively few steroid conjugates that are commercially available, and not on the steroid conjugates that cichlids actually release, a situation which is important to address because cichlids could theoretically produce more than 500 sulfated and/or glucuronated steroids, and only approximately 70 were used in EOG screening studies. The 3G-1 receptor appears to be the only olfactory receptor for conjugated sex steroid in the basal clades Chromidotilapiini, Hemichromini, Pelmatochromini, and Etiini (Fig. 3.18).

To the extent that it can be determined with the limited number of available steroid conjugates, the 3G-1 receptor appears to have retained its ligand specificity throughout the more than 50 myr since its origin. All tested Africans (other than *Heterochromis*) have been found to be most sensitive to 17P-3α-g, a metabolite of 17α-hydroxyprogesterone, the immediate precursor of 17,20β-P, and the typical relative potencies of the four pregnan and androstan compounds that interacted with the 3G-1 receptor are as depicted in Figure 3.18.

Of the remaining four steroid conjugate receptors that I have documented in *A. burtoni* (17-G, 3-S, 17-S, 3,17-S; Cole and Stacey, 2006), two appear apparently simultaneously about 40 myr after

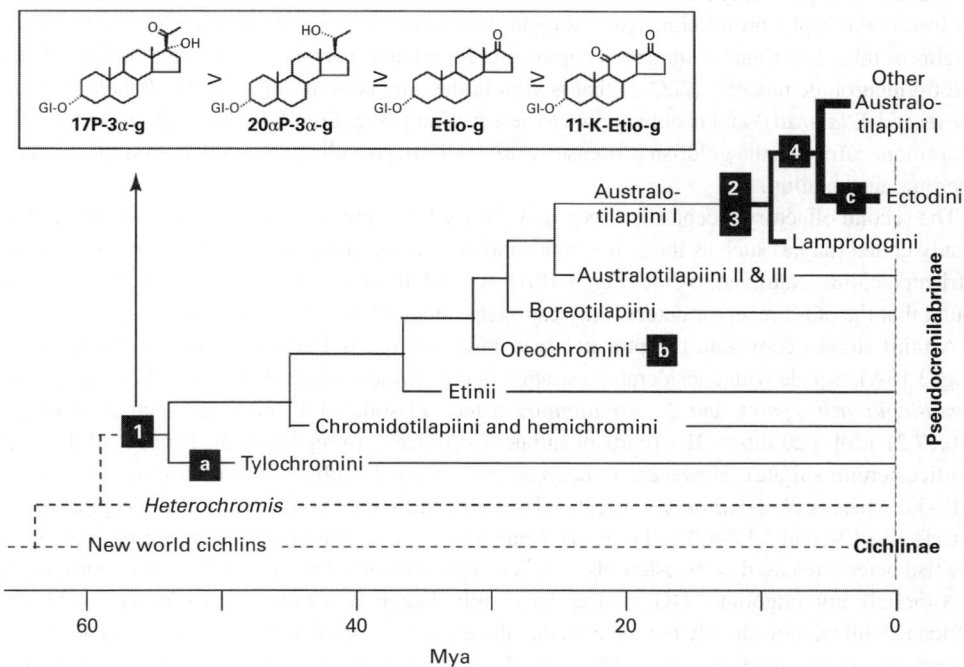

Figure 3.18. Cichlid phylogeny and the origins of olfactory sensitivity to steroid conjugates. *Astatotilapia burtoni* detects a range of conjugated steroids (Cole and Stacey 2006) that appear to be detected by five olfactory receptor mechanisms responding to 3-glucuronides, 3-sulfates, 17-glucuronides, 17-sulfates, and 3,17-disulfates. EOG studies of more than 90 cichlid species suggest detection of steroid conjugates arose near the base of the African pseudocrenilabrin clade with the 3-glucuronide receptor (3G-1; [1]). Approximately 40 myr later, two receptors appear in groups such as the lamprologins: a very specific 17-glucuronide receptor (17G-1; 2) and a nonspecific 17-sulfate receptor (17S-1; 3) that detects both 17-sulfates and 3,17-disulfates. In more highly derived pseudocrenilabrins (4), sulfate receptors are more specific, as in *A. burtoni*, although the situation is complex and unresolved. In addition to the olfactory receptors observed in *A. burtoni*, three additional receptors have been found: a 20-glucuronide receptor (20G-1; [a]) in tylochromins, a second 3-glucuronide receptor (3G-2; [b]), and a 21-sulfate receptor (21S-1; [3]); see text for details. Cichlid phylogeny and age estimates from Sparks and Smith (2004), Genner *et al.* (2007), and Schwarzer *et al.* (2009).

the 3G-1 receptor appears (Fig. 3.18), a very specific glucuronide receptor (17G-1) and a surprisingly nonspecific sulfate receptor (17S-1). As in *A. burtoni*, the 17G-1 receptor appears to be an "estrogen" receptor, because it detects E2-17-g, as does *A. burtoni*, and E3-17-g, which was not tested in *A. burtoni* (Cole and Stacey, 2006), but does not detect 17-glucuronated androgens (21-carbon, progesterone-like steroids cannot be glucuronated at carbon 17).

I have found that the 17S-1 receptor in lamprologins detects disulfates such as E2-3,17-di-s and 5-androstene-3β,17β-di-s as well as a variety of 17-sulfates including E2-17-s, T-s and 5α-androstan-3β,17β-diol-17β-s, a testosterone metabolite. However, in other more derived clades in the Austrotilapiin I lineage, there are, as in *A. burtoni*, two or more sulfate receptors specific to conjugates with the 3-s, 17-s, and 3,17-s configuration, but the situation is complex and as yet unresolved.

In addition to providing a rough outline of when and in what sequence steroid conjugate detection evolved in the haplochromin lineage leading to *A. burtoni*, EOG screening studies have revealed the origins of three additional conjugate receptors not found in *A. burtoni*. Likely, the first to evolve was a 20β-glucuronide receptor (20G-1) that is seen in the only tylochromins tested (*Tylochromis sudanensis* and *T. lateralis*) and is only known to detect 17,20β-P-g. Significantly, 17,20β-P-g has primer pheromone effects in the goldfish (Sorensen *et al.*, 1995b), providing an excellent example of parallel pheromone evolution.

The second olfactory receptor not seen in *A. burtoni* is found in some oreochromin species (commonly called tilapia) such as the Nile tilapia, and is a glucuronide receptor (3G-2) not seen in other African cichlids examined to date (Fig. 3.18A). As with the 17G-1 receptor discussed earlier, it was found that the 3G-2 receptor detects only estrogens such as E1-3-G and E2-3-G.

A third steroid conjugate receptor not seen in *A. burtoni* is found in some species in Ectodini (Fig. 3.18A), a clade with considerable variation in the steroid conjugates detected. For example, both *Xenotilapia ochrogenys* and *X. ornatipinnis* detect 21-sulfated compounds such as 4-pregnen-11β,17,21-triol-3,20-dione-21-s (cortisol sulfate; Cort-s) and 4-pregnen-11β-21-diol-3,20-dione-21-s (corticosterone sulfate). However, *X. ochrogenys* detects 21-sulfates via a specific 21-S receptor (21S-1), whereas *X. ornatipinnis* detects 21-sulfates via a relatively nonspecific receptor that also detects 3β-, 17-, and 3,17-sulfated steroids. Nonetheless, these ectodins provide the first evidence that any fish detects released corticosteroids, which might be involved in signaling stress to conspecifics.

Although my ongoing EOG studies have only begun to examine pheromone evolution in African cichlids, they already have shown that the evidently complex steroidal pheromone system of *A. burtoni* did not arise *de novo* in the haplochromine lineage, but rather originated at the base of pseudocrenilabrin diversification (Genner *et al.*, 2007; Koblmüller, Sefc, and Sturmbauer, 2008; Schwarzer *et al.*, 2009). The pattern of olfactory receptor accumulation during pseudocrenilabrin evolution raises the key question of whether novel receptors are indicative of novel pheromone functions, or whether they simply indicate enhanced ability to discriminate a pre-existing function. In addition, it is striking that during the first 40 myr of pseudocrenilabrin evolution, all hormonal pheromones might have been steroid glucuronides which, if released via bile to the feces, might not have provided the efficient signaling systems that appear to have arisen through releasing steroid sulfates in urine (Barata *et al.*, 2007; Scott and Ellis, 2007; Maruska and Fernald, 2012).

3.5 PHYSIOLOGICAL REGULATION OF RESPONSIVENESS TO HORMONAL PHEROMONES

A remarkable aspect of hormonal pheromones is that the reproductive endocrine system is not only their source, but it also regulates their effects, thus synchronizing periods of pheromonal responsiveness with appropriate life history stages. Although relatively little is known of this aspect of hormonal pheromone function, these endocrine effects likely occur both in senders and receivers.

In senders, for example, the endocrine system is not simply the hormonal pheromone source, but it also exerts indirect effects on pheromone function. A male's androgenic hormones that attract females by enhancing his visual (Pankhurst, 2008) or acoustic signals (Bass and Zakon, 2005), for example, will thereby increase the likelihood that females will enter his pheromonal active space. Similarly, in receivers, hormones that increase a female's attraction to territorial male signals will increase the likelihood that she subsequently encounters his pheromones. Although these indirect endocrine effects on hormonal pheromone function have received little attention, other studies clearly indicate that reproductive hormones increase pheromonal responsiveness of receiving individuals through direct actions on both the olfactory epithelium (Cardwell *et al.*, 1995) and unknown central sites (Murphy and Stacey, 2002).

EOG studies indicating that adult gender differences in olfactory organ sensitivity are androgen dependent provide the clearest evidence that pheromonal responsiveness can be regulated by hormonal actions at the peripheral sensory level. For example, in a variety of cypriniforms such as goldfish, tinfoil barb, redtail sharkminnow, and zebrafish (Sorensen and Goetz, 1993; Cardwell *et al.*, 1995; Stacey *et al.*, 2003; Belanger, Pachkowski, and Stacey, 2010), and in several salmoniforms (lake whitefish and brown trout; Laberge and Hara, 2003), mature males have both lower olfactory detection thresholds to prostaglandins and larger prostaglandin-induced EOG responses, than do females. At least in the cyprinids, it is likely that androgenic hormones induce such gender differences in adult fish, because androgen treatment of juveniles greatly enhances response to F-prostaglandin (Cardwell *et al.*, 1995; Stacey *et al.*, 2003; Belanger, Pachkowski, and Stacey, 2010). Although nothing is known about how androgen might exert these changes in sensitivity to prostaglandins, the findings are consistent with reports of nuclear androgen receptors in fish olfactory epithelium (Pottinger and Moore, 1997).

Evidence that androgens can exert central hormonal effects that regulate both primer and releaser responses to hormonal pheromones comes from studies of fishes in which steroid odorants induce sexually dimorphic biological responses when steroid-induced EOG responses are sexually isomorphic. In goldfish, mature males and females exhibit equivalent EOG responses to the steroid pheromone 17,20β-P (Sorensen *et al.*, 1987), even though this odorant induces distinctly sexually dimorphic endocrine primer responses. In females, for example, 17,20β-P exposure appears able to trigger an LH increase only at night (Kobayashi, Sorensen, and Stacey, 2002), the normal time of the preovulatory LH surge (Stacey *et al.*, 1979a,b); whereas, it triggers an LH surge in male goldfish at any time of day (Dulka, Sorensen, and Stacey, 1987b), a male-typical response pattern that is exhibited by females following androgen treatment (Kobayashi *et al.*, 1997b).

The clearest evidence that hormonal androgens can act centrally to influence releaser responses to hormonal pheromones is seen in the round goby (*N. melanostomus*), in which males and females exhibit equivalent EOG responses to three steroids (Etio, E2-3-G, and E1) that are known from EOG cross-adaptation studies to act through separate olfactory receptor mechanisms (Murphy, Stacey, and Corkum, 2001; see Section 3.4.1). Males respond to all three steroid odorants by briefly increasing their ventilation rate, whereas females exhibit this response only to Etio, a gender difference that is unlikely due to sensory mechanisms given the equivalent steroid-induced EOG responses of males and females (Murphy, Stacey, and Corkum, 2001; Fig. 3.17A). The male-typical pattern of steroid odor-induced ventilation in round gobies evidently is androgen-dependent because, following androgen implant, female gobies hyperventilate in response to all three steroid odorants and also exhibit male-typical discrimination among the three odorants (Murphy, Stacey, and Corkum, 2001, Murphy and Stacey, 2002; Fig. 3.17E). The proposed central effects of androgens in round gobies may simply involve a reduced response threshold to sensory input, because in nonreproductive males the behavioral threshold for the E1-induced ventilation response is 1 nM, whereas in reproductive males it falls to 10 pM (Belanger, Corkum, and Zielinski, 2007), the threshold for E1-induced EOG response in males and females (Murphy, Stacey, and Corkum, 2001).

In masu salmon, precociously maturing male parr become attracted to the urine of ovulated females (Yambe *et al.*, 1999) and androgenic hormones appear to mediate not only this behavioral change (Yambe and Yamazaki, 2001a) but also a steroidal primer response to urine (Yambe *et al.*, 2003). Also in the cyprinid *Barilius bendelisis*, androgen treatment increases behavioral responsiveness to 15k-PGF$_{2\alpha}$ (Bhatt, Kandwal, and Nautiyal, 2002). In neither masu salmon nor *Barilius*, however, is there any information on the site of androgen action.

In summary, hormonal androgens clearly regulate male-typical responses to hormonal pheromones in a variety of fishes through actions at both the peripheral sensory and central levels. Although the mechanisms mediating these effects are unknown, the clear androgen effects on EOG responses in many cyprinids suggest it should be straightforward at least to determine whether androgen effects on sensitivity are mediated by altering the function of existing olfactory receptor neurons or stimulating new ones to develop. Similarly, the finding that bulbar units in crucian carp (*C. carassius*) display dramatic gender differences in discrimination of hormonal pheromones (Lastein, Hamdani, and Døving, 2006) provides a clue as to where androgen might act centrally to effect sexually dimorphic pheromone response. Although it seems likely that these mechanisms first evolved to restrict pheromonal responsiveness to appropriate stages of reproduction, it is possible that they now also might function to reduce heterospecific interactions, as discussed in Section 3.6.

3.6 HORMONAL PHEROMONES AND SPECIES SPECIFICITY

Successful sexual reproduction requires the evolution of species recognition systems that ensure sexual interactions are confined to conspecifics. Such recognition systems need not be unique, but simply distinct from those of sympatric heterospecifics. Accordingly, lack of specificity will not be problematic between such distinctly allopatric species as the goldfish and *T. sudanensis*, a Congo River cichlid, which have independently evolved the ability to detect 17,20β-P-g (Sorensen *et al.*, 1995b; Stacey, 2010). By contrast, specificity is expected to be problematic when incipient or recently diverged sister species retain a common suite of hormonal odorants and share similarities in morphology, mating system, habitat use, and nonchemosensory species recognition systems.

It is relatively easy to understand how species recognition systems that utilize visual or auditory information can operate to synchronize the activities of conspecifics, while avoiding interactions with sympatric heterospecifics, because visual and auditory information can be easily observed, and its transmission from sender to receiver can be readily determined. However, understanding possible specificity in hormonal pheromone systems is much more complex and difficult. For example, it is difficult to know exactly when hormonal pheromones are being released, what path they might take from sender to receiver, and how they might be diluted and degraded during this process (Sorensen *et al.*, 2000; Hubbard, Barata, and Canário, 2002; Mesquite, Canário, and Melos, 2003; Ellis *et al.*, 2004; Fabian *et al.*, 2007). In addition, hormonal pheromones such as the goldfish preovulatory steroid pheromone appear to be unspecialized gonadal by-products that change rapidly in composition and concentration (Figs. 3.4 and 3.6) and are released by multiple routes, making it difficult to understand how such a variable hormonal bouquet might encode species specificity. Finally, there is the question of whether and how other sensory systems (e.g., vision) might complement olfaction.

It is for these reasons that there is almost no generic information that bears directly on the issue of the species specificity of hormonal pheromones, despite considerable information about the kinds of hormonal odorants that some fishes release, detect, and respond to. This is not to say there is no evidence for species specificity in fish pheromones; numerous studies have indicated that fishes can clearly discriminate behaviorally between conspecific odor and the odor of other, even closely related, species (Liley, 1982; Stacey *et al.*, 1986; Sisler and Sorensen, 2008; Levesque *et al.*, 2011; Lim and Sorensen, 2011; Rosenthal *et al.*, 2011). It is important to note, however, that such studies typically use behavioral choice tests to show that conspecific odor is more attractive than heterospecific odor, a test protocol that has little relevance to natural situations, where fish would rarely if ever face such choices. More convincing evidence for specificity would be experiments showing that reproductive responses are induced only by conspecific odors in natural settings, although unfortunately this is rarely done in any pheromone studies (e.g., McKinnon and Liley, 1986; McLennan and Ryan, 1997; Olsén *et al.*, 2000; Yambe and Yamazaki, 2001b).

There are two key reasons why it has been difficult to understand how discrimination of conspecific from heterospecific odors might at times be based on pheromones composed solely of hormonal compounds. First, sex steroids and prostaglandins have been very highly conserved throughout vertebrate evolution, making them seemingly poor candidates for species-specific odorants. However, given the large number of sex steroid metabolites that are known to be detected by fish (Fig. 3.2), and the very great likelihood that many more remain to be discovered, the conserved nature of reproductive hormones may be less of a problem than once thought. Second, although EOG studies clearly indicate that distantly related fishes typically detect different hormonal odorants, they also indicate that the range of hormonal odorants can be remarkably similar (or identical) among closely related species, such as the goldfish and related carps, where maladaptive heterospecific interaction is most likely to occur (Irvine and Sorensen, 1993; Stacey and Sorensen, 2009; Stacey, 2010).

Given the obvious evolutionary importance of avoiding wasteful, fitness-reducing sexual interactions with heterospecifics, it is both unfortunate and surprising that so little can be said definitively about the issue of hormonal pheromone specificity. In the hope of stimulating future research, this chapter ends by discussing three key questions: (1) Is there always a biological need for hormonal pheromones to be species-specific? (2) How might species specificity evolve? (3) What is the evidence for species-specificity of hormonal pheromones? These and other issues in this complex topic have been discussed in earlier reviews (Stacey and Sorensen, 1991; Sorensen and Stacey, 1999; Stacey and Sorensen, 2008; Stacey, 2009, 2010), and related discussions on this topic are found in Sorensen and Baker (2014).

3.6.1 Is There Always a Biological Need for Hormonal Pheromones to be Species-Specific?

The basic concept of a species-specific pheromone is that an individual will benefit if it cannot detect, or does not respond to, the odors of sympatric heterospecifics. However, just as there is no need for specificity between allopatric species, neither is there a need for specificity between sympatric species that do not enter each other's pheromonal active space. Thus, to answer the question of the need for specificity, we first need to know how large hormonal pheromone active spaces are in relation to interspecific distances.

Although interspecific shoaling distances are unique to each set of species, there are some general features of hormonal pheromone active space derived from the nature of the hormones that produce them. Naively, we might expect that active spaces of hormonal pheromones would be small, given that hormones and their precursors and metabolites are in low concentrations even prior to release. To estimate active space, we require information on the amount of pheromone released, the olfactory detection threshold, and whether the pheromone is released continuously, as free steroids are across the gills, or intermittently, as steroid sulfates are in urine. This information is available for a number of goldfish hormonal odorants, and the volumes of the estimated active spaces differ considerably (Table 3.2).

For example, AD is passively released by spawning male goldfish and can be estimated to produce an active space of ~$250\,l\,h^{-1}$; such high AD release is presumably simply a result of high plasma AD levels, although these have not been measured. In contrast to the spawning male goldfish, the female goldfish releases peaks of preovulatory AD (~$100\,ng\cdot h^{-1}$) and 17,20β-P (~$5\,ng\cdot h^{-1}$) estimated to generate active spaces of only about $30\,l\,h^{-1}$, and that therefore should be detectable only in the female's immediate vicinity. The female's pulsatile release of 17,20β-P-s and 15k-PGF$_{2\alpha}$, on the other hand, would create urinary patches detectable at greater distances (Table 3.2).

Although the small estimated active spaces of goldfish hormonal pheromones are unlikely to be encountered by heterospecifics, this is probably a reflection of the goldfish social system rather than a common feature of all hormonal pheromones. For goldfish, living in bisexual aggregations likely has obviated the necessity of evolving specializations for detecting conspecifics at a distance. In other fishes, however, two key factors are expected to increase the active space of hormonal pheromones and thus to create selective pressures for species specificity.

First, a species' mating system can operate to separate the sexes. In the gobies, for example, females during the spawning season are separated from males, which guard spawning nests and eggs. Here, it appears likely that the female goby's chemically mediated searching for males has exerted sexual selection resulting in the evolution of a specialized testicular structure, the mesorchial gland, that amplifies production of male steroidal odorants and thus increases pheromonal active space (Colombo *et al.*, 1980; Locatello, Mazzoldi, and Rasotto, 2002; Gammon *et al.*, 2005). Thus, unlike the situation in goldfish, the active space of the male goby's attractant pheromone might well routinely encompass sympatric heterospecifics, potentially creating selective pressures for species-specific pheromones.

Second, active space is affected by release route, and hormonal odorants that are accumulated and released in urine pulses have the potential to create larger active spaces than their unconjugated precursors that are released continuously across the gills. Thus, for species such as cichlids, in which all steroidal odorants appear to be conjugates that would be released in urine (Cole and Stacey, 2006; Stacey, 2009 and unpublished results), selection for specificity is likely to be greater than for those species such as mochokid catfishes (Narayanan and Stacey, 2003) in which steroid odorants are likely all unconjugated. This effect of urinary release should be particularly pronounced in marine species, because reduced urine production results in elevated urinary steroid conjugates (Scott *et al.*, 1991). Unfortunately, almost nothing is known about hormonal pheromones in marine species, although the euryhaline round goby detects steroids (Murphy, Stacey, and Corkum, 2001) and the Indo-Pacific tarpon, *M. cyprinoides*, detects both PGFs and sex steroids (Stacey and Sorensen 2005).

3.6.2 How Might Species Specificity Evolve?

Species-specificity likely arises through two very different scenarios. First, specificity obviously can arise passively, either through divergence of allopatric sister species or by independent evolution of distinct hormonal pheromones. For example, mochokid catfishes, which evidently detect only unconjugated steroids (Narayanan and Stacey 2003), and cichlids, which evidently detect only conjugated steroids (Cole and Stacey 2006; Stacey 2010), are sympatric in many African rivers and lakes, but their species-specific olfactory responses to steroids could be considered to be passive because they undoubtedly arose independently.

In the second and more biologically interesting scenario, specificity could arise through active processes that occur when sympatric species using the same or similar odorants experience maladaptive interspecific responses that create selection pressure for increased conspecific discrimination. However, given the special nature of hormonal pheromones, it is unclear how these active processes might operate.

For example, most hormonal pheromones do not conform to the traditional concept that pheromones are specialized chemical signals, because their synthesis and release appear to be simply the inevitable consequence of reproductive endocrine activity, rather than being specialized to enhance communication to conspecifics. Such use of relatively unspecialized pheromones, which has been termed *chemical spying* to distinguish it from *chemical communication* (Fig. 3.19), has been proposed to be the primitive condition in hormonal pheromone evolution, and to have originated when receivers evolved the ability to detect and respond adaptively to (i.e., to benefit from) conspecifics' released hormones and hormonal metabolites, which at this evolutionary stage would be termed pheromonal *cues* (Sorensen and Stacey 1999; Stacey and Sorensen 2002, 2009; Stacey 2009, 2011). Conspecifics that release a pheromone cue may or may not benefit from the receivers' responses but, critical to the definition of chemical spying, remain unspecialized with respect to production and release of the cue. The primitive spying stage could then progress to chemical *communication* if the response of receivers can exert selective pressure for specialization in production and/or release of the hormonal odorant, which now would be termed a pheromonal *signal*, released by a *signaler*. Notably, while this definition of a pheromone is consistent with that suggested by this book (Sorensen, 2014), it is not consistent with all definitions (Wisenden, 2014). Semantics can be important.

The preovulatory steroid pheromone of the goldfish is likely an example of spying, because there is no evidence that steroid production or release is specialized for pheromonal function. In addition,

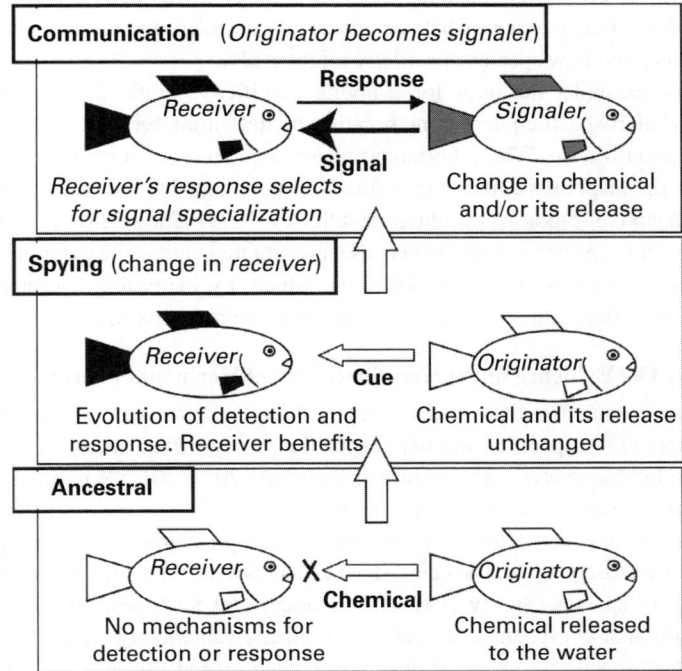

Figure 3.19. Proposed stages in the evolution of pheromonal communication. Theory suggested that all fishes are, or have been, at a prepheromonal *ancestral* stage, where hormones and metabolites are released to the water in an unspecialized manner, but are not detected by conspecifics. This stage progresses to the *spying* stage if receivers evolve specializations for detecting and responding to hormonal compounds but does not lead to the communication stage unless receivers exert selective pressure for specializations in pheromone production and release. Stacey, N.E. & Sorensen, P.W. (2002) Hormonal Pheromones in Fish. In: Hormones, Brain, and Behavior, Vol. 2, (eds D.W. Pfaff, A.P. Arnold, A.M. Etgen, et al.). pp. 375–435. Academic Press, New York. Copyright © 2002, Elsevier.

there are several reasons why the goldfish mating system should preclude specialization of the preovulatory pheromone. First, spawning aggregations typically involve multiple males and females in close proximity, making it unlikely that males could effectively exert mate preference based on individual variation in preovulatory pheromone release. Second, mistaken responses to this pheromone would have little consequence. Third, and most importantly, any mate preference exerted by males undoubtedly should be based on differential release of the postovulatory prostaglandin pheromone rather than the preovulatory steroid pheromone (i.e., there is little evolutionary pressure to specialize pheromone production and release).

Unlike this situation in goldfish, however, in species such as gobies, whose females evidently use olfactory cues to locate individual nesting males, it is reasonable to assume that female mate selection based on differential steroid release by males has driven the evolution of the specialized testicular mesorchial gland synthesizing pheromonal steroids and resulted in true chemical communication (e.g., Locatello, Mazzoldi, and Rasotto, 2002; Rasotto and Mazzoldi 2002; Arbuckle *et al.*, 2005). The proposed distinction between chemical spying and chemical communication, which has been discussed more extensively elsewhere (Sorensen and Stacey, 1999, Stacey and Sorensen, 2002,

2009; Wisenden and Stacey, 2005; Stacey, 2011), is far from trivial because, by focusing on the fundamental nature of the interaction between pheromone senders and receivers, it leads to key questions as to whether and how species-specificity might evolve.

In spying, for example, the onus for evolving specificity should fall entirely on the species responding to a heterospecific pheromone because, by definition, receivers do not act as a selective force for cue specialization. Where hormonal pheromones serve a communicatory function, however, heterospecific responses could reduce fitness of both signaler and receiver, thereby leading to more complex adaptive responses (Sorensen and Stacey, 1999). The spying–communication distinction also is relevant to species specificity because, in communication, one potential result of selection by receivers on signalers is signal amplification, which, by increasing pheromonal active space, potentially increases the number of species in functional reproductive sympatry.

3.6.3   What is The Evidence for Species Specificity of Hormonal Pheromones?

As noted earlier, despite abundant evidence that fishes can discriminate behaviorally between the whole body odors of conspecifics and heterospecifics (Liley, 1982; Stacey *et al.*, 1986; Sisler and Sorensen, 2008; Levesque *et al.*, 2011; Lim and Sorensen, 2011, 2012; Rosenthal *et al.*, 2011), there is little definitive evidence that hormonal pheromones *per se* are species-specific. In particular, where EOG studies have used suites of sex steroids and prostaglandins to compare the olfactory responsiveness of closely related species (Irvine and Sorensen, 1993; Essington and Sorensen, 1996) or groups of species (Stacey *et al.*, 1995; Stacey and Sorensen, 2009; Stacey, unpublished results), the hormonal compounds detected are typically remarkably similar between congeners, suggesting that, if specificity exists, it does not depend on the use of a few unique compounds. Another possible mechanism that could underlie species specificity in some cases would be for fish to release specific mixtures or blends of hormonal products (Sorensen and Stacey, 1999; Lim and Sorensen, 2011, 2012); but here, evidence is mixed and the rationale complicated by the existence of different release routes. Another possibility (see below; also see Sorensen and Baker, 2014) is that pheromone mixtures are complimented by mixtures of other nonhormonal products and perhaps other sensory cues.

The goldfish has been one of the few models available to address the role of pheromone mixtures. However, this model is complicated by the fact that although both genders release a bouquet of odorous steroids (Scott and Sorensen, 1994; Sorensen and Scott, 1994; Sorensen, Pinillos, and Scott, 2005), exposure to only 17,20β-P (Dulka *et al.*, 1987a) or 17,20β-P-s (Sorensen *et al.*, 1995b) is sufficient to induce apparently normal primer endocrine effects in males. These findings raise the important question of whether these primer effects would also be induced by heterospecific odor that contained 17,20β-P, a simple experiment that unfortunately has not been conducted in goldfish. However, male crucian carp, which respond to 17,20β-P exposure (Bjerselius, Olsén, and Zheng, 1995) or ovulatory conspecifics (Olsén, Sawisky, and Stacey, 2006) with the same, likely homologous, LH and milt responses seen in goldfish, do not exhibit these responses when held with ovulatory heterospecific cyprinids (white bream, *Blicca bjoerkna*; rudd, *Scardinius erythrophthalmus*) (Stacey *et al.*, 2012); unfortunately, steroid release by bream and rudd was not determined in these studies.

Even though single hormonal odorants can elicit biological responses (Defraipont and Sorensen, 1993; Sorensen *et al.*, 1989; Poling, Fraser, and Sorensen, 2001) in the goldfish, hormonal pheromones might often be mixtures that can be discriminated by the nature and ratios of their components. EOG and pheromone bioassay studies indicate that fish can discriminate a variety of hormonal products (goldfish; Kobayashi, Sorensen, and Stacey, 2002: round goby; Murphy, Stacey, and Corkum, 2001: *A. burtoni*; Cole and Stacey, 2006), potentially enabling discrimination between conspecific and heterospecific mixtures. Recently, Lim and Sorensen (2011, 2012) reported that even though ovulated common carp and goldfish release essentially the same postovulatory prostaglandin pheromones that stimulate male courtship, male common carp discriminate between the odor of ovulated conspecifics and ovulated goldfish in response to unknown polar metabolites that are unlikely to be hormonal products. These and related findings (Levesque *et al.*, 2011) are significant because they suggest that, even though their prostaglandin and steroid components are highly conserved and likely ubiquitous among at least the cyprinids (Stacey, 2010 and unpublished results), hormonal pheromones might readily become species-specific through the addition of nonhormonal components. These types of pheromones have been termed "complexes" and are described by Sorensen and colleagues (Lim and Sorensen, 2011, 2012; Sorensen and Baker, 2014). Of course, fish are a diverse and ancient group; therefore, species-specificity may be encoded in various ways. This important issue warrants considerable future research effort.

3.7 SUMMARY

The discovery that many fishes have evolved pheromonal functions for their released reproductive hormones has both significantly broadened our perspective of fish reproductive biology and raised important and interrelated issues that invite serious further study. The first is simply that sex steroids and prostaglandins—the only fish hormones known to act as pheromones—function both endogenously as hormones and exogenously as pheromones. Thus, unlike the situation with nonhormonal pheromones, the evolutionary specialization of hormonal pheromones is likely constrained by endocrine function. Likely solutions to this problem could entail specialization in hormonal precursors and metabolites, whose production and release could be regulated independently of the hormones *per se*. Although many fishes have evolved to use hormonal precursors and metabolites as pheromonal components (and cichlids may do this exclusively), there is as yet no evidence that this is the result of selection to separate pheromone synthesis from hormone synthesis.

The second issue is that it is not intuitively obvious how hormonal pheromones might function as species-specific odors, given that sex steroids and prostaglandins have been so highly conserved during vertebrate evolution, and that closely related species commonly detect very similar hormonal compounds. Although recent research on prostaglandin pheromones in cyprinids indicates specificity is achieved by the addition of nonhormonal odorants to the pheromonal mixture, the situation with steroidal pheromones remains to be explored.

Third, it has been proposed in this chapter and in earlier reviews (Sorensen and Stacey, 1999; Stacey and Sorensen, 2002, 2009; Wisenden and Stacey, 2005; Stacey, 2010) that, because their synthesis and release often appear to be unspecialized, and simply the inevitable consequence of endocrine activity, many hormonal pheromones may not conform to the traditional concept that pheromones are always specialized chemical signals. Although it may be difficult to find definitive proof that pheromonal specialization has not occurred, strong evidence might come from studies comparing pheromonal systems of related species where differences in mating systems might be mirrored by differences in hormonal pheromones. Good models for such studies might be the North American cyprinids (Tribe Leuciscinae), which detect hormonal compounds (Sorensen *et al.*, 1992) and exhibit diverse mating strategies including nonterritorial broadcast spawning (*Phoxinus*, *Ptychocheilus* spp.), where pheromonal specialization might not be predicted, as well as paternal territoriality and nest building (*Pimephales*, *Nocomis*, *Campostoma* spp.), where a situation analogous to that of the gobies might have evolved.

Last, EOG studies in groups such as the cichlids demonstrate that olfactory sensitivity to hormonal compounds can be highly conserved through evolutionary history, enabling us to understand when and how complex hormonal pheromone systems have evolved. Such studies should allow us to select model species that can shed light on hormonal pheromone function in large numbers of related species.

ACKNOWLEDGMENTS

The author is grateful to the Natural Sciences and Engineering Research Council of Canada (NSERC) for providing consistently generous support throughout his career.

REFERENCES

Almeida, O.G., Barata, E.N., Hubbard, P.C. *et al.* (2003) Urination rate of male tilapia (*Oreochromis mossambicus*) is highly dependent on social context. *Comparative Biochemistry and Physiology A*, **134**, S27–S28.

Amorim, M.C.P., Simões, J.M., Fonseca, P.J., and Turner, G.F. (2008) Species differences in courtship acoustic signals among Lake Malawi cichlid species (*Pseudotropheus spp.*). *Journal of Fish Biology*, **72**, 1355–1368.

Andersen, Ø. and Døving, K.B. (1991) Gonadotropin releasing hormone (GnRH)—a novel olfactory stimulant in fish. *Neuroreport*, **2**, 458–460.

Appelt, C.W. and Sorensen, P.W. (2007) Female goldfish signal spawning readiness by altering when and where they release a urinary pheromone. *Animal Behavior*, **74**, 1329–1338.

Arbuckle, W.J., Belanger, A.J., Corkum, L.D. *et al.* (2005) *In vitro* biosynthesis of novel 5ß-reduced steroids by the testis of the round goby, *Neogobius melanostomus*. *General and Comparative Endocrinology*, **140**, 1–13.

Barata, E.N., Hubbard, P.C., Almeida, O.G. *et al.* (2007) Male urine signals social rank in the Mozambique tilapia (*Oreochromis mossambicus*). *BMC Biology*, **5**, 54–64.

Barata, E.N., Fine, J.M., Hubbard, P.C. *et al.* (2008) A sterol-like odorant in the urine of Mozambique tilapia males likely signals social dominance to females. *Journal of Chemical Ecology*, **34**, 438–449.

Bass, A.H. and Zakon, H.H. (2005) Sonic and electric fish: at the crossroads of neuroethology and behavioral neuroendocrinology. *Hormones and Behavior*, **48**, 360–372.

Bélanger, A.J., Arbuckle, W.J., Corkum, L.D. *et al.* (2004) Behavioral and reproductive responses by reproductive female *Neogobius melanostomus* to odours released by conspecific males. *Journal of Fish Biology*, **65**, 933–946.

Belanger, R.M., Corkum, L.D., Li, W., and Zielinski, B.S. (2006) Olfactory sensory input increases gill ventilation in male round gobies (*Neogobius melanostomus*) during exposure to steroids. *Comparative Biochemistry and Physiology A*, **144**, 196–202.

Belanger, R.M., Corkum, L.D., and Zielinski, B.S. (2007) Differential behavioral responses by reproductive and non-reproductive male round gobies (*Neogobius melanostomus*) to the putative pheromone estrone. *Comparative Biochemistry and Physiology A*, **147**, 77–83.

Belanger, R.M., Pachkowski, M.D., and Stacey, N.E. (2010) Methyltestosterone-induced changes in electro-olfactogram responses and courtship behaviors of cyprinids. *Chemical Senses*, **35**, 65–74.

Bhatt, J.P., Kandwal, J.S. and Nautiyal, R. (2002) Water temperature and pH influence olfactory sensitivity to preovulatory and postovulatory ovarian pheromones in male *Barilius bendelisis*. *Journal of Biosciences*, **27**, 273–281.

Bjerselius, R. and Olsén, K.H. (1993) A study of the olfactory sensitivity of crucian carp (*Carassius carassius*) and goldfish (*Carassius auratus*) to 17α,20β-dihydroxyprogesterone and prostaglandin F2α. *Chemical Senses*, **18**, 427–436.

Bjerselius, R., Olsén, K.H., and Zheng, W. (1995) Endocrine, gonadal and behavioral responses of male crucian carp (*Carassius carassius*) to the hormonal pheromone 17α,20β-dihydroxy-4-pregnen-3-one. *Chemical Senses*, **20**, 221–230.

Borg, B. (1994) Androgens in teleost fishes. *Comparative Biochemistry and Physiology C*, **108**, 219–245.

Burnard, D., Gozlan, R.E., and Griffiths, S.W. (2008) The role of pheromones in freshwater fishes. *Journal of Fish Biology*, **73**, 1–16.

Canosa, L.M., Stacey, N., and Peter, R.E. (2008) Changes in brain mRNA levels of GnRH, PACAP and SS during ovulatory LH and GH surges in goldfish. *American Journal of Physiology, Regulative, Integrative, and Comparative Physiology*, **295**, R1815–R1821.

Cardwell, J.R. and Stacey, N.E. (1995) Hormonal sex pheromones in characiform fishes: An evolutionary case study, in *Fish Pheromones: Origins and Modes of Action* (eds A.V.M. Canário and D.M. Power), University of Algarve Press, Faro, Portugal, pp. 47–55.

Cardwell, J.R., Dulka, J.G., and Stacey, N.E. (1992) Acute olfactory sensitivity to prostaglandins but not to gonadal steroids in two sympatric species of *Catostomus* (Pisces: Cypriniformes). *Canadian Journal of Zoology*, **70**, 1897–1803.

Cardwell, J.R., Stacey, N.E., Tan, E.S.P. *et al.* (1995) Androgen increases olfactory receptor response to a vertebrate sex pheromone. *Journal of Comparative Physiology A*, **176**, 55–61.

Carolsfeld, J., Scott, A.P., and Sherwood, N.M. (1997) Pheromone-induced spawning of Pacific herring. 2. Plasma steroids distinctive to fish responsive to spawning pheromone. *Hormones and Behavior*, **31**, 269–276.

Chung-Davidson, Y.W., Rees, C.B., Bryan, M.B., and Li, W. (2008) Neurogenic and neuroendocrine effects of goldfish pheromones. *Journal of Neuroscience*, **28**, 14492–14499.

Cole, T.B. and Stacey, N.E. (2006) Olfactory responses to steroids in an African mouth brooding cichlid, *Haplochromis burtoni* (Günther). *Journal of Fish Biology*, **68**, 661–680.

Colombo, L., Belvedere, P.C., and Pilati, A. (1977) Biosynthesis of free and conjugated 5β-reduced androgens by the testis of the black goby. *Gobius jozo* L. *Bollettino di Zoologia*, **44**, 131–144.

Colombo, L., Marconato, A., Belvedere, P.C., *et al.* (1980) Endocrinology of teleost reproduction: a testicular steroid pheromone in the black goby, *Gobius jozo* L. *Bollettino di Zoologia*, **47**, 355–364.

Corkum, L.D., Sapota, M.R., and Skora, K.E. (2004) The round goby, *Neogobius melanostomus*, a fish invader on both sides of the Atlantic Ocean. *Biological Invasions*, **6**, 173–181.

Corkum, L.D., Meunier, B., Moscicki, M. *et al.* (2008) Behavioural responses of female round gobies (*Neogobius melanostomus*) to putative steroidal pheromones. *Behaviour*, **145**, 1347–1365.

Crespi, B.J. and Fulton, M.J. (2004) Molecular systematics of Salmonidae: combined nuclear data yields a robust phylogeny. *Molecular Phylogenetics and Evolution*, **31**, 658–679.

DeFraipont, M. and Sorensen, P.W. (1993) Exposure to the pheromone 17α,20β-dihydroxy-4-pregnen-3-one enhances the behavioural spawning success, sperm production, and sperm motility of male goldfish. *Animal Behavior*, **46**, 245–256.

Demski, L.S. and Northcutt, R.G. (1983) The terminal nerve: A new chemosensory system in the vertebrates. *Science*, **202**, 435–437.

Dittman, A.H. and Quinn, T.P. (1994) Avoidance of a putative pheromone, 17α,20β-dihydroxy-4-pregnen-3-one, by precociously mature Chinook salmon (*Oncorhynchus tshawytscha*). *Canadian Journal of Zoology*, **72**, 215–219.

Døving, K. (1976). Evolutionary trends in olfaction, in *The Structure-Activity Relationships in Olfaction* (ed. G. Benz), IRL Press, London, pp. 149–159.

Dulka, J.G. and Demski, L.S. (1986) Sperm duct contractions mediate centrally evoked sperm release in goldfish. *Journal of Experimental Zoology*, **237**, 271–279.

Dulka, J.G. and Stacey, N.E. (1990) Effects of olfactory tract lesions on gonadotropin and milt responses to the female sex pheromone, 17α,20β-dihydroxy-4-pregnen-3-one, in male goldfish. *Journal of Experimental Zoology*, **257**, 223–229.

Dulka, J.G., Stacey, N.E., Sorensen, P.W., and Van Der Kraak G.J. (1987a) A sex steroid pheromone synchronizes male-female spawning readiness in goldfish. *Nature*, **325**, 251–253.

Dulka, J.G., Sorensen, P.W., and Stacey, N.E. (1987b) Socially-stimulated gonadotropin release in male goldfish: differential circadian sensitivities to a steroid pheromone and spawning stimuli, in *Proceedings of the Third International Symposium on the Reproductive Physiology of Fish* (eds D.R. Idler, L.W. Crim, and J.M. Walsh), Memorial University Press, St. John's, Nfld, p. 160.

Dulka, J.G., Sloley, B.D., Stacey, N.E., and Peter, R.E. (1992) A reduction in pituitary dopamine turnover is associated with sex pheromone-induced gonadotropin secretion in male goldfish. *General and Comparative Endocrinology*, **86**, 496–505.

Ellis, T., James, J.D., Stewart, C., and Scott, A.P. (2004) A non-invasive stress assay based upon measurement of free cortisol released into the water by rainbow trout. *Journal of Fish Biology*, **65**, 1233–1252.

Ellis, T., James, J.D., and Scott, A.P. (2005) Branchial release of free cortisol and melatonin by rainbow trout. *Journal of Fish Biology*, **67**, 535–540.

Essington, T.E. and Sorensen, P.W. (1996) Overlapping sensitivities of brook trout and brown trout to putative hormonal pheromones. *Journal of Fish Biology*, **48**, 1027–1029.

Fabian, N.J., Albright, L.B., Gerlach, G., *et al.* (2007). Humic acid interferes with species recognition in zebrafish (*Danio rerio*). *Journal of Chemical Ecology*, **33**, 2090–2096.

Frade, P., Hubbard, P.C., Barata, E.N., and Canario, A.V.M. (2002) Olfactory sensitivity of the Mozambique tilapia to conspecific odours. *Journal of Fish Biology*, **61**, 1239–1254.

Fraser, E.J. and Stacey, N.E. (2002) Isolation increases milt production in goldfish. *Journal of Experimental Zoology*, **293**, 511–524.

Freidrich, R.W. and Korsching, S.I. (1998) Chemotopic, combinatorial, and noncombinatorial odorant representations in the olfactory bulb revealed using a voltage-sensitive axon tracer. *Journal of Neuroscience*, **18**, 9977–9988.

Fujita, I., Sorensen, P.W., Stacey, N.E., and Hara, T.J. (1991) The olfactory system, not the terminal nerve, functions as the primary chemosensory pathway mediating responses to sex pheromones in goldfish. *Brain Behavior and Evolution*, **38**, 313–321.

Gammon, D.B., Li, W., Scott, A.P. *et al.* (2005) Behavioural responses of female *Neogobius melanostomus* to odours of conspecifics. *Journal of Fish Biology*, **67**, 615–626.

Garcia-Vazquez, E., Perez, J., Ayllon, F. *et al.* (2004) Asymmetry of post-F1 interspecific reproductive barriers among brown trout (*Salmo trutta*) and Atlantic salmon (*Salmo salar*). *Aquaculture*, **234**, 77–84.

Genner, M.J., Seehausen, O., Lunt, D.H. *et al.* (2007) Age of cichlids: new dates for ancient lake fish radiations. *Molecular Biology and Evolution*, **24**, 1269–1282.

Hamdani, E.H., Kasumyan, A., and Døving, K.B. (2001) Is feeding behaviour in crucian carp mediated by the lateral olfactory tract? *Chemical Senses*, **26**, 1133–1138.

Hänfling, B., Bolton, P., Harley, M., and Carvalho, G.R. (2005) A molecular approach to detect hybridisation between crucian carp (*Carassius carassius*) and non-indigenous carp species (*Carassius* spp. and *Cyprinus carpio*). *Freshwater Biology*, **50**, 403–417.

Hara, T.J. and Zhang, C. (1997) Topographic bulbar projections and dual neural pathways of the primary olfactory neurons in salmonid fishes. *Neuroscience*, **82**, 301–313.

Hoysak, D.J. and Stacey, N.E. (2008) Large and persistent effect of a female steroid pheromone on ejaculate size in goldfish. *Journal of Fish Biology*, **73**, 1573–1584.

Hubbard, P.C., Barata, E.N., and Canário, A.V.M. (2002) Possible disruption of pheromonal communication by humic acid in the goldfish, *Carassius auratus*. *Aquatic Toxicology*, **60**, 169–183.

Immler, S., Mazzoldi, C., and Rasotto, M.B. (2004) From sneaker to parental male: change of reproductive traits in the black goby, *Gobius niger* (Teleostei, Gobiidae). *Journal of Experimental Zoology A*, **301**, 177–185.

Irvine, I.A.S. and Sorensen, P.W. (1993) Acute olfactory sensitivity of wild common carp, *Cyprinus carpio*, to goldfish sex pheromones is influenced by gonadal maturity. *Canadian Journal of Zoology*, **71**, 2199–2210.

Jasra, S.K., Arbuckle, W.J., Corkum, L.D. *et al.* (2007) The seminal vesicle synthesizes steroids in the round goby, *Neogobius melanostomus*. *Comparative Biochemistry and Physiology A*, **148**, 117–123.

Katare, Y.K., Scott, A.P., Laframboise, A.J. *et al.* (2011) Release of free and conjugated forms of the putative pheromonal steroid11-oxo-etiocholanolone by reproductively mature male round goby (Neogobius melanostomus Pallas, 1814). *Biology of Reproduction*, **84**, 288–298.

Kawabata, K. (1993) Induction of sexual behavior in male fish (*Rhodeus ocellatus ocellatus*) by amino acids. *Amino Acids*, **5**, 323–327.

Kitamura, S., Ogata, H., and Takashima, F. (1994a) Activities of F-type prostaglandins as releaser sex pheromones in cobitide loach, *Misgurnus anguillicaudatus*. *Comparative Biochemistry and Physiology A*, **107**, 161–169.

Kitamura, S., Ogata, H., and Takashima, F. (1994b) Olfactory responses of several species of teleost to F-prostaglandins. *Comparative Biochemistry and Physiology A*, **107**, 463–467.

Kobayashi, M., Aida, K., and Hanyu, I. (1986a) Gonadotropin surge during spawning in male goldfish. *General and Comparative Endocrinology*, **62**, 70–79.

Kobayashi, M., Aida, K., and Hanyu, I. (1986b) Pheromone from ovulatory female goldfish induces gonadotropin surge in males. *General and Comparative Endocrinology*, **63**, 451–455.

Kobayashi, M., Aida, K., and Hanyu, I. (1988) Hormone changes during the ovulatory cycle in goldfish. *General and Comparative Endocrinology*, **69**, 301–307.

Kobayashi, M., Amano, M., Kim, M.-H. *et al.* (1994) Gonadotropin-releasing hormones of terminal nerve origin are not essential to ovarian development and ovulation in goldfish. *General and Comparative Endocrinology*, **95**, 192–200.

Kobayashi, M., Amano, M., Kim, M.-H. *et al.* (1997a) Gonadotropin-releasing hormone and gonadotropin in goldfish and masu salmon. *Fish Physiology and Biochemistry*, **17**, 1–8.

Kobayashi, M., Furukawa, K., Kim, M.-H., and Aida, K. (1997b) Induction of male-type gonadotropin secretion by implantation of 11-ketotestosterone in female goldfish. *General and Comparative Endocrinology*, **108**, 434–445.

Kobayashi, M., Sorensen, P.W., and Stacey, N.E. (2002) Hormonal and pheromonal control of spawning behavior in the goldfish. *Fish Physiology and Biochemistry*, **26**, 71–84.

Kobayashi, M., Kuroyanagi, H., Otomo, S., and Hayakawa, Y. (2008) Involvement of aquatic plants in the spawning behaviour of goldfish and crucian carp. *Cybium*, **32** (*Supplement*), 310–311.

Koblmüller, S., Sefc, K.M., and Sturmbauer, C. (2008) The Lake Tanganyika species assemblage: recent advances in molecular genetics. *Hydrobiologia*, **615**, 5–20.

Kyle, A.L., Stacey, N.E., Peter, R.E., and Billard, R. (1985) Elevations in gonadotrophin concentrations and milt volumes as a result of spawning behavior in the goldfish. *General and Comparative Endocrinology*, **57**, 10–22.

Laberge, F. and Hara, T.J. (2003) Behavioral and electrophysiological responses to F-prostaglandins, putative spawning pheromones, in three salmonid fishes. *Journal of Fish Biology*, **62**, 206–221.

Laframboise, A.J. and Zielinski, B.S. (2011) Responses of round goby (*Neogobius melanostomus*) olfactory epithelium to steroids released by reproductive males. *Journal of Comparative Physiology A*, **197**, 999–1008.

Lastein, S., Hamdani, E.H. and Døving, K.B. (2006) Gender distinction in neural discrimination of sex pheromones in the olfactory bulb of crucian carp, *Carassius carassius*. *Chemical Senses*, **31**, 69–77.

Lastein, S., Hamdani, E.H. and Døving, K.B. (2014) Olfactory discrimination of pheromones, in *Fish Pheromones and Related Cues* (eds Peter W. Sorensen and Brian D. Wisenden), John Wiley & Sons, Inc., Hoboken.

Levesque, H.M., Scaffidi, D., Polkinghorne, C.N. *et al.* (2011) A multi-component species identifying pheromone in the goldfish. *Journal of Chemical Ecology*, **37**, 219–227.

Liley, N.R. (1982) Chemical communication in fish. *Canadian Journal of Fisheries and Aquatic Sciences*, **39**, 22–35.

Liley, N.R. and Tan, E.S.P. (1985) The induction of spawning behavior in *Puntius gonionotus* (Bleeker) by treatment with prostaglandin $F_{2\alpha}$. *Journal of Fish Biology*, **26**, 491–502.

Li, W., Scott, A.P., Siefkes, M.J. *et al.* (2002). Bile acid secreted by male sea lamprey that acts as a sex pheromone. *Science*, **296**, 138–141.

Lim, H. and Sorensen P.W. (2011) Polar metabolites synergize the activity of prostaglandin $F_{2\alpha}$ in a species-specific hormonal sex pheromone released by ovulated common carp. *Journal of Chemical Ecology*, **37**, 695–704.

Lim, H.K. and Sorensen, P.W. (2012) Common carp implanted with prostaglandin $F_{2\alpha}$ release a sex pheromone complex that attracts conspecific males in both the laboratory and field. *Journal of Chemical Ecology*, **38**, 127–134.

Locatello, L., Mazzoldi, C., and Rasotto, M.B. (2002) Ejaculate of sneaker males is pheromonally inconspicuous in the black goby, *Gobius niger* (Teleostei, Gobiidae). *Journal of Experimental Zoology*, **293**, 601–605.

Lower, N., Scott, A.P., and Moore, A. (2004) Release of sex steroids into the water by roach. *Journal of Fish Biology*, **64**, 16–33.

Marentette, J.R., Fitzpatrick, J.L., Berger, R.G., and Sigal, B. (2009) Multiple male reproductive morphs in the invasive round goby (*Apollonia melanostoma*). *Journal of Great Lakes Research*, **35**, 302–308.

Maruska, K.P. and Fernald, R.D. (2012) Contextual chemosensory urine signaling in an African cichlid fish. *Journal of Experimental Biology*, **215**, 68–74.

Maunder, R.J., Matthiessen, P., Sumpter, J.P., and Pottinger, T.G. (2007) Rapid bioconcentration of steroids in the plasma of sticklebacks (*Gasterosteus aculeatus*) exposed to water-borne testosterone and 17β-oestradiol. *Journal of Fish Biology*, **70**, 678–690.

McGregor, P.K. and Peake, T.M. (2000) Communication networks: social environments for receiving and signaling behavior. *Acta Ethologica*, **2**, 71–81.

McKinnon, J.F. and Liley, N.R. (1986) Asymmetric species-specificity in responses to female sexual pheromone by males of two species of *Trichogaster* (Pisces: Belontiidae). *Canadian Journal of Zoology*, **65**, 1129–1134.

McLennan, D.A. and Ryan, M.J. (1997) Responses to conspecific and heterospecific olfactory cues in the swordtail *Xiphophorus cortezi*. *Animal Behaviour*, **54**, 1077–1088.

Mesquite, R.M.R.S., Canário, A.V.M., and Melos, E. (2003) Partition of fish pheromones between water and aggregates of humic acid. Consequences for sexual signaling. *Environmental Science and Technology*, **37**, 742–746.

Meunier, B., Yavno, S., Ahmed, S., and Corkum, L.D. (2009). First documentation of spawning and nest guarding in the laboratory by the invasive fish, the round goby (*Neogobius melanostomus*). *Journal of Great Lakes Research*, **35**, 608–612.

Miranda, A., Almeida, O.G., Hubbard, P.C. *et al.* (2005) Olfactory discrimination of female reproductive status by male tilapia (*Oreochromis mossambicus*). *Journal of Experimental Biology*, **208**, 2037–2043.

Miura, T., Yamauchi, K., Takahashi, H., and Nagahama, Y. (1992) The role of hormones in the acquisition of sperm motility in salmonid fish. *Journal of Experimental Zoology*, **261**, 359–363.

Moore, A. (1991) Behavioral and physiological responses of precocious male Atlantic salmon (*Salmo salar* L.) parr to testosterone, in *Proceedings of the 4th International Symposium on Reproductive Physiology of Fish* (ed. A.P. Scott, J.P. Sumpter, D.E. Kime, *et al.*). FishSymp 91, Sheffield, pp. 194–196.

Moore, A. and Scott, A.P. (1991) Testosterone is a potent odorant in precocious male Atlantic salmon (*Salmo salar* L.) parr. *Philosophical Transactions of the Royal Society of London B*, **332**, 241–244.

Moore, A. and Scott, A.P. (1992) 17α,20β-dihydroxy-4-pregnen-3-one-20-sulfate is a potent odorant in precocious male Atlantic salmon parr which have been pre-exposed to the urine of ovulated females. *Proceedings of the Royal Society of London B*, **249**, 205–209.

Moore, A. and Waring, C.P. (1995) Seasonal changes in olfactory sensitivity of mature male Atlantic salmon (*Salmo salar* L.) parr to prostaglandins, in *Proceedings of the Fifth International Symposium on the Reproductive Physiology of Fish* (eds F.W. Goetz and P. Thomas), Fish Symposium 95, Austin Texas, p. 273.

Moore, A. and Waring, C.P. (1996) Electrophysiological and endocrinological evidence that F-series prostaglandins function as priming pheromones in mature male Atlantic salmon (*Salmo salar*) parr. *Journal of Experimental Biology*, **199**, 2307–2316.

Moore, A., Olsén, K.H., Lower, N., and Kindahl, H. (2002) The role of F-series prostaglandins as reproductive priming pheromones in the brown trout. *Journal of Fish Biology*, **60**, 613–624.

Murphy, C.A. and Stacey, N.E. (2002) Methyl-testosterone induces male-typical behavioral responses to putative steroidal pheromones in female round gobies (*Neogobius melanostomus*). *Hormones and Behavior*, **42**, 109–115.

Murphy, C.A., Stacey, N., and Corkum, L.D. (2001) Putative steroidal pheromones in the round goby, *Neogobius melanostomus*: olfactory and behavioral responses. *Journal of Chemical Ecology*, **27**, 443–470.

Murthy, C.K., Zheng, W., Trudeau, V.L. *et al.* (1994) *In vivo* actions of a gonadotropin-releasing hormone (GnRH) antagonist on gonadotropin-II and growth hormone secretion in goldfish, *Carassius auratus*. *General and Comparative Endocrinology*, **96**, 427–437.

Nagahama, Y., Yoshikuni, M., Yamashita, M. *et al.* (1994) Regulation of oocyte maturation in fish, in *Fish Physiology*, vol. **13** (eds N.M. Sherwood and C.L. Hew), Academic Press, San Diego, pp. 393–439.

Narayanan, A. and Stacey, N.E. (2003) Olfactory responses to putative steroidal pheromones in allopatric and sympatric species of mochokid catfish. *Fish Physiology and Biochemistry*, **28**, 275–276.

Nelson, J.S. (2006) *Fishes of the World*, 4th edn, John Wiley and Sons, Hoboken.

Nevitt, G.A. (1991) Do fish sniff? A new mechanism of olfactory sampling in pleuronectid flounders. *Journal of Experimental Biology*, **157**, 1–18.

Ogata, H., Kitamura, S., and Takashima, F. (1994) Release of 13,14-dihydro-15-keto-prostaglandin $F_{2\alpha}$, a sex pheromone, to water by cobitid loach following ovulatory stimulation. *Fisheries Science*, **60**, 143–148.

Oliveira, R.F. (2006) Neuroendocrine mechanisms of alternative reproductive tactics in fish, in *Fish Physiology: Behavior and Physiology of Fish*, vol. **24** (eds K.A. Sloman, R.W. Wilson and S. Balshine), Elsevier, New York, pp. 297–357.

Oliveira, R.F., Almada, V.C., and Canario, A.V.M. (1996) Social modulation of the sex steroid concentrations in the urine of male cichlid fish *Oreochromis mossambicus*. *Hormones and Behavior*, **30**, 2–12.

Oliveira, R.F., Lopes, M., Carneiro, L.A., and Canário, A.V.M. (2001) Watching fights raises fish hormone levels. *Nature*, **409**, 475.

Olsén, K.H., Bjerselius, R., Petersson, E., *et al.* (2000) Lack of species-specific primer effects of odours from female Atlantic salmon, *Salmo salar*, and brown trout. *Salmo trutta. Oikos*, **88**, 213–220.

Olsén, K.H., Bjerselius, R., Mayer, I., *et al.* (2001) Both ovarian fluid and female urine increase sex steroid levels in mature Atlantic salmon (*Salmo salar* L.) male parr. *Journal of Chemical Ecology*, **27**, 2337–2349.

Olsén, K.H., Johanssen, A.-K., Bjerselius, R., *et al.* (2002) Mature Atlantic salmon (*Salmo salar* L.) male parr are attracted to ovulated female urine but not ovarian fluid. *Journal of Chemical Ecology*, **28**, 29–40.

Olsén, K.H., Sawisky, G.R., and Stacey, N.E. (2006) Endocrine and milt responses of male crucian carp (*Carassius carassius* L.) to periovulatory females under field conditions. *General and Comparative Endocrinology*, **149**, 294–302.

Pankhurst, N.W. (2008) Gonadal steroids: functions and patterns of change, in *Fish Reproduction* (eds M. J. Rocha, A. Arukwe, and B. G. Kapoor), Science Publishers, Enfield, pp. 67–111.

Patiño, R. and Sullivan, C.G. (2002) Ovarian follicle growth, maturation, and ovulation in teleost fish. *Fish Physiology and Biochemistry*, **26**, 57–70.

Pinheiro, M.F.M., Souza, S.M.G., and Barcellos, L.J.G. (2003) Exposure to 17α,20β-dihydroxy-4-pregnen-3-one changes seminal characteristics in Nile tilapia, *Oreochromis niloticus*. *Aquaculture Research*, **34**, 1047–1052.

Pinillos, M.L., Guijarro, A.I., Delgado, M.J. *et al.* (2002) Production, release and olfactory detection of sex steroids by the tench (*Tinca tinca* L.). *Fish Physiology and Biochemistry*, **26**, 197–210.

Poling, K.R., Fraser, E.J., and Sorensen, P.W. (2001) The three steroidal components of the goldfish preovulatory pheromone signal evoke different behaviors in males. *Comparative Biochemistry and Physiology B*, **129**, 645–651.

Pottinger, T.G. and Moore, A. (1997) Characterization of putative steroid receptors in the membrane, cytosol and nuclear fractions from the olfactory tissue of brown and rainbow trout. *Fish Physiology and Biochemistry*, **16**, 45–63.

Rasotto, M.B. and Mazzoldi, C. (2002) Male traits associated with alternative reproductive tactics in *Gobius niger*. *Journal of Fish Biology*, **61**, 173–184.

Resink, J.W., Voorthuis, P.K., Van Den Hurk, R. *et al.* (1989) Steroid glucuronides of the seminal vesicle as olfactory stimuli in African catfish, *Clarias gariepinus*. *Aquaculture*, **83**, 153–166.

Rollo, A., Andraso, G., Janssen, J., and Higgs, D. (2007) Attraction and localization of round goby (*Neogobius melanostomus*) to conspecific calls. *Behaviour*, **144**, 1–21.

Rosenthal, G.G., Fitzsimmons, J.N., Woods, K.U. *et al.* (2011) Tactical release of a sexually-selected pheromone in a swordtail fish. *PLoS One*, **6**, e16994.

Saitoh, K., Miya, M., Inoue, J.G. *et al.* (2003) Mitochondrial genomics of ostariophysan fishes: perspectives on phylogeny and biogeography. *Journal of Molecular Evolution*, **56**, 464–472.

Salzburger, W., Mack, T., Verheyen, E., and Meyer, A. (2005). Out of Tanganyika: genesis, explosive speciation, key-innovations and phylogeography of the haplochromine cichlid fishes. *BMC Evolutionary Biology*, **5**, 17.

Santos, A.J.G., Furukawa, K., Kobayashi, M., *et al.* (1986) Plasma gonadotropin and steroid hormone profiles during ovulation in the carp *Cyprinus carpio. Bulletin of the Japanese Society of Scientific Fisheries*, **52**, 1159–1166.

Schwarzer, J., Misof, B., Tautz, D., and Schliewen, U.K. (2009) The root of the East African cichlid radiations. *BMC Evolutionary Biology*, **9**, 186.

Scott, A.P. and Ellis, T. (2007) Measurement of fish steroids in water - a review. *General and Comparative Endocrinology*, **153**, 392–400.

Scott, A.P. and Sorensen, P.W. (1994) Time course of release of pheromonally active steroids and their conjugates by ovulatory goldfish. *General and Comparative Endocrinology*, **96**, 309–323.

Scott, J.W. and Scott-Johnson, P.E. (2002) The electroolfactogram: a review of its history and uses. *Microscopic Research Techniques*, **58**, 152–160.

Scott, A.P., Canário, A.V.M., Sherwood, N.M., and Warby, C.M. (1991). Levels of steroids, including cortisol and 17α,20β-dihydroxy-4-pregnen-3-one, in plasma, seminal fluid, and urine of Pacific herring (*Clupea harengus pallasi*) and North Sea plaice (*Pleuronectes platessa* L.). *Canadian Journal of Zoology*, **69**, 111–116.

Scott, A.P., Liley, N.R., and Vermeirssen, E.L.M. (1994) Urine of reproductively mature female rainbow trout, *Oncorhynchus mykiss* (Walbaum), contains a priming pheromone which enhances plasma levels of sex steroids and gonadotrophin II in males. *Journal of Fish Biology*, **44**, 131–147.

Scott, A.P., Pinillos, M., and Huertas, M. (2005) The rate of uptake of sex steroids from water by tench *Tinca tinca* L. is influenced by their affinity for sex steroid binding protein in plasma. *Journal of Fish Biology*, **67**, 182–200.

Scott, A.P., Sumpter, J.P., and Stacey, N. (2010) The role of the maturation-inducing steroid, 17,20β-dihydroxypregn-4-en-3-one, in male fishes: a review. *Journal of Fish Biology*, **76**, 183–224.

Seehausen, O., Terai, Y., Magalhaes, I.S. *et al.* (2008) Speciation through sensory drive in cichlid fish. *Nature*, **455**, 620–626.

Sisler, S.P. and Sorensen, P.W. (2008) Common carp and goldfish discern conspecific identity using chemical cues. *Behaviour*, **145**, 1409–1425.

Sorensen, P.W. (2014) Introduction to pheromones and related chemical cues in fishes, in *Fish Pheromones and Related Cues* (eds Peter W. Sorensen and Brian D. Wisenden), John Wiley & Sons, Inc., Hoboken

Sorensen, P.W. and Baker, C. (2014) Species-specific pheromones and their roles in shoaling, migration and reproduction: a critical review and synthesis, in *Fish Pheromones and Related Cues* (eds Peter W. Sorensen and Brian D. Wisenden), John Wiley & Sons, Inc., Hoboken.

Sorensen, P.W. and Goetz, F.W. (1993) Pheromonal and reproductive function of F-prostaglandins and their metabolites in teleost fish. *Journal of Lipid Mediators*, **6**, 385–393.

Sorensen P.W. and Hoye, T.R. (2007) A critical review of the discovery and application of a migratory pheromone in an invasive fish, the sea lamprey *Petromyzon marinus* L. *Journal of Fish Biology*, **71** (Supplement. D), 100–114.

Sorensen, P.W. and Scott, A.P. (1994) The evolution of hormonal sex pheromones in teleost fish: Poor correlation between the pattern of steroid release by goldfish and olfactory sensitivity suggests that these cues evolved as a result of chemical spying rather than signal specialization. *Acta Physiologica Scandinavica*, **152**, 191–205.

Sorensen, P.W. and Stacey, N.E. (1999) Evolution and specialization of fish hormonal pheromones, in *Advances in Chemical Signals in Vertebrates* (eds R.E. Johnston, D. Müller-Schwarze, and P.W. Sorensen), Kluwer Academic/Plenum Publishers, New York, pp. 15–47.

Sorensen, P.W. and Stacey, N.E. (2004) Brief review of fish pheromones and discussion of their possible uses in the control of non-indigenous teleost fishes. *New Zealand Journal of Marine and Freshwater Research*, **38**, 399–417.

Sorensen, P.W., Hara, T.J., and Stacey, N.E. (1987) Extreme olfactory sensitivity of mature and gonadally-regressed goldfish to a potent steroidal pheromone, 17α,20β-dihydroxy-4-pregnen-3-one. *Journal of Comparative Physiology A*, **160**, 305–313.

Sorensen, P.W., Hara, T.J., Stacey, N.E., and Goetz, F.W. (1988) F prostaglandins function as potent olfactory stimulants that comprise the postovulatory female sex pheromone in goldfish. *Biology of Reproduction*, **39**, 1039–1050.

Sorensen, P.W., Stacey, N.E., and Chamberlain, K.J. (1989) Differing behavioral and endocrinological effects of two female sex pheromones on male goldfish. *Hormones and Behavior*, **23**, 317–332.

Sorensen, P.W., Hara, T.J., Stacey, N.E., and Dulka, J.G. (1990) Extreme olfactory specificity of the male goldfish to the preovulatory steroidal pheromone 17α,20β-dihydroxy-4-pregnen-3-one. *Journal of Comparative Physiology A*, **166**, 373–383.

Sorensen, P.W., Hara, T.J., and Stacey, N.E. (1991) Sex pheromones selectively stimulate the medial olfactory tracts of male goldfish. *Brain Research*, **558**, 343–347.

Sorensen, P.W., Irvine, I.A.S., Scott, A.P., *et al.* (1992) Electrophysiological measures of olfactory sensitivity suggest that goldfish and other fish use species-specific mixtures of hormones and their metabolites as sex pheromones, in *Chemical Signals in Vertebrates*, vol. **6** (eds R. Doty and D. Muller-Schwarze), Plenum Press, New York, pp. 357–364.

Sorensen, P.W., Brash, A.R., Goetz, F.W., *et al.* (1995a) Origins and functions of F prostaglandins as hormones and pheromones in the goldfish, in *Proceedings of the Fourth International Symposium on the Reproductive Physiology of Fish* (eds F.W. Goetz and P. Thomas), FishSymp 95, Austin, pp. 252–254.

Sorensen, P.W., Scott, A.P., Stacey, N.E., and Bowdin, L. (1995b) Sulfated 17,20β-dihydroxy-4-pregnen-3-one functions as a potent and specific olfactory stimulant with pheromonal actions in the goldfish. *General and Comparative Endocrinology*, **100**, 128–142.

Sorensen, P.W., Cardwell, J.R., Essington, T., and Weigel, D.E. (1995c) Reproductive interactions between brook and brown trout in a small Minnesota stream. *Canadian Journal of Fisheries and Aquatic Sciences*, **52**, 1958–1965.

Sorensen, P.W., Scott, A.P., and Kihslinger, R.L. (2000) How common hormonal metabolites function as relatively specific pheromonal signals in goldfish, in *Proceedings of the Sixth International Symposium on the Reproductive Physiology of Fish* (eds B. Norberg, O.S. Kjesbu, G.L. Taranger, *et al.*), John Grieg AS, Bergen, pp. 125–128.

Sorensen, P.W., Murphy, C.A., Loomis, K. *et al.* (2004) Evidence that 4-pregnen-17,20β,21-triol-3-one functions as a maturation-inducing hormone and pheromone precursor in the percid fish, *Gymnocephalus cernuus*. *General and Comparative Endocrinology*, **139**, 1–11.

Sorensen, P.W., Pinillos, M., and Scott, A.P. (2005) Sexually mature male goldfish release large quantities of androstenedione into the water where it functions as a pheromone. *General and Comparative Endocrinology*, **140**, 164–175.

Sorensen, P.W., Fine, J.M., Dvornikovs, V. *et al.* (2005) Mixture of new sulfated steroids functions as a migratory pheromone in the sea lamprey. *Nature Chemical Biology*, **1**, 324–328.

Souza, S.M.G., Lucion, A.B., and Wassermann, G.F. (1998) Influence of 17α,20β-dihydroxy-4-pregnen-3-one injected into a post-ovulatory female on the reproductive behavior of male Nile tilapia (*Oreochromis niloticus*). *Comparative Biochemistry and Physiology A*, **119**, 759–763.

Sparks, J.S. and Smith, W.L. (2004) Phylogeny and biogeography of cichlid fishes (Teleostei: Perciformes: Cichlidae). *Cladistics*, **20**, 501–517.

Stacey, N.E. (1976) Effects of indomethacin and prostaglandins on spawning behavior of female goldfish. *Prostaglandins*, **12**, 113–128.

Stacey, N.E. (1991) Hormonal pheromones in fish: status and prospects, in *Proceedings of the Fourth International Symposium on the Reproductive Physiology of Fish* (eds A.P. Scott, J.P. Sumpter, and D.S. Kime, *et al.*), FishSymp 91, Sheffield, pp. 177–181.

Stacey, N.E. (2003) Hormones, pheromones, and reproductive behavior. *Fish Physiology and Biochemistry*, **28**, 229–235.

Stacey, N.E. (2009) Pheromones and reproduction, in *Reproductive Biology and Phylogeny of Fishes,* vol. **8B** (ed. B.G.M. Jamieson), Science Publishers, Enfield, pp. 94–137.

Stacey, N.E. (2010) Hormonally derived sex pheromones in fishes, in *Hormones and Reproduction of Vertebrates*, vol. **1** (eds D.O. Norris and K.H. Lopez), Elsevier, San Diego, pp. 169–192.

Stacey, N.E. and Cardwell, J.R. (1995) Hormones as sex pheromones in fish: widespread distribution among freshwater species, in *Proceedings of the 5th International Symposium on Reproductive Physiology of Fish* (eds F.W. Goetz and P. Thomas). Fish Symposium 95, University of Texas at Austin.

Stacey, N.E. and Kyle, A.L. (1983) Effects of olfactory tract lesions on sexual and feeding behavior of goldfish. *Physiology and Behavior*, **30**, 621–628.

Stacey, N.E. and Liley, N.R. (1974) Regulation of spawning behavior in the female goldfish. *Nature*, **247**, 71–72.

Stacey, N.E. and Peter, R.E. (1979) Central action of prostaglandins in spawning behavior of female goldfish. *Physiology and Behavior*, **22**, 1191–1196.

Stacey, N.E. and Sorensen, P.W. (1991) Function and evolution of fish hormonal pheromones, in *Biochemistry and Molecular Biology of Fishes*, vol. **1** (eds P.W. Hochachka and T.P. Mommsen), Elsevier, Amsterdam, pp. 109–135.

Stacey, N.E. and Sorensen, P.W. (2002) Hormonal pheromones in fish, in *Hormones, Brain, and Behavior*, vol. **2** (eds D.W. Pfaff, A.P. Arnold, A.M. Etgen, *et al.*), Academic Press, New York, pp. 375–435.

Stacey, N.E. and Sorensen, P.W. (2005) Reproductive pheromones, in *Fish Physiology: Behavior and Physiology of Fish*, vol. **24** (eds K.A. Sloman, R.W. Wilson, and S. Balshine), Elsevier, San Diego pp. 359–412.

Stacey, N.E. and Sorensen P.W. (2008) Hormonally derived sex pheromones in fish, in *Fish Reproduction* (eds. M.J. Rocha, A. Arukwe, and B.G. Kapoor), Oxford-IBH Publishers, pp. 201–244.

Stacey, N.E. and Sorensen, P.W. (2009) Fish hormonal pheromones, in *Hormones, Brain, and Behavior*, 2nd edn vol. **2** (eds D.W. Pfaff, A.P. Arnold, A.M. Etgen, *et al.*), Elsevier, San Diego, pp. 639–681.

Stacey N.E. and Sorensen P.W. (2011) Hormonal pheromones, in *Encyclopedia of Fish Physiology: From Genome to Environment*, vol. **1** (ed. A.P. Farrell), Elsevier, San Diego, pp. 1553–1562.

Stacey, N.E., Cook, A.F., and Peter, R.E. (1979a) Spontaneous and gonadotropin-induced ovulation in the goldfish, *Carassius auratus* L.: effects of external factors. *Journal of Fish Biology*, **15**, 349–361.

Stacey, N.E., Cook, A.F., and Peter, R.E. (1979b) Ovulatory surge of gonadotropin in the goldfish *Carassius auratus*. *General and Comparative Endocrinology*, **37**, 246–249.

Stacey, N.E., Kyle, A.L., and Liley, N.R. (1986) Fish reproductive pheromones, in *Chemical Signals in Vertebrates*, vol. **4** (eds. D. Duvall, D. Müller-Schwarze, and R.M. Silverstein), Plenum Press, New York. pp. 117–133.

Stacey, N.E., Chamberlain, K.J., Sorensen, P.W., and Hara. T.J. (1987) Milt volume increase in goldfish: interaction of pheromonal and behavioral stimuli, in *Proceedings of the Third International Symposium on the Reproductive Physiology of Fish* (eds D.R. Idler, L.W. Crim, and J.M. Walsh), St. John's University Press, St. John's, Newfoundland, p. 165.

Stacey, N.E., Sorensen, P.W., Van Der Kraak, G.J., and Dulka J.G. (1989) Direct evidence that 17α,20β-dihydroxy-4-pregnen-3-one functions as a goldfish primer pheromone: preovulatory release is closely associated with male endocrine responses. *General and Comparative Endocrinology*, **75**, 62–70.

Stacey, N.E., Zheng, W.B., and Cardwell, J.R. (1994) Milt production in common carp (*Cyprinus carpio*): stimulation by a goldfish steroid pheromone. *Aquaculture*, **127**, 265–276.

Stacey, N.E., Cardwell, J.R., and Murphy, C. (1995) Hormonal pheromones in freshwater fishes: preliminary results of an electro-olfactogram survey, in *Fish Pheromones: Origins and Modes of Action* (eds A.V.M. Canário and D.M. Power), University of Algarve Press, Faro, Portugal, pp. 47–55.

Stacey, N.E., Fraser, E.J., Sorensen, P.W., and Van Der Kraak, G. (2001) Milt production in goldfish: regulation by multiple social stimuli. *Comparative Biochemistry and Physiology C*, **130**, 467–476.

Stacey, N.E., Chojnacki, A., Narayanan, A., *et al.* (2003) Hormonally-derived sex pheromones in fish: exogenous cues and signals from gonad to brain. *Canadian Journal of Physiology and Pharmacology*, **81**, 329–341.

Stacey, N.E., Van Der Kraak, G.J., and Olsen, K.H. (2012) Male primer endocrine responses to preovulatory female cyprinids under natural conditions in Sweden. *Journal of Fish Biology*, **80**, 147–165.

Sveinsson, T. and Hara, T.J. (2000) Olfactory sensitivity and specificity of arctic char, *Salvelinus alpinus*, to a putative male pheromone, prostaglandin F_{2a}. *Physiology and Behavior*, **69**, 301–307.

Taborsky, M. (2001) The evolution of bourgeois, parasitic, and cooperative reproductive behaviors in fishes. *Journal of Heredity*, **92**, 100–110.

Tavolga, W.N. (1956). Visual, chemical and sound stimuli as cues in the sex discriminatory behavior of the gobiid fish, *Bathygobius soporator*. *Zoologica*, **41**, 49–64.

Taylor, J. and Mahon, R. (1977) Hybridization of *Cyprinus carpio* and *Carassius auratus*, the first two exotic species in the lower Laurentian Great Lakes. *Environmental Biology of Fishes*, **1**, 205–208.

Taylor, M.I. and Knight, M.E. (2008) Mating systems in fishes, in *Fish Reproduction* (eds M.J. Rocha, A. Arukwe, and B.G. Kapoor), Science Publishers, Enfield, pp. 277–309.

Thomas, P. (2003) Rapid, nongenomic steroid actions initiated at the cell surface: lessons from studies with fish. *Fish Physiology and Biochemistry*, **28**, 3–12.

Thomas, P. (2012) Rapid steroid hormone actions initiated at the cell surface and the receptors that mediate them with an emphasis on recent progress in fish models. *General and Comparative Endocrinology*, **175**, 367–383.

Van Den Hurk, R. and Lambert, J.G.D. (1983) Ovarian steroid glucuronides function as sex pheromones for male zebrafish, *Brachydanio rerio*. *Canadian Journal of Zoology*, **61**, 2381–2387.

Van Den Hurk, R. and Resink, J.W. (1992) Male reproductive system as sex pheromone producer in teleost fish. *Journal of Experimental Zoology*, **261**, 204–213.

Van Staaden, M.J. and Smith, A.R. (2011) Cutting the Gordian knot: complex signaling in African cichlids is more than multimodal. *Current Zoology*, **57**, 237–252.

Vermeirssen, E.L.M. and Scott, A.P. (1996) Excretion of free and conjugated steroids in rainbow trout (*Oncorhynchus mykiss*): evidence for branchial excretion of the maturation-inducing steroid, 17,20β-dihydroxy-4-pregnen-3-one. *General and Comparative Endocrinology*, **101**, 180–194.

Vermeirssen, E.L.M. and Scott, A.P. (2001) Male priming pheromone is present in bile, as well as urine, of female rainbow trout. *Journal of Fish Biology*, **58**, 1039–1045.

Vermeirssen, E.L.M., Scott, A.P., and Liley, N.R. (1997) Female rainbow trout urine contains a pheromone which causes a rapid rise in plasma 17,20β-dihydroxy-4-pregnen-3-one levels and milt amounts in males. *Journal of Fish Biology*, **50**, 107–119.

Waring, C. and Moore, A. (1995) F-series prostaglandins have a pheromonal priming effect on mature male Atlantic salmon (*Salmo salar*) parr, in *Proceedings of the Fifth International Symposium on the Reproductive Physiology of Fish* (eds F.W. Goetz and P. Thomas), Fish Symposium 95, Austin Texas, pp. 255–257.

Waring, C.P., Moore, A., and Scott, A.P. (1996) Milt and endocrine responses of mature male Atlantic salmon (*Salmo salar* L.) parr to water-borne testosterone, 17,20β-dihydroxy-4-pregnen-3-one-20-sulfate, and the urines from adult female and male salmon. *General and Comparative Endocrinology*, **103**, 142–149.

Wilson, E.O. and Bossert, W.H. (1963) Chemical communication among animals, in *Recent Progress in Hormone Research*, vol. **19** (ed. G. Pincus), Academic Press, New York, pp. 673–716.

Wisenden, B.D. (2014) The Cue-Signal Continuum: A hypothesized evolutionary trajectory for chemical communication in fishes, in *Fish Pheromones and Related Cues* (eds Peter W. Sorensen and Brian D. Wisenden), John Wiley & Sons, Inc., Hoboken.

Wisenden, B.D. and Stacey, N.E. (2005) Fish semiochemicals and the evolution of communication networks, in *Animal Communication Networks* (ed. P. McGregor), Cambridge University Press, London, pp. 540–567.

Yamamoto, N., Oka, Y., and Kawashima, S. (1997) Lesions of gonadotropin-releasing hormone-immunoreactive terminal nerve cells: effects on the reproductive behavior of male dwarf gouramis. *Neuroendocrinology*, **65**, 403–412.

Yamazaki, F. (1990) The role of urine in sex discrimination in the goldfish *Carassius auratus*. *Bulletin of the Faculty of Fisheries, Hokkaido University*, **41**, 155–161.

Yamazaki, F. and Watanabe, K. (1979) The role of sex hormones in sex recognition during spawning behavior of the goldfish, *Carassius auratus* L. *Proceedings of the Indian National Academy of Sciences*, Part B, **45**, 505–511.

Yambe, H. and Yamazaki, F. (2000) Urine of ovulated female masu salmon attracts immature male parr treated with methyltestosterone. *Journal of Fish Biology*, **57**, 1058–1064.

Yambe, H. and Yamazaki, F. (2001a) A releaser pheromone that attracts methyltestosterone-treated immature fish in the urine of ovulated female trout. *Fisheries Science*, **67**, 214–220.

Yambe, H. and Yamazaki, F. (2001b) Species-specific releaser effect of urine from ovulated female masu salmon and rainbow trout. *Journal of Fish Biology*, **59**, 1455–1464.

Yambe, H., Shindo, M., and Yamazaki, F. (1999) A releaser pheromone that attracts males in the urine of mature female masu salmon. *Journal of Fish Biology*, **55**, 158–171.

Yambe, H., Munakata, A., Kitamura, S., *et al.* (2003) Methyltestosterone induces male sensitivity to both primer and releaser pheromones in the urine of ovulated female masu salmon. *Fish Physiology and Biochemistry*, **28**, 279–280.

Yambe, H., Kitamura, S., Kamio, M., *et al.* (2006) l-Kynurenine, an amino acid identified as a sex pheromone in the urine of ovulated female masu salmon. *Proceedings of the National Academy of Sciences*, **103**, 15370–15374.

Yambe, H., Konno, S., and Kitamura, S. (2008) Urine of ovulated female masu salmon (*Oncorhynchus masou*) contains a priming pheromone which increases plasma 17,20β-dihydroxy-4-pregnen-3-one level in mature male parr. *Cybium*, **32** (Supplement), 307.

Yavno, S. and Corkum, L. D. (2010) Reproductive female round gobies (*Neogobius melanostomus*) are attracted to visual male models at a nest rather than to olfactory stimuli in urine of reproductive males. *Behaviour*, **147**, 121–132.

Zheng, W. and Stacey, N.E. (1996) Two mechanisms for increasing milt volume in male goldfish. *Journal of Experimental Zoology*, **276**, 287–295.

Zheng, W. and Stacey, N.E. (1997) A steroidal pheromone and spawning stimuli act *via* different neuroendocrine mechanisms to increase gonadotropin and milt volume in male goldfish (*Carassius auratus*). *General and Comparative Endocrinology*, **105**, 228–235.

Zheng, W., Strobeck, C., and Stacey, N.E. (1997) The steroid pheromone 4-pregnen-17α,20β-diol-3-one increases fertility and paternity in goldfish. *Journal of Experimental Biology*, **200**, 2833–2840.

Zohar, Y., Munoz-Cueto, J.A., Elizur, A., and Kah, O. (2010) Neuroendocrinology of reproduction in teleost fish. *General and Comparative Endocrinology*, **165**, 438–455.

Chapter 4
Conspecific Odors as Sexual Ornaments with Dual Functions in Fishes

Lynda D. Corkum[1] and Karen M. Cogliati[2]

[1]University of Windsor, Windsor, Canada
[2]Oregon State University, Corvallis, USA

4.1 INTRODUCTION

Chemical signals enable species to discriminate among conspecifics, maintain dominance hierarchies, form shoals (or other types of social organizations), and find mates with which to spawn (Liley and Stacey, 1983; Stacey and Sorensen, 2006; Ward *et al.*, 2008). Chemical communication occurs when both originator (signaler) and one or more receivers possess specializations for chemical exchange of information (Corkum and Belanger, 2007). The release and detection of water-borne pheromones are a major means of intraspecific communication in fishes because visibility in water is often limited.

Sex pheromones play a significant role in coordinating reproductive behavior within species. Typically, releaser pheromones induce behavioral responses in a receiver, whereas primer pheromones induce physiological changes without necessarily triggering behavioral displays (Wilson and Bossert, 1963). Interestingly, there are pheromones that have both functions (Sorensen *et al.*, 1990). The task of identifying putative pheromones is challenging (Fine and Sorensen, 2008); yet, behavioral interactions are fundamental to understanding chemical communication between and among individuals. In this chapter, we examine how fishes perceive and selectively respond to conspecific chemical stimuli rather than how fishes produce chemical signals.

Locating a mate is critical to reproductive success. Odors released and received by females or males in the context of sexual selection may serve the same function as elaborate visual ornaments or displays that enhance mating success (Wyatt, 2003). The use of pheromones of one sex to attract members of the opposite sex during the breeding season could lead to a shift in the operational sex ratio of a species, resulting in intra- and intersexual selection.

Many studies have documented physiological responses of ripe (spermiated) males or sexually receptive females to chemical cues released by the opposite sex (Sorensen and Stacey, 1999; Corkum and Belanger, 2007), but behavioral responses between conspecifics are less commonly reported.

Fish Pheromones and Related Cues, First Edition. Edited by Peter W. Sorensen and Brian D. Wisenden.
© 2015 John Wiley & Sons, Inc. Published 2015 by John Wiley & Sons, Inc.

Moreover, olfactory signals may be involved in cross-modality ornamentation of visual and mecha-nosensory signals in both courtship and agonistic interactions.

We consider conspecific odors to be a chemical or mixture of chemicals released by an individual and detected by conspecific olfactory system that elicits differential physiological changes and behavioral responses. Wisenden (2000) discussed the evolution of olfactory ornamentation with respect to alarm cues, suggesting that chemicals released by prey served as "public information" that warned off predators.

We explore the behavioral responses to sex pheromones (evolved stimuli whose production are adaptive; see Chapter 1) that enhance reproductive success in fishes. We postulate a novel view that in some cases a sex pheromone (released either by a male or female) may also function as an ornament, that is, a chemical signal that attracts a conspecific to mate. Thus, chemical ornamentation reinforces the effectiveness of information that a signaler emits in other modalities such as vision or sound (B. Wisenden, personal communication). However, not all attractants are necessarily ornaments because a recipient may have evolved to recognize and be attracted to conspecific cues because of benefits without specialized production.

The idea of an odor as a sexual ornament was originally suggested in the avian literature where crested auklet (*Aethia cristatella*) odors advertise mate quality (identified by parasite resistance), serving as a basis for mate choice (Douglas *et al.*, 2001; Hagelin, Jones, and Rasmussen, 2003; Jones *et al.*, 2004). The sources and uses of avian odors are varied (Hagelin and Jones, 2007), but several findings support the idea that some odors may serve as olfactory ornaments (Hagelin, 2007), although not pheromones *per se*, facilitating social cohesion within flocks of crested auklets (Jones *et al.*, 2004; Sorensen and Hoye, 2010).

Ornaments may serve a dual function in that signals may be either attractants and/or threats. For example, a male's odor may induce aggression or repulsion among rival males that may ultimately influence mate choice by female onlookers. Variation in androgen levels may explain the inability of some adult males, with similarly high gonadosomatic index levels, to initiate courtship and spawning behaviors. Such variability suggests that there is a cost associated with odor production that may be related to male quality.

4.2 ORNAMENTS OVERVIEW

We assume that time spent by a reproductive female near the odor of a reproductive male (and *vice versa*) is an indicator of mate attraction.

Sexual selection arises from competition among males for access to females, or mate choice by females selecting among candidate males, both of which result in differential reproduction (Andersson, 1994; Jones and Ratterman, 2009). Before mating occurs, males often advertise their quality, enabling females to select among signalers (Searcy and Nowicki, 2005). Variation in reproductive success among males is typically high (Bateman, 1948). Because some males mate with many females, the risk for males that they will not find a single female is quite high, whereas the risk that a female will not find a male to donate sperm is quite low. Thus, there is selection on males to advertise, but females do not typically face this competitive pressure (Andersson, 1994). In general, females are less orna-mented than males because resources that are invested in a female ornament take away resources allocated to reproduction (Fitzpatrick, Berglund, and Rosenqvist, 1995). Nevertheless, some females (e.g., two-spotted goby, *Gobius flavescens*) may exhibit elaborate ornaments (orange bellies during the breeding season) and compete for mates (Svensson *et al.*, 2010). In convict (and other) cichlids, females also have color-enhanced abdomens to advertise fecundity (Nuttall and Keenleside, 1993; Wisenden, 1995; Beeching *et al.*, 1998).

Olfactory information about major histocompatibility complex (MHC) genotypes appears to be important for mate choice in at least one species of stickleback that reduces mating with kin, and therefore plays a significant role in sexual selection (Milinski *et al.*, 2010). Milinski *et al.* (2010) report that female three-spined stickleback, *Gasterosteus aculeatus*, assess males on the basis of odors that relay information about their MHC diversity as well as a maleness signal. Moreover, these costly signals (produced only by males) depend on the reproductive state and health of the male (Milinski *et al.*, 2010). Specifically, signals are presented only when male sticklebacks are in a reproductive state, and stronger signals are presented by healthy males. These findings suggest that healthy males emit more/stronger pheromones and that females may be attracted to the odors (ornaments) based on the MHC of the males.

Amundsen (2003) defines an ornament as any trait in either males or females that is conspicuous, and "costly in utilitarian (natural selection) terms." The expression of elaborate secondary sexual traits is typically regulated by sex steroids (Partridge, Boettcher, and Jones, 2010). Rarely, the same trait is selected for in both the sexes (Jones and Hunter, 1993). If the expression of an ornament (a signal) is a measure of individual quality, there should be a relationship between the efficacy of sex attractants released in fishes and mate quality (e.g., levels of parasitism). Barber and Wright (2005) suggest that fish should be able to identify parasite incidence through their odors; yet, research on parasitism and chemosensory abilities has not been explored. However, see below (section 4.3.1 Male Odors) where anal gland secretions by the peacock blenny, *Salaria pavo*, inhibit bacterial growth (Pizzolon *et al.*, 2010). Lastly, ornaments that advertise quality should be costly so that they cannot be faked by individuals unfit to produce them (Zahavi, 1975; Getty, 2006).

Metabolic costs of pheromone signals are often lower than visual or acoustic signals because only small amounts of water-borne pheromones need to be released into the water for the signal to be dispersed over long distances (Thornhill and Alcock, 1983; Wyatt, 2003, Johnson and Li, 2010). For example, a trap deployed in a river with a concentration of 1.5×10^{-12} M 7α, 12α, 24-trihydroxy-5α-cholan-3-one 24-sulfate (3kPZS) from sea lamprey male washings was sufficient to attract and capture ovulating conspecific females over a distance of more than 650 m (Johnson *et al.*, 2009).

4.3 ODORS AS ORNAMENTS

If/when conspecific odors serve as ornaments, they should (by our definition) have the following properties: (i) stimulate receiver's sensory biology in a way that serves to attract attention; (ii) be innate and not learned (if the odor is truly pheromonal); (iii) show evidence of metabolic cost so that they meet the expectation of some kind of handicap or honest signal; and (iv) show variation among individuals that can serve as a measure of individual quality in mate choice.

Pheromonal secretions that meet the above criteria are good candidates for the label "sexual ornament." To date, there is insufficient evidence available in which all of these conditions are met. Nevertheless, we attempt to identify some likely candidates among those organisms where compounds (androgens or estrogens) are secreted from one individual with benefits to both sender and conspecific receiver (see Corkum and Belanger, 2007).

Døving (1976) suggested that fishes may have evolved to release hormones as sex pheromones. Since then, researchers have shown that pheromones produced in the final stages of maturation function in the synchronization of mating in many species. Examples include Cypriniformes (e.g., goldfish, *Carassius auratus*; Sorensen and Stacey, 1999), Siluriformes (African catfish, *Clarias gariepinus*; Van den Hurk and Resink, 1992), Salmoniformes (e.g., Atlantic salmon, *Salmo salar*; Waring and Moore, 1997; brook trout, *Salvelinus fontinalis*; Young, Micek, and Rathbun, 2003), and Perciformes (black gobies, *Gobius jozo* (*G. niger*); Colombo, Colombo Belvedere, and Marconato, 1979; round gobies, *Neogobius melanostomus*; Murphy, Stacey, and Corkum, 2001).

Androstenedione (AD), testosterone (T) and 11-ketotestosterone (11-KT) are examples of important androgens used in the reproduction of fishes. AD, the precursor of testosterone, functions as a male pheromone in goldfish (a scramble spawner) with greatly increased amounts being released by reproductively stimulated, sexually active males (Sorensen *et al.*, 2005). AD increases aggressive interactions (pushing and nudging) among males competing for access to females (Poling, Fraser, and Sorensen, 2001) and attracts sexually receptive females (Levesque and Sorensen, personal communication), and thus appears to meet the criteria of an ornament.

Testosterone is more important for the initial development of gonadal tissue (e.g., Kindler *et al.*, 1989), whereas 11-KT is associated with the production of secondary sexual characteristics in males, including aggressive, territorial, and vocalizing behaviors (e.g., Kindler *et al.*, 1989; Oliveira, Almada, and Canario, 1996; Hay and Pankhurst, 2005; Remage-Healey and Bass, 2005; Desjardins *et al.*, 2006; Parikh, Clement, and Fernald, 2006).

Testosterone, the precursor to 11-KT, may lead to increased fitness through mate attraction, but is also known to act as an immunosuppressant (Folstad and Karter, 1992; McKean and Nunney, 2001; Getty, 2002). According to Folstad and Karter (1992), this immunological detriment suggests that testosterone-dependent signals are viability indicators, which are costly to the male and act as honest indicators of quality as outlined by the handicap hypothesis (Zahavi, 1975; Zahavi and Zahavi, 1977; Grafen, 1990; Folstad and Karter, 1992; Irving, 1996). Thus, those individuals with more T have better long-term health with good genes as they are able to handle the detrimental effects of T and still express larger secondary sex characteristics through 11-KT.

4.3.1 Odors as Attractants

FEMALE ODORS

Depending on the species, male or female odors can be used to attract members of the opposite sex (Tables 4.1 and 4.2). Commercial fishers were known to seed traps with sexually mature female channel catfish, *Ictalurus punctatus*, to increase catches of mature males (Timms and Kleerekoper, 1972). Adult males respond to hormonal pheromones released by receptive females in Eurasian ruffe, *Gymnocephalus cernuus* (Sorensen *et al.*, 2004), frillfin goby, *Bathygobius soporator* (Tavolga, 1956); goldfish (Sorensen and Stacey, 1999); crucian carp, *Carassius carassius* (Olsén, Sawisky, and Stacey, 2006); guppy, *Poecilia reticulata* (Crow and Liley, 1979; Guevara-Fiore, Skinner, and Watt, 2009); and other species (Table 4.1).

The goldfish is the most thoroughly studied fish with respect to hormonal pheromones (Sorensen and Stacey, 1999). Females release a mixture of several hormonal pheromones, which attract males that compete for access to females. Sexually receptive females increase their rate of urination when placed with males (sexually active or inactive), but not females, indicating that females release a pheromone to signal their spawning readiness to males (Appelt and Sorensen, 2007). In addition, females appear to mark spawning sites (vegetation) with urine, and this serves as an advertising signal to males, luring them to a spawning site (Appelt and Sorensen, 2007).

There can be significant costs when individuals search for receptive mates. Sorensen *et al.* (2004) showed that swimming activity in males of Eurasian ruffe increased in response to water from conspecific pre-ovulatory females when presented with females injected with 4-pregnen-17, 20-ß, 21-triol-3-one (20-ß-S) and when presented with urine from females injected with 20-ß-S. Urine from ovulating female masu salmon, *Oncorhynchus masou*, attracted spermiating conspecific males, but not immature males (Yambe *et al.*, 2006).

Fishes should conserve energy when seeking mates. Small live-bearing male guppies serve as an example of this by seeking receptive rather than unreceptive females (Guevara-Fiore *et al.*, 2010). The operational sex ratio (OSR) is male biased in guppies and males (with higher multiple mating rates than females) constantly search for receptive females (Guevara-Fiore, Skinner, and Watt, 2009). Female guppies spend more time than males schooling, a defense used against predators (Magurran and Seghers, 1994). By nature, males continuously move between shoals to find both receptive females and areas with female-biased OSR (Guevara-Fiore, Skinner, and

Table 4.1. The Effects of Female Odors on Behaviors of Conspecifics.

Family	Species	Common Name	Substance	Behavior Observed	References
Cichlidae	*Oreochromis mossambicus*	Mozambique tilapia	Males in isolation presented with either pre- or postovulatory females	Male urination frequency significantly increased in the presence of pre-ovulatory females	Barata *et al.* (2008a)
Cyprinidae	*Carassius auratus*	Goldfish	Males exposed to water with concentrated (5×10^{-10}M) 17,20P	Males that were exposed to 17,20P for a given time had higher blood GtH concentrations than control males	Dulka *et al.* (1987); Sorensen and Stacey (1999)
Cyprinidae	*Carassius auratus*	Goldfish	Female urinary prostaglandins (15K-PGF$_{2\alpha}$)	Initiated rapid swimming by males increasing spawning substrate inspection	Appelt and Sorensen (2007)
Cyprinidae	*Carassius carasius*	Crucian carp	Ovaprim™ & steroid, 17,20ßP	Field study where mature males increased in lutinizing hormone and stripped milt in response to injected females	Olsén, Sawisky, and Stacey (2006)
Gasterosteidae	*Culaea inconstans*	Brook stickleback	CW	Males were attracted to the cues from ovulated conspecific females	McLennan (2004)
Gobiidae	*Bathygobius soporator*	Frillfin goby	CW and ovarian fluid	Within 5–10 s of exposure to gravid female CW (0.5 mL), the male "exploded into courtship." Ovarian fluid also induced courtship behavior by males. No responses by females were noted	Tavolga (1956)
Gobiidae	*Neogobius melanostomus*	Round goby	CW	Nonreproductive females (NRF) spent significantly more time near reproductive female (RF) CW odors than to control odors. NRF exhibited directed movement to RF odors.	Gammon *et al.* (2005)

(continued)

Table 4.1. The Effects of Female Odors on Behaviors of Conspecifics. (*Continued*)

Family	Species	Common Name	Substance	Behavior Observed	References
Gobiidae	*Neogobius melanostomus*	Round goby	Free steroids (11-kT, AD, 11β-OH-AD, ETIO, and 11-oxo-ETIO)	Non-reproductive females spent significantly more time in the arm of a Y-maze that received the free-steroid blend vs. the control (ethanol carrier) treatment	Corkum *et al.* (2008)
Percidae	*Gymnocephalus cernuus*	Eurasian ruffe	CW	Water from pre-ovulatory females (odors) causes male swimming activity to increase	Sorensen *et al.* (2004)
Percidae	*Gymnocephalus cernuus*	Eurasian ruffe	Female injected with 20 β-S	Males stimulated to swim more actively when presented with females injected with 20 β-S	Sorensen *et al.* (2004)
Percidae	*Gymnocephalus cernuus*	Eurasian ruffe	Female injected with 20 β-S	Male swimming activity increased when presented with the urine of females that were injected with 20 β-S	Sorensen *et al.* (2004)
Poeciliidae	*Poecilia reticulata*	Guppy	CW	Males were attracted to and increased their courtship activity when exposed to female CW	Crow and Liley (1979)
Poeciliidae	*Poecilia reticulata*	Guppy	Virgin vs. mated females	Males associated more with the smell of virgin than mated females	Guevara-Fiore, Skinner, and Watt (2009)
Poeciliidae	*Poecilia reticulata*	Guppy	Receptive vs. nonreceptive female shoals	Searching activity of males increases when exposed to olfactory cues from shoaling receptive females	Guevara-Fiore *et al.* (2010)
Salmonidae	*Oncorhynchus masou*	Masu salmon	Ovulated female urine	In Y-maze preference tests, ovulated female urine contains a male-attracting pheromone. Spermiating males clearly responded to female urine, whereas immature males did not	Yambe *et al.* (2006)

CW, conditioned water (i.e., washings from a fish).
Family names are listed in alphabetical order.

Table 4.2. The Effects of Male Odors on Female Behaviors.

Family	Species	Common Name	Substance	Behavior Observed	Reference
Blenniidae	*Blennius pavo* (*Salaria pavo*)	Peacock blenny	CW	Females were attracted to water from males with well developed anal fins, and not to water of other males or females	Laumen, Pern, and Blöm (1974)
Blenniidae	*Salaria fluviatilis*	Freshwater blenny	Pairs of males presented, one with anal gland removed and one with 2 anal fins rays removed	Males with anal glands received more eggs in their nests. However, the two types of males (with and without anal glands) did not differ in attractiveness to females or frequency of visits by females	Barata *et al.* (2008b)
Blenniidae	*Salaria pavo*	Peacock blenny	CW; anal gland and anal fin ray macerate	In Y-maze experiment, females preferred options with male CW and anal gland macerate	Barata *et al.* (2008b)
Blenniidae	*Salaria pavo*	Peacock blenny	Males with either 2 anal fin rays or anal glands removed	Females visited and courted males that still had their anal glands intact more often	Barata *et al.* (2008b)
Blenniidae	*Salaria pavo*	Peacock blenny	CW and anal gland macerate from *S. pavo* and *S. fluviatilis*	Females were attracted to conspecific odors, both from CW and anal gland macerate, and were not attracted to heterospecific odors	Serrano *et al.* (2008)
Blenniidae	*Salaria pavo*	Peacock blenny	CW from males without anal glands	Female attraction to anal gland odors is ephemeral. Females were still attracted to whole odors from males without anal glands	Serrano *et al.* (2008)
Blenniidae	*Salaria pavo*	Peacock blenny	Mature male anal gland macerate; food macerate	Nonreproductive females were not attracted to conspecific anal gland odor	Serrano *et al.* (2008)

(*continued*)

Table 4.2. The Effects of Male Odors on Female Behaviors. (*Continued*)

Family	Species	Common Name	Substance	Behavior Observed	Reference
Cichlidae	*Pseudotropheus emmiltos*	Lake Malawi cichlid	Female choice tests based on full contact with males, male visual stimuli only, and male visual and olfactory stimuli	When females were given a choice between two heterospecific males, they chose their conspecific only when full contact or when visual and olfactory cues were available. Females did not differentiate between the two male species when given visual signals alone.	Plenderleith *et al.* (2005)
Clariidae	*Clarias gariepinus*	African sharptooth catfish	Seminal vesicle fluid (3α, 17α-dihydroxy 5β- pregnan-20-one-3α glucuronide	Attraction of females	Lambert and Resink (1991)
Cyprinidae	*Brachydanio rerio*	Zebrafish	CW	Attraction of females to male water	Bloom and Perlmutter (1977)
Gobiidae	*Gobius jozo* (*Gobius niger*)	Black goby	Etiocholanolone glucuronide (ETIO-g) injected in nests	Reproductive females only showed interest in nests where the chemical was injected	Colombo, Colombo Belvedere, and Marconato (1979)
Gobiidae	*Neogobius melanostomus*	Round goby	CW	Reproductive male water attracted reproductive females	Gammon *et al.* (2005)
Gobiidae	*Neogobius melanostomus*	Round goby	Synthesized steroid blends	Reproductive females showed a tendency toward the conjugated steroid blend, but results were not significant	Corkum *et al.* (2008)
Petromyzontidae	*Petromyzon marinus*	Sea lamprey	CW	Females were attracted to conditioned (male odorants) rather than unconditioned water in a two-choice maze	Li *et al.* (2002)
Petromyzontidae	*Petromyzon marinus*	Sea lamprey	Spermiating males in traps	Field study in the Ocqueoc River, MI, showed that 74% of released ovulating females were caught in traps (with spermiating males) compared to empty traps and traps with nonspermiating males in which no females were caught	Johnson, Siefkes, and Li (2005)

Petromyzontidae	*Petromyzon marinus*	Sea lamprey	Spermiating males in traps	Continuous male washings caught more females than pulsed washings in the Ocqueoc River, MI. Females with occluded olfactory organs could not locate males	Johnson *et al.* (2006)
Petromyzontidae	*Petromyzon marinus*	Sea lamprey	Synthesized 3-keto petromyzonol sulfate (3kPZS)	Synthesized 3-kPZS released into a stream captured ovulating females, travelling distances of up to 650 m to the trap	Johnson *et al.* (2009)
Petromyzontidae	*Petromyzon marinus*	Sea lamprey	3kPZS and CW with spermiating males	Ovulating females swam to spermiating male CW and to synthetic 3kPZS	Siefkes, Winterstein, and Li (2005)
Poeciliidae	*Poecilia reticulata*	Guppy	CW	Females were found to significantly associate with other females when the choice was between females and control water. They did not associate significantly more often with male water when paired with control water.	Shohet and Watt (2004)
Poeciliidae	*Poecilia reticulata*	Guppy	CW	Females were found to significantly associate first with other females, but then would reverse their association to the male odor	Shohet and Watt (2004)
Poeciliidae	*Poecilia reticulata*	Guppy	CW	Females were found to associate preferentially with certain males based on odors, but this choice was exactly opposite to the preferences females exerted with they received visual cues from the same males	Shohet and Watt (2004)

(continued)

Table 4.2. The Effects of Male Odors on Female Behaviors. (*Continued*)

Family	Species	Common Name	Substance	Behavior Observed	Reference
Salmonidae	*Salvelinus alpinus*	Arctic char	Prostaglandins (PGF$_{2\alpha}$)	Ripe females were attracted to odor. The odor induced some ripe females to exhibit digging (a typical spawning behavior)	Sveinsson and Hara (1995)
Salmonidae	*Salmo gairdneri* (*O. mykiss*)	Rainbow trout	CW	Females were attracted to water from males and to water from spawning fish	Newcombe and Hartman (1973)

CW, conditioned water (i.e., washings from a fish).
References are listed by alphabetical order, according to family name.

Watt, 2009). Male guppies associated equally with visual and odor cues, but only when odor was from receptive females.

Female odors may serve as reproductive ornaments to enhance courtship behavior in males and as lures to attract nonreproductive females to aggregate. Reproductive male frillfin gobies, were induced to court when exposed to water that had previously contained ovulating females (suggesting a chemical ornament), but males did not respond to water that had contained other males (Tavolga, 1956). Odor signals from reproductive female round gobies as well as a blend of free synthesized steroids attract nonreproductive females (NRF) significantly more than control odors (Gammon *et al.*, 2005; Corkum *et al.*, 2008). These NRF also exhibited elevated electro-olfactogram values when exposed to reproductive females conditioned water but not to water from NRF or to males (reproductive or not), suggesting that NRF are excited by odors of reproductive females. Female intrasexual pheromones may be linked to aggregation or shoaling behavior as suggested for zebrafish, *Danio rerio* (Bloom and Perlmutter, 1977) and guppies (Shohet and Watt, 2004), species in which females are attracted to conspecific female reproductive odors verus control water. Round gobies have high site fidelity and aggregate to such a degree that they are typically the dominant benthivore in lakes (Ray and Corkum, 2001; Dopazo, Corkum, and Mandrak, 2008). Pheromone communication is important to round gobies that occupy deep, dark regions of lakes and turbid rivers; these fish are night active (Schaeffer *et al.*, 2005; Dopazo, Corkum, and Mandrak, 2008).

MALE ODORS

Pheromones produced by reproductive males attract females to spawning sites in several fishes that provide paternal care of eggs (Table 4.2). Parental care occurs in 21% of teleost fish families and 61% of these families have male care (Gross and Sargent, 1985). Species (typically benthic fishes) in which parental males guard territories or nests appear to have specialized structures for production of sex pheromones (Serrano *et al.*, 2008). Examples include (i) mesorchial glands (black goby, Colombo *et al.*, 1980; and round goby, Arbuckle *et al.*, 2005, Gammon *et al.*, 2005), (ii) anal glands (peacock blenny, Barata *et al.*, 2008b), (iii) seminal vesicles (African catfish, Lambert and Resink, 1991) and (iv) testes (*combtooth blennies, Hypsoblennius* spp., Losey, 1969; round gobies, Arbuckle *et al.*, 2005). Parental males that occupy nests and exhibit high nest fidelity benefit from advertising their location to ripe females.

The peacock blenny is a bottom-dwelling fish in which the male defends and guards embryos in a nest (cavity) where reproductive females have deposited eggs. The species has a promiscuous mating system and breeds in the Mediterranean and nearshore Atlantic Ocean (Gonçalves *et al.*, 2002). During the breeding season, the first two rays of the anal fin of parental males develop into a pair of anal glands from which hydrophilic odorants are released (Serrano *et al.*, 2008). Reproductive females were attracted to male odor and anal gland macerates; yet, female responses to males without anal glands were more variable (Serrano *et al.*, 2008). In another experiment, there were more female visits and more eggs within nests of parental males with anal glands than in nests without anal glands (i.e., the glands were excised), indicating that putative pheromones from anal glands of males attract females and increase reproductive success (Barata *et al.*, 2008b). Interestingly, anal gland secretions also inhibit the growth of both gram-negative and gram-positive bacteria (Pizzolon *et al.*, 2010). Together, these studies on the peacock blenny provide support for a male odor ornament that attracts reproductive females to high-quality males that exhibit antimicrobial defense.

Understanding chemical communication and identification of pheromone structure has been explored for the management of the invasive sea lamprey, *Petromyzon marinus* (Sorensen *et al.*, 2005; Fine and Sorensen, 2008; Johnson and Li, 2010). Spermiating males release a conjugated bile acid 3-keto-petromyzonal sulfate (3kPZS) to which ovulating females are attracted (Li *et al.*, 2002;

Siefkes, Winterstein, and Li, 2005). Siefkes, Winterstein, and Li (2005) showed that ovulating females, but not males or preovulating females, were attracted to 3kPZS and washings from spermiating males. Results of experiments in a natural stream that flowed on either side of an island (mimicking a Y-maze lab experiment) showed that spermiating male washings attracted ovulating females to move upstream to the odor source (Johnson, Siefkes, and Li, 2005; Johnson *et al.*, 2006). A synthesized bile acid (3kPZS) from spermiating sea lamprey males lured conspecific ovulating females to a trap over a distance of 650 m (Johnson *et al.*, 2009).

The evidence of female attraction to odors from spermiating male sea lamprey is compelling and suggestive of an odor ornament. Concentrations of 3kPZS released into conditioned water samples by spermiating male lampreys are highly variable and range from 21.5 to 785.2 ng/ml (or 53.7 µg to 1.9 mg/fish/h) (Yun *et al.*, 2002). This variability in release rate is potential evidence for an odor cost. Given the odor release rate by males and the response by reproductive females to concentrations of 10^{-14} M 3kPZS (Johnson *et al.*, 2009), the odors can be considered to be conspicuous. Lastly, the compound (3kPZS) appears to be actively released across the gills of spermiating males from glandular cells with secretory papillae, suggesting that males are "active signalers" (Li *et al.*, 2002; Siefkes *et al.*, 2003; Li, 2005).

A series of papers published by the Colombo lab demonstrated the importance of a steroid-secreting structure in the mesorchial gland that specialized in the biosynthesis of 5β reduced androgens in the black goby a bottom-dwelling marine fish that exhibits parental care. Colombo and Burighel (1974) identified a glandular mass in the black goby, and reported that the cells lying along the mesorchium were specialized for steroid hormonal production. Later, Colombo, Colombo Belvedere, and Pilati (1977) identified the testicular synthesis of etiocholanolone in its conjugated form (etiocholanolone glucuronide, ETIO-g), speculating that the testes of the black goby release water-soluble sex pheromones into the urine. When ripe females were exposed to synthetic ETIO-g, females were attracted to the odor and some females deposited eggs (Colombo, Colombo Belvedere, and Marconato, 1979). Also, females (between ovulation and deposition) responded to low concentrations (<2 µm) of ETIO-g steroid, demonstrating the effectiveness of this sex attractant in the absence of a male (Colombo *et al.*, 1980). Because the male black goby is territorial and remains in a nest, it was expected that females would move toward male odor. These findings support the notion that this pheromone (ETIO-g) represents an ornament.

The round goby, another Gobiidae with male parental care, is a colonial breeder. MacInnis and Corkum (2000) determined that 15 reproductive female round gobies deposited eggs in a nest of a single paternal male. The round goby is an invasive species and in nonnative regions, the OSR is male biased (Corkum, Sapota, and Skora, 2004), indicating that some males are chosen much more frequently than others. The round goby has clusters of steroid producing, Leydig-like cells, in both testes (Arbuckle *et al.*, 2005) and seminal vesicles (Jasra *et al.*, 2007). In laboratory experiments, Gammon *et al.* (2005) demonstrated that ripe round goby females spent significantly more time near odors from conspecific reproductive males' washings compared with control water. Interestingly, when specific blends of synthetic conjugated pheromones (previously identified in the gonads of reproductive round goby males, Arbuckle *et al.*, 2005) were compared with a control in a Y-maze, there was no significant attraction by reproductive females to the blends tested (Corkum *et al.*, 2008). This is reminiscent of the studies on sea lamprey (Johnson and Li, 2010) where other factors (e.g., water quality and the behavioral activities of the parental male (Meunier *et al.*, 2009) or (in the case of the round goby) compounds not yet identified) are responsible for initiating courtship and/or spawning behaviors in reproductive females. Once female round gobies are attracted to a nest, the flow emanating from the nest by tail-fanning parental males is suggestive of male quality (Meunier *et al.*, 2009; Meunier, White, and Corkum, 2013).

The sticklebacks (Gasterosteidae) also have male-only care. Males build nests and perform a zigzag dance to attract females. Results from laboratory experiments have shown that reproductive male odors are effective in luring ripe females to a nest. Ripe female three-spined stickleback prefer odors from reproductive males in nests over control water, odors from males over females, and odors from males in nests over males without nests (Haberli and Aeschlimann, 2004). The source of odors in three-spined stickleback is unknown, but may be associated with "spiggin", the substance used to glue nest materials (Jones *et al.*, 2001) or the androgen, 11-KT (Pa'll, Mayer, and Borg, 2002a, 2002b). Also, an increase in pH (associated with eutrophication in nature) resulted in an increase of female attraction to male odors in three-spined stickleback (Heuschele and Candolin, 2007). The enhanced response by females was not merely in response to an increase in the pH of water, but was in response to signals from male odors in water with raised pH. This enhanced response by females has been attributed to pH-dependent binding of male olfactory signals to olfactory receptors in females (Heuschele and Candolin, 2007). These findings highlight the importance of olfaction in seeking mates when vision is impaired in eutrophic environments.

In brook stickleback, *Culaea inconstans*, the response of females to male odors depends on the stage of the female's ovarian cycle (McLennan, 2004, 2005). Females on the day of ovulation were strongly attracted to male odor and exhibited courtship displays; yet, these courtship responses declined after 24 h and stopped (or females avoid odors) after 48 h only to peak again at the next ovulation (McLennan, 2005), thereby providing further support for odors as ornaments.

In a laboratory study, ripe (spermiated and ovulated) male and female Pacific herring, *Clupea harengus pallasi*, responded to a sexual pheromone in the testis filtrate/milt (but not to food) by increasing swimming speed, rising, milling (temporary cessation of schooling), papilla extension, and substrate spawning (Stacey and Hourston, 1982).

4.3.2   Ornaments as Deterrents

Although it is known that some male fishes use conspecific odors for aggressive displays (Table 4.3), it is unknown if the same suite of odors used in mate attraction also is used in aggressive displays (with the exception of the goldfish for which androstenedione serves both attraction and repulsion (Sorensen *et al.*, 2005).

Male three-spined sticklebacks prefer to nest near one another (Assem, 1967), but their response to stimuli depends on the context. For example, some male odors attract conspecifics from distant locations to breeding areas, but repel nearby male neighbors if they are perceived to be trespassers (Waas and Colgan, 1992). Nonterritorial three-spined sticklebacks display aggressive behaviors (bumps and bites) when presented with male-conditioned water (Waas and Colgan, 1992), suggesting that the signal intensified agonistic behavior.

Locatello, Mazzoldi, and Rasotto (2002) showed that parental male black gobies responded aggressively by tail-beating and biting the pheromone-containing ejaculate of other parental males, but not to the ejaculate of sneaker males. Because the sneaker males are chemically "silent" (i.e., the fish do not respond to odors), behavioral responses by males to odors appear to be a reproductive strategy exhibited by parental males (Locatello, Mazzoldi, and Rasotto, 2002). However, in another experiment with this species, the presence of parental males inhibited the change of reproductive tactics in sneakers, a physiological effect that may be mediated through chemical signals (Immler, Mazzoldi, and Rasotto, 2004). Interestingly, a male may switch from a sneaker to a parental male, depending on social context (Immler, Mazzoldi, and Rasotto, 2004). In contrast to aggressive interactions displayed by the black goby, Marentette and Corkum (2008) showed a lack of response by male round goby to odors from males (or females).

Dominance hierarchies among males of other fishes may be established to reduce the intensity of costly antagonistic encounters (Smith, 1974). Male dominance has been linked to body size and

Table 4.3. The Effects of Pheromones on Behavior during Male–male Interactions.

Family	Species	Common Name	Substance	Behavior Observed	References
Blennidae	*Hypsoblennius* spp.	Combtooth blennies	CW from sexually ripe nonparental males	Ripe, nonparental males were strongly attracted to conspecific courting males, but not females or egg-guarding males, suggesting social facilitation	Losey (1969)
Cichlidae	*Oreochromis mossambicus*	Mozambique tilapia	Resident male exposed to an intruder male	Resident males markedly increased their urination frequency when acting aggressive, but not when submissive	Barata *et al.* (2008a)
Cichlidae	*Oreochromis mossambicus*	Mozambique tilapia	In community tank with 13 males, 5 submissive, 7 more dominant	Mean urine volume was lower in subordinate males than dominant males. The dominant males released urine in short pulses	Barata *et al.* (2008a)
Cyprinidae	*Brachydanio rerio*	Zebrafish	CW	Attraction of males to other male water	Bloom and Perlmutter (1977)
Gasterosteidae	*Culaea inconstans*	Brook stickleback	CW	Males were attracted to the scent of territorial males	McLennan (2004)
Gobiidae	*Neogobius melanostomus*	Round goby	CW	No behavioral responses exhibited by males to male or female CW	Marentette and Corkum (2008)
Salmonidae	*Salvelinus alpinus*	Arctic charr	0.1 nm PGF$_{2\alpha}$ (released by males) tested in U-shaped maze	Ripe males and females (but not nonripe females) were attracted to odor	Sveinsson and Hara (1995)

CW, conditioned water (i.e., washings from a fish).

to secondary sexual traits such as breeding colors or weapon size (Kortet *et al.*, 2003; Jacob *et al.*, 2009). Jacob *et al.* (2009) showed that male dominance in European minnow, *Phoxinus phoxinus* (a broadcast spawner), was related to body size and the number of breeding tubercles (although body size and tubercle number were not correlated). Specifically, large males and males with more tubercles were most dominant. Moreover, there was an exponential positive relationship between dominance score and reproductive success in the European minnow (Jacob *et al.*, 2009). Kortet *et al.* (2003) showed that a positive relationship existed between 11-KT and breeding tubercles in male roach, *Rutilus rutilus*.

4.4   VARIATION IN STEROID LEVELS AMONG INDIVIDUALS

Large variation in circulating steroid levels exist among individuals in a population (Maruska, Korzan, and Mesinger, 2009), explaining why some parental males appear to be more successful in attracting females throughout the spawning period. Interindividual variation in 11-KT among male oyster toadfish, *Opsanus tau*, was high and may be explained by differences in spermatogenic stage, dominance hierarchy, parental care or intrinsic genetic differences (Maruska, Korzan, and Mesinger, 2009). This variation in steroid levels explains why some adult males (even if females are available) do not engage in courtship and nesting behavior during peak spawning season.

The round goby has both parental and sneaker males (Marentette *et al.*, 2009), yet not all adult parental males breed (Fig. 4.1). Qureshi (2007) explored the production of steroid secretion of 11-oxo-etiocholanolone (11-oxo-ETIO) among reproductive male round gobies using gonadotropin-releasing hormone (GnRH), which stimulates steroidogenesis in the gonads. Results from an enzyme-linked immunosorbent assay (ELISA) showed that the concentration of 11-oxo-ETIO (ng/ml) in fish washings released post GnRH injection to pre-injection increased by a factor, ranging from 0.291 to 5.307 for five reproductive males (GSI values >1.3%) all of which had secondary sexual morphological traits (black coloration, slimy coating, enlarged cheeks). On average, GnRH injections increased concentration of 11-oxo-ETIO by 2.29. Two of the five males exhibited a decrease in androgen concentration after GnRH injection, and therefore were considered to be "duds" compared with the other three males ("studs") in which 11-oxo-ETIO increased after injection. This variability shows evidence of an ornament with respect to both cost

Figure 4.1.   Reproductive male (dark fish on left) and female (right) round goby on rocky bottom of the St. Lawrence River at Morrisburg, Ontario. Morrisburg was partially flooded during the construction of the St. Lawrence Seaway. Photograph was taken by Lonny Howard, September 5, 2010. (*See insert for color representation of the figure.*)

and quality in the release of 11-oxo-ETIO, indicating that the highest quality males likely release the highest concentrations of the androgen.

Sisneros *et al.* (2004) examined how seasonal variation of steroid hormone levels was related to gonadal development (measured using GSI) and reproductive behavior in the plainfin midshipman, *Porichthys notatus*. Both T and 11-KT increased during the pre-nesting (courtship) period in males, but declined over the breeding season when males shifted to parental care (Sisneros *et al.*, 2004). Earlier, Knapp, Wingfield, and Bass (1999) showed that T and 11-KT levels were higher in male plainfin midshipman when there were no or few eggs in a nest than when nests had developed embryos. In plainfin midshipman, GSI levels begin to increase rapidly during the prenesting/courtship period, and it is at this stage when GSI variation is highest (Sisneros *et al.*, 2004). Thus, individual males with the highest GSI and hormone levels during prenesting period may be the "studs" to which females are attracted. 11-KT is responsible for secondary sexual characteristics, including the induction of sonic muscle development (Brantley, Wingfield, and Bass, 1993; Sisneros *et al.*, 2004). High levels of 11-KT are associated with male humming behavior (Knapp, Marchaterre, and Bass, 2001). Parental males that produce a low-frequency "hum" are attractive to females (Brantley and Bass, 1994). Clearly, variation in steroid and GSI levels exist among males such that females deposit eggs in nests of males with the highest steroid levels, which may in part be from the odors released into the water.

Males may tradeoff costs of pheromone ornaments early in the reproductive season in favor of increased growth and mating opportunities later in the season. Frischknecht (1993) suggested that male three-spined stickleback decide early in the year to invest energy into ornaments (in this case, color) or size. An early investment into an ornament is an advantage, resulting in the attraction of high-quality mates and higher reproductive success early in the season. However, an investment in size yields higher survivorship late in the season, resulting in higher overwinter survivorship (Goodman, 1982) with males having the earliest mating advantage in the following year. In general, this tradeoff in size versus reproduction likely accounts for an extended reproductive season found in so many species.

4.5 SUMMARY

We suggest that sex pheromones, which may be released by males or females, can at times be specialized and function as ornaments in mate attraction in certain species. Given that ornaments reflect individual quality, there may often be a cost associated with their production so that stronger signals are presented by healthier individuals (Amundsen, 2003; Milinski *et al.*, 2010).

Ornaments may attract a mate to spawn directly or lead potential mates to spawning locations. Odors from reproductive females attract conspecific males in many fishes, including Cyprinidae and Poeciliidae. Female odors also attract other females to shoal or aggregate. Female shoals may facilitate male searching behavior and result in reduced predation risk (Guevara-Fiore, Skinner, and Watt, 2009; Guevara-Fiore *et al.*, 2010). Studies on many species of fish (e.g., Blenniidae, Gasterosteidae, Gobiidae, Petromyzontidae) showed that male pheromones attract conspecific females to mate.

We also review the literature, reporting cases where male fishes use odors as deterrents for aggressive displays to warn off social dominance or quality (Locatello, Mazzoldi, and Rasotto, 2002). In some species of Cyprinidae, dominance score is related to reproductive success (Kortet *et al.*, 2003; Jacob *et al.*, 2009).

Adult males can differ in reproductive success. Importantly, seasonal variation in steroid levels that relate to gonad development (indicated by GSI values) help explain why only some males engage in courtship and spawning behavior during peak spawning periods (Sisneros *et al.*, 2004). Tradeoff costs, which occur between sex attractants and increased growth (Frischknecht, 1993), may account for variation in reproductive success and in spawning time over the season.

Most odorant-mediated behavioral studies have been conducted in laboratory settings. In a literature review of more than 400 papers where pheromones played a role in chemical ecology of fishes and crustaceans, 80% of the studies were conducted in a laboratory (Johnson and Li, 2010). Johnson and Li (2010) state that contradictory results often occur between laboratory experiments and field investigations for various reasons and urge researchers to validate lab results in the field, a conclusion with which we concur.

4.6 FUTURE APPROACHES

We have attempted to present a case for both female and male odors as sexual ornaments in fishes. However, the evidence is gleaned from studies where experiments focused on the broader topic of pheromone communication. We feel that future experiments should aim at investigating mate preference decisions based on direct comparisons of conspecific odors. Finally, we feel it is important to establish a more definitive link between circulating hormone levels and odors released into the water by the signaler in species aside from the goldfish. We hope this background information will stimulate others to test hypotheses that odors may be ornaments that are attractive or threatening among fishes that exhibit competition for mates.

ACKNOWLEDGMENTS

We thank Drs Rachelle Belanger, Ekatrina Kovalenko, and especially Peter Sorensen and Brian Wisenden for their constructive comments on this chapter. We also thank Dr Nicholas S. Johnson for his discussions on sea lamprey pheromones. Lonny Howard kindly contributed the round goby photograph.

REFERENCES

Amundsen, T. (2003) Fishes as models in studies of sexual selection and parental care. *Journal of Fish Biology*, **63** (Suppl A), 17–52.

Andersson, M. (1994) *Sexual Selection*. Princeton University Press, Princeton.

Appelt, C.W. and Sorensen, P.W. (2007) Female goldfish signal spawning readiness by altering when and where they release a urinary pheromone. *Animal Behaviour*, **74**, 1329–1338.

Arbuckle, W.J., Belanger, A.J., Corkum, L.D. *et al.* (2005) In vitro biosynthesis of novel 5β-reduced steroids by the testis of the round goby, *Neogobius melanostomus*. *General and Comparative Endocrinology,* **140**, 1–13.

Assem van den, J. (1967) Territory in the three-spined stickleback *Gasterosteus aculeatus* L. An experimental study in intra-specific competition. *Behavioural Supplement*, **16**, 1–64.

Barata, E.N., Fine, J.M., Hubbard, P.C. *et al.* (2008a) A sterol-like odorant in the urine of Mozambique tilapia males likely signals social dominance to females. *Journal of Chemical Ecology*, **34**, 438–449.

Barata, E.N., Serrano, R.M. Miranda, A. *et al.* (2008b) Putative pheromones from the anal glands of male blennies attract females and enhance reproductive success. *Animal Behavior*, **75**, 379–389.

Barber, I. and Wright, H.A. (2005) Effects of parasites on fish behaviour: interactions with host physiology, in *Behaviour and Physiology of Fish*, vol. **24** (eds K.A. Sloman, R.W. Wilson, and S. Balshine), Academic Press, Elsevier, pp. 109–149.

Bateman, A.J. (1948) Intra-sexual selection in Drosophila. *Heredity*, **2**, 349–368.

Beeching, S.C., Gross, S.H., Bretz, H.S., and Hariatis, E. (1998) Sexual dichromatism in a cichlid fish: the ethological significance of female ventral colouration in the convict cichlid, *Cichlasoma nigrofasciatum*. *Animal Behaviour*, **56**, 1021–1026.

Bloom, H.D. and Perlmutter, A. (1977) A sexual aggregating pheromone system in the zebrafish, *Brachydanio rerio* (Hamilton-Buchanan). *Journal of Experimental Zoology*, **199**, 215–226.

Brantley, R.K. and Bass, A.H. (1994) Alternative male spawning tactics and acoustic signals in the plainfin midshipman fish, *Porichthys notatus* (Teleostei, Batrachoididae). *Ethology*, **96**, 213–232.

Brantley, R.K., Wingfield, J.C., and Bass, A.H. (1993) Sex steroid levels in *Porichthys notatus*, a fish with alternative reproductive tactics, and a review of the hormonal bases for male dimorphism among teleost fishes. *Hormones and Behavior*, **27**, 332–347.

Colombo, L. and Burighel, P. (1974) Fine structure of the testicular gland of the black goby, *Gobius jozo* L. *Cell Tissue Research*, **154**, 39–49.

Colombo, L., Colombo Belvedere, P., and Pilati, A. (1977) Biosynthesis of free and conjugated 5β-reduced androgens by the testis of the black goby, *Gobius jozo* L. *Bollettino di Zoologia*, **44**, 131–134.

Colombo, L., Colombo Belvedere, P., and Marconato, A. (1979) Biochemical and functional aspects of gonadal biosynthesis of steroid hormones in teleost fish. *Proceedings of the Indian National Science Academy*, **B45**, 226–234.

Colombo, L., Marconato, A., Colombo Belvedere, P.C., and Frisco, C. (1980) Endocrinology of teleost reproduction: a testicular steroid pheromone in the black goby, *Gobius jozo* L. *Bollettino di Zoologia*, **47**, 355–364.

Corkum, L.D. and Belanger, R.M. (2007) Use of chemical communication in the management of freshwater aquatic species that are vectors of human diseases or are invasive. *General and Comparative Endocrinology*, **153**, 401–417.

Corkum, L.D., Sapota, M.R., and Skora, K.E. (2004) The round goby, *Neogobius melanostomus*, a fish invader on both sides of the Atlantic Ocean. *Biological Invasions*, **6**, 173–181.

Corkum, L.D., Meunier, B., Moscicki, M., *et al.* (2008) Behavioural responses of female round gobies (*Neogobius melanostomus*) to putative steroidal pheromones. *Behaviour*, **145**, 1347–1365.

Crow, R.T. and Liley, N.R. (1979) Sex pheromone in the guppy, *Poecilia reticulata* (Peters). *Canadian Journal of Zoology*, **57**, 184–188.

Desjardins, J., Hazelden, M., Van der Kraak, G., and Balshine, S. (2006) Male and female cooperatively breeding fish provide support for the "Challenge Hypothesis". *Behavioral Ecology*, **17**, 149–154.

Dopazo, S.N., Corkum, L.D., and Mandrak, N.E. (2008) Fish assemblages and environmental variables associated with gobiids in nearshore areas of the lower Great Lakes. *Journal of Great Lakes Research*, **34**, 450–460.

Douglas, H.D., III, Co, J.E., Jones, T.H., and Conner, W.E. (2001) Heteropteran chemical repellents identified in the citrus odor of a seabird (Crested Auklet: *Aethia cristatella*): evolutionary convergence in chemical ecology. *Naturwissenschaften*, **88**, 330–332.

Døving, K. (1976) Evolutionary trends in olfaction, in *The Structure Activity Relationships in Olfaction* (ed. G. Benz), IRL Press, London, pp. 149–159.

Dulka, J.G., Stacey, N.E., Sorensen, P.W., and Van Der Kraak, G.J. (1987) Sex steroid pheromone synchronizes male-female spawning readiness in the goldfish. *Nature Lond*, **325**, 251–253.

Fine, J.M. and Sorensen, P.W. (2008) Isolation and biological activity of the multi-component sea lamprey migratory pheromone. *Journal of Chemical Ecology*, **34**, 1259–1267.

Fitzpatrick, S., Berglund, A., and Rosenqvist, G. (1995) Ornaments or offspring: costs to reproductive success restrict sexual selection processes. *Biological Journal of the Linnean Society*, **55**, 251–260.

Folstad, I. and Karter, A.J. (1992) Parasites, bright males, and the immunocompetence handicap. *American Naturalist*, **139**, 603–622.

Frischknecht, M. (1993) The breeding colouration of male three-spined sticklebacks (*Gasterosteus aculeatus*) as an indicator of energy investment in vigour. *Evolutionary Ecology*, **7**, 439–450.

Gammon, D.B., Li, W., Scott, A.P. *et al.* (2005) Behavioral responses of female *Neogobius melanostomus* to odors of conspecifics. *Journal of Fish Biology*, **67**, 615–626.

Getty, T. (2002) Signalling health versus parasites. *The American Naturalist*, **159**, 363–371.

Getty, T. (2006) Sexually selected signals are not similar to sports handicaps. *Trends in Ecology & Evolution*, **21**, 83–88.

Gonçalves, D.M., Barata, E.N., Oliveira, R.F., and Canario, A.V.M. (2002) The role of male visual and chemical cues on the activation of female courtship behaviour in the sex-role reversed peacock blenny. *Journal of Fish Biology*, **61**, 96–105.

Goodman, D. (1982) Optimal life histories, optimal notation, and the value of reproductive value. *American Naturalist*, **119**, 803–829.

Grafen, A. (1990) Biological signals as handicaps. *Journal of Theoretical Biology*, **144**, 517–546.

Gross, M.R. and Sargent, R.C. (1985) The evolution of male and female parental care in fishes. *American Zoologist*, **25**, 807–822

Guevara-Fiore, P., Skinner A., and Watt, P.J. (2009) Do male guppies distinguish virgin females from recently mated ones? *Animal Behaviour*, **77**, 425–431.

Guevara-Fiore, P., Stapley, P.J., Krause, J. *et al.* (2010) Male mate-searching strategies and female cues: how do male guppies find receptive females? *Animal Behaviour*, **79**, 1191–1197.

Haberli, M.A. and Aeschlimann, P.B. (2004) Male traits influence odour-based mate choice in the three-spined stickleback. *Journal of Fish Biology*, **64**, 702–710.

Hagelin, J.C. (2007) The citrus scent of crested auklets: reviewing the evidence for an avian olfactory ornament. *Journal of Ornithology*, **148** (Suppl 2), S195–S201.

Hagelin, J.C. and Jones, I.L. (2007) Bird odors and other chemical substances: a defense mechanism or overlooked mode of intraspecific communication? *The Auk*, **124**, 741–761.

Hagelin, J.C., Jones, I.L., and Rasmussen, L.E.L. (2003) A tangerine-scented social odor in a monogamous seabird. *Proceedings of the Royal Society of London Series B*, **270**, 1323–1329.

Hay, A.C. and Pankhurst, N.W. (2005) Effect of paired encounters on plasma androgens and behaviour in males and females of the spiny damselfish *Acanthochromis polyacanthus*. *Marine and Freshwater Behaviour and Physiology*, **38**, 127–138.

Heuschele, J. and Candolin, U. (2007) An increase in pH boosts olfactory communication in sticklebacks. *Biology Letters*, **3**, 411–413.

Immler, S., Mazzoldi, C., and Rasotto, M.B. (2004) From sneaker to parental male: change of reproductive traits in the black goby, *Gobius niger* (Teleostei, Gobiidae). *Journal of Experimental Zoology*, **301a**, 177–185.

Irving, P.W. (1996) Sexual dimorphism in club cell distribution in the European minnow and immunocompetence signaling. *Journal of Fish Biology*, **48**, 80–88.

Jacob, A., Evanno, G., Renai, E. *et al.* (2009) Male body size and breeding tubercles are both linked to intrasexual dominance and reproductive success in the minnow. *Animal Behaviour*, **77**, 823–829.

Jasra, S.K., Arbuckle, W.J., Corkum, L.D. *et al.* (2007) The seminal vesicle synthesizes steroids in the round goby. *Neogobius melanostomus*. *Comparative Biochemistry and Physiology, Part A*, **148**, 117–123.

Johnson, N.S. and Li, W. (2010) Understanding behavioral responses of fish to pheromones in natural freshwater environments. *Journal of Comparative Physiology A*, **196**, 701–711.

Johnson, N.S., Siefkes, M.J., and Li, W. (2005) Capture of ovulating female sea lampreys in traps baited with spermiating male sea lampreys. *North American Journal of Fisheries Management*, **25**, 67–72.

Johnson, N.S., Luehring, M.A., Siefkes, M.J. *et al.* (2006) Mating pheromone reception and induced behavior in ovulating female sea lampreys. *North American Journal of Fisheries Management*, **26**, 88–96.

Johnson, N.S., Yun, S.-S., Thompson, H.T. *et al.* (2009) A synthesized pheromone induces upstream movement in female sea lamprey and summons them into traps. *Proceedings of the National Academy of Science*, **106**, 1021–1026.

Jones, A.G. and Ratterman, N.L. (2009) Mate choice and sexual selection: what have we learned since Darwin? *Proceedings of the National Academy of Sciences*, **106** (Suppl 1), 10001–10008.

Jones, I.L. and Hunter F.M. (1993) Mutual sexual selection in a monogamous seabird. *Nature*, **362**, 238–239.

Jones, I.L., Lindberg, C., Jakobsson, S. *et al.* (2001) Molecular cloning and characterization of spiggin. *Journal of Biological Chemistry*, **276**, 17857–17869.

Jones, I.L., Hagelin, J.C. Major, H.L., and Rasmussen, L.E.L. (2004) An experimental field study of the function of crested auklet feather odor. *The Condor*, **106**, 71–78.

Kindler, P.M, Philipp, D.P, Gross, M.R., and Bahr, J.M. (1989) Serum 11-ketotestosterone and testosterone concentrations associated with reproduction in male bluegill (*Lepomis macrochirus*: Centrarchidae). *General and Comparative Endocrinology*, **75**, 446–453.

Knapp, R., Wingfield, J.C., and Bass, A.H. (1999) Steroid hormones and parental care in the plainfin midshipman (*Porichthys notatus*). *Hormones and Behavior*, **35**, 81–89.

Knapp, R., Marchaterre, M.A., and Bass, A.H. (2001) The relationship between courtship behavior and steroid hormone levels in parental male midshipman fish. *Hormones and Behavior*, **39**, 335 (abstract).

Kortet, R., Vainikka, A., Rantala, M.J. *et al.* (2003) Sexual ornamentation, androgens and papillomatosis in male roach (*Rutilus rutilus*). *Evolutionary Ecology Research*, **5**, 411–419.

Lambert, J.G.D. and Resink, J.W. (1991) Steroid glucuronides as male pheromones in the reproduction of the African catfish *Clarias gariepinus*—a brief review. *Journal of Steroid Biochemistry and Molecular Biology*, **40**, 549–556.

Laumen, J., Pern, U., and Blum, V. (1974) Investigations on the function and hormonal regulation of the anal appendices in *Blennius pavo* (Risso). *Journal of Experimental Zoology*, **190**, 47–56.

Li, W. (2005) Potential multiple functions of a male sea lamprey pheromone. *Chemical Senses*, **30** (Suppl 1), 307–308.

Li, W., Scott, A.P., Siefkes, M.J. *et al.* (2002) Bile acid secreted by male sea lamprey that acts as a sex pheromone. *Science*, **296**, 138–141.

Liley, N.R. and Stacey, N.E. (1983) Hormones, pheromones and reproductive behavior in fish, in *Fish Physiology*, vol. **9b** (eds W.S. Hoar, D.J. Randall, and E.M. Donaldson), Academic Press, New York, pp. 1–63.

Locatello, L., Mazzoldi, C., and Rasotto, M.B. (2002) Ejaculate of sneaker males is pheromonally inconspicuous in the black goby, *Gobius niger* (Teleostei, Gobiidae). *Journal of Experimental Zoology*, **293**, 601–605.

Losey, G.S., Jr. (1969) Sexual pheromone in some fishes of the genus *Hypsoblennius* Gill. *Science*, **163**, 181–183.

MacInnis, A.J. and Corkum, L.D. (2000) Fecundity and reproductive season of the round goby, *Neogobius melanostomus*, in the upper Detroit River. *Transactions of the American Fisheries Society*, **129**, 136–144.

Magurran, A.E. and Seghers, B.H. (1994) Sexual conflict as a consequence of ecology: evidence from guppy, *Poecilia reticulata,* populations in Trinidad. *Proceedings of the Royal Society B*, **255**, 31–36.

Marentette, J.R. and Corkum, L.D. (2008) Does the reproductive status of male round gobies (*Neogobius melanostomus*) influence their response to conspecific odours? *Environmental Biology of Fishes*, **81**, 447–455.

Marentette, J.R., Fitzpatrick, J.L., Berger, R.G., and Balshine, S. (2009) Multiple male reproductive morphs in the invasive round goby (*Apollonia melanostoma*). *Journal of Great Lakes Research*, **35**, 302–308.

Maruska, K.P., Korzan, W.J., and Mesinger, A.F. (2009) Individual, temporal, and population-level variations in circulating 11-ketotestosterone and 17β-estradiol concentrations in the oyster toadfish *Opsanus tau*. *Comparative Biochemistry and Physiology, Part A*, **152**, 569–578.

McKean, K.A. and Nunney, L. (2001) Increased sexual activity reduces male immune function in *Drosophila melanogaster*. *Proceedings of the National Academy of Sciences of the USA*, **98**, 7904–7909.

McLennan, D.A. (2004) Male brook sticklebacks' (*Culaea inconstans*) response to olfactory cues. *Behaviour*, **141**, 1411–1424.

McLennan, D.A. (2005) Changes in response to olfactory cues across the ovulatory cycle in brook sticklebacks, *Culaea inconstans*. *Animal Behaviour*, **69**, 181–188.

Meunier, B., Yavno, S., Ahmed, S., and Corkum, L.D. (2009) First documentation of spawning and nest guarding in the laboratory by the invasive fish, the round goby (*Neogobius melanostomus*). *Journal of Great Lakes Research*, **35**, 608–612.

Meunier, B., White, B., and Corkum, L.D. (2013) The role of fanning behavior in water exchange by a nest-guarding benthic fish before spawning. *Limnology and Oceanography: Fluids and Environments*, **3**, 198–209.

Milinski, M., Griffiths, S.W., Reusch, B.H., and Boehm, T. (2010) Costly major histocompatibility complex signals produced only by reproductively active males, but not females, must be validated by a 'maleness signal' in three-spined sticklebacks. *Proceedings of the Royal Society B*, **277**, 391–398.

Murphy, C.A., Stacey, N.E., and Corkum, L.D. (2001) Putative steroidal pheromones in the round goby, *Neogobius melanostomus*: olfactory and behavioural responses. *Journal of Chemical Ecology*, **27**, 443–469.

Newcombe, C. and Hartman, G. (1973) Some chemical signals in the spawning behavior of rainbow trout (*Salmo gairdneri*). *Journal of Fisheries Research Board, Canada*, **30**, 995–997.

Nuttall, D.B. and Keenleyside, M.H.A. (1993) Mate choice by the male convict cichlid (*Cichlasoma nigrofasciatum*; Pisces, Cichlidae). *Ethology*, **95**, 247–256.

Oliveira, R.F., Almada, V.C., and Canario, A.V.M. (1996) Social modulation of sex steroid concentrations in the urine of male cichlid fish *Oreochromis mossambicus*. *Hormones and Behavior*, **30**, 2–12.

Olsén, K.H., Sawisky, G.R., and Stacey, N.E. (2006) Endocrine and milt responses of male crucian carp (*Carassius carassius* L.) to periovulatory females under field conditions. *General and Comparative Endocrinology*, **149**, 294–302.

Pa'll, M.K., Mayer, I., and Borg, B. (2002a) Androgen and behavior in the male three-spined stickleback, *Gasterosteus aculeatus*. I. Changes in 11-ketotestosterone levels during the nesting cycle. *Hormones and Behavior*, **41**, 377–383.

Pa'll, M.K., Mayer, I., and Borg, B. (2002b) Androgen and behavior in the male three-spined stickleback, *Gasterosteus aculeatus*. II. Castration and 11-ketotestosterone effects on courtship and parental care during the nesting cycle. *Hormones and Behavior*, **42**, 337–344.

Parikh, V.N., Clement, T.S., and Fernald, R.D. (2006) Androgen level and male social status in the African cichlid. *Astatotilapia burtoni*. *Behavioural Brain Research*, **166**, 291–295.

Partridge, C., Boettcher, A., and Jones, A.G. (2010) Short-term exposure to a synthetic estrogen disrupts mating dynamics in a pipefish. *Hormones and Behavior*, **58**, 800–807.

Pizzolon, M., Giacomello, E., Marri, L. *et al.* (2010) When fathers make the difference: efficacy of male sexually selected antimicrobial glands in enhancing fish hatching success. *Functional Ecology*, **24**, 141–148.

Plenderleith, M., van Oosterhout, C., Robinson, R.L., and Turner, G.F. (2005) Female preference for conspecific males based on olfactory cues in a Lake Malawi cichlid fish. *Biology Letters*, **1**, 411–414.

Poling, K.R., Fraser, E.J., and Sorensen, P.W. (2001) The three steroidal components of the goldfish pre-ovulatory pheromone signal evoke different behaviors in males. *Comparative Biochemistry & Physiology Part A*, **129**, 645–651.

Qureshi, N.A. 2007. Gravid female round goby (*Neogobius melanostomus*) response to odours of conspecific males injected with Gonadotropin Releasing Hormone (GnRH). B.Sc. Thesis. Department of Biological Sciences, University of Windsor, Windsor, ON.

Ray W.J. and Corkum L.D. (2001) Habitat and site affinity of the round goby. *Journal of Great Lakes Research*, **27**, 329–334.

Remage-Healey, L. and Bass, A.H. (2005) Rapid elevations in both steroid hormones and vocal signaling during playback challenge: a field experiment in Gulf toadfish. *Hormones and Behavior*, **47**, 297–305.

Schaeffer, J.S., Bowen, A., Thomas, M. *et al.* (2005) Invasion history, proliferation and offshore diet of the round goby *Neogobius melanostomus* in western Lake Huron, USA. *Journal of Great Lakes Research*, **31**, 414–425.

Searcy, W.A. and Nowicki, S. (2005) *The Evolution of Animal Communication. Reliability and Deception in Signaling Systems.* Princeton University Press, Princeton.

Serrano, R.M., Barata, E.N., Birkett, M.A. *et al.* (2008) Behavioral and olfactory responses of female *Salaria pavo* (Pisces: Blenniidae) to a putative multi-component male pheromone. *Journal of Chemical Ecology*, **34**, 647–658.

Shohet, A.J. and Watt, P.J. (2004) Female association preferences based on olfactory cues in the guppy, *Poecilia reticulata*. *Behavioral Ecology and Sociobiology*, **55**, 363–369.

Siefkes, M.J., Bergstedt, R.A., Twohey, M.B., and Li, W. (2003) Chemosterilization of male sea lampreys (*Petromyzon marinus*) does not affect sex pheromone release. *Journal of Fisheries and Aquatic Sciences*, **60**, 23–31.

Siefkes, M.J., Winterstein, S.R., and Li, W. (2005) Evidence that 3-keto petromyzonol sulfate specifically attracts ovulating female sea lamprey, *Petromyzon marinus*. *Animal Behaviour*, **70**, 1037–1045.

Sisneros, J.A., Forlano, P.M., Knapp, R., and Bass, A.H. (2004) Seasonal variation of steroid hormone levels in an inter-tidal nesting fish, the vocal plainfin midshipman. *General and Comparative Endocrinology*, **136**, 101–116.

Smith, J.M. (1974) Theory of games and evolution of social conflicts. *Journal of Theoretical Biology*, **47**, 209–221.

Sorensen, P.W. and Hoye, T.H. (2010) Pheromones in vertebrates, in *Chemical Ecology, Comprehensive Natural Products Chemistry II: Chemistry and Biology*, vol. **4** (ed. K. Mori), Elsevier Press, Oxford, pp. 225–262.

Sorensen, P.W. and Stacey, N.E. (1999) Evolution and specialization of fish hormonal pheromones, in *Advances in Chemical Signals in Vertebrates* (eds R.E. Johnston, D. Müller-Schwwarze, and P.W. Sorensen), Kluwer Academic, Plenum Publishers, New York, pp. 15–47.

Sorensen, P.W., Hara, T.J., Stacey, N.E., and Dulka, J.G. (1990) Extreme olfactory specificity of male goldfish to the preovulatory pheromone 17a,20b-dihydroxy-4-pregnen-3-one. *Journal of Comparative Physiology A*, **166**, 373–385.

Sorensen, P.W., Murphy, C.A., Loomis, K. *et al.* (2004) Evidence that 4-pregnen-17,20β,21-triol-3-one functions as a maturation-inducing hormone and pheromonal precursor in the percid fish, *Gymnocephalus cernuus. General and Comparative Endocrinology*, **139**, 1–11.

Sorensen, P.W., Fine, J.M., Dvornikovs, V. *et al.* (2005) Mixture of new sulfated steroid functions as a migratory pheromone in the sea lamprey. *Nature Chemical Biology*, **1**, 324–328.

Stacey, N.E. and Hourston, A.S. (1982) Spawning and feeding behavior of captive Pacific herring, *Clupea harengus pallasi. Canadian Journal of Fisheries and Aquatic Sciences*, **39**, 489–498.

Stacey, N.E. and Sorensen, P.W. (2006) Reproductive pheromones, in *Behaviour and Physiology of Fish*, vol. **24** (eds K.A. Sloman, R.W. Wilson, and S. Balshine). Academic Press, Amsterdam, pp. 359–412.

Sveinsson, T. and Hara, T. (1995) Mature males of Arctic charr, *Salvelinus alpinus*, release F-type prostaglandins to attract conspecific mature females and stimulate their spawning behaviour. *Environmental Biology of Fishes*, **42**, 253–266.

Svensson, P.A., Blount, J.D., Forsgren, E. *et al.* (2010) Female ornamentation and egg carotenoids of six sympatric gobies. *Journal of Fish Biology*, **75**, 2777–2787.

Tavolga, W.N. (1956) Visual, chemical and sound stimuli as cues in the sex discriminatory behaviour of the gobiid fish *Bathygobius soporator. Zoologica*, **41**, 49–64.

Thornhill, R. and Alcock, J. (1983) *The Evolution of Insect Mating Systems*, Harvard University Press, Cambridge.

Timms, A.M. and Kleerekoper, H. (1972) The locomotor responses of male *Ictalurus punctatus*, the channel catfish, to a pheromone released by the ripe female of the species. *Transactions of the American Fisheries Society*, **101**, 302–310.

Van den Hurk, R. and Resink, J.W. (1992) Male reproductive system as a sex pheromone producer in teleost fish. *Journal of Experimental Biology*, **261**, 204–213.

Waas, J.R. and Colgan, P.W. (1992) Chemical cues associated with visually elaborate aggressive displays of three-spined sticklebacks. *Journal of Chemical Ecology*, **18**, 2277–2284.

Ward, A.J.W, Sumpter, D.J.T., Couzins, I.D. *et al.* (2008) Quorum decision-making facilitates information transfer in fish shoals. *Proceedings of the National Academy of Science*, **105**, 6948–6953.

Waring, C.P. and Moore, A. (1997) Sublethal effects of a carbamate pesticide on pheromonal mediated endocrine function in mature male Atlantic salmon (*Salmo salar* L.) parr. *Fish Physiology and Biochemistry*, **17**, 203–211.

Wilson, E.O. and Bossert, W.H. (1963) Chemical communication among animals, in *Recent Progress in Hormone Research*, vol. **19** (ed. G. Pincus), Academic Press, New York, pp. 673–716.

Wisenden, B.D. (1995) Reproductive behaviour of free-ranging convict cichlids, *Cichlasoma nigrofasciatum. Environmental Biology of Fishes*, **43**, 121–134.

Wisenden, B.D. (2000) Scents of danger: the evolution of olfactory ornamentation in chemically-mediated predator-prey interactions, in *Signalling and Signal Design in Animal Communication* (eds Y. Espmark, T. Amundsen, and G. Rosenqvist), Tapir Academic Press, Trondheim, pp. 365–386.

Wyatt, T.D. (2003) *Pheromones and Animal Behaviour*, Cambridge University Press, Cambridge.

Yambe, H., Kitamura, S., Kamio, M. *et al.* (2006) L-Kynurenine, an amino acid identified as a sex phero-mone in the urine of ovulated female masu salmon. *Proceedings of the National Academy of Science*, **103**, 15370–15374.

Young, M.K., Micek B.K., and Rathbun M. (2003) Probable pheromone attraction of sexually mature brook trout to mature male conspecifics. *North American Journal of Fisheries Management*, **23**, 276–282.

Yun, S.-S., Siefkes, M.J., Scott, A.P., and Li, W. (2002) Development and application of an ELISA for a sex pheromone released by the male sea lamprey (*Petromyzon marinus* L.). *General and Comparative Endocrinology*, **129**, 163–170.

Zahavi, A. (1975) Mate selection—a selection for a handicap. *Journal of Theoretical Biology*, **53**, 205–214.

Zahavi, A. and Zahavi, A. (1977) *The Handicap Principle*, Oxford University Press, Oxford.

Chapter 5
Intraspecific Social Recognition in Fishes via Chemical Cues

Ashley J.W. Ward

University of Sydney, Sydney, Australia

5.1 INTRODUCTION

Research using fish models has made a major contribution to our understanding of animal communication and recognition mechanisms. A wide range of sensory channels are available to fish to gather information about their physical and social environments; and of these, chemical communication is of major importance for various reasons, including constraints on the visual environment caused by turbidity and light absorption and the property of water as the "universal solvent" and medium for transmitting pheromones and chemical cues.

Chemical communication in fishes, or at least the transfer of chemical information, has been studied in a scientific context for the better part of a century. In 1932, Wrede reported that minnows (*Phoxinus phoxinus*) were attracted by the mucus of conspecifics, therefore, providing a basis for the formation of shoals using chemical cues (Wrede, 1932). Building on this, Keenleyside (1955) showed that rudd (*Scardinius erythrophthalmus*) that had been temporarily blinded were able to locate and maintain proximity to conspecifics on the basis of their detection of chemical cues. In subsequent years, numerous studies based on a diverse range of species have shown the importance of chemical cues to fish in recognizing and discriminating among conspecifics (reviewed in Solomon, 1977; Hara 1992; Wisenden and Stacey, 2005; Rosenthal and Lobel, 2006).

When one animal detects another in its environment, it may make an assessment by comparing the cues that it detects with a "recognition template" (Mateo, 2004). Depending on which cues are gathered, the quality of those cues and the sophistication of the recognition template, the animal may achieve some level of recognition. The specificity of this recognition may range from identification to the species level, particularly discrimination of conspecific from heterospecific, to the more complex discernment of subsets of the local conspecific population, including the ability to determine kin from nonkin, and even individual recognition. The recognition template may be genetically determined, providing the receiver with an innate ability to recognize the cue, such as is the case with the detection and recognition of most conspecific pheromones. Alternatively, the recognition template may be more flexible, reflecting the dynamic nature of many of the chemical cues involved in recognition. In such cases, the template may be generated by learning, or through self-referencing where a

Fish Pheromones and Related Cues, First Edition. Edited by Peter W. Sorensen and Brian D. Wisenden.
© 2015 John Wiley & Sons, Inc. Published 2015 by John Wiley & Sons, Inc.

fish matches its own odor to that of a conspecific. The kinds of patterns of social organization that are observed in fish populations in nature demonstrate that fishes are capable of discriminating between conspecifics to a fairly sophisticated level and that this ability to discriminate mediates their subsequent social behavior. The precise mechanisms are not always well understood, however.

In this chapter, the role of chemical cues in conspecific recognition in fishes is reviewed. The ability of fish to recognize related individuals, kin, in the population as well as their capacity to discriminate between unrelated conspecifics in a social context will be considered. Further, what is known about the chemical cues that facilitate these forms of recognition will be discussed. Last, future directions for this field of research will be suggested.

5.2 KIN RECOGNITION IN FISHES VIA CHEMICAL CUES

In the early-to-mid-twentieth century, a number of evolutionary biologists, including R.A. Fisher and J.B.S. Haldane, noted that the simple fact that related individuals share genes may give rise to evolutionary strategies that promote nepotism, or the favoring of kin over nonkin. These insights were formalized by W.D. Hamilton into what became known as Hamilton's rule (Hamilton, 1963, 1964a, 1964b), which asserts that probability of cooperation between two individuals is likely to be increased if the two are related. An individual can increase its own fitness, albeit indirectly, simply by promoting the fitness of a relative. That an individual may increase its fitness indirectly as well as directly forms the basis for the concept of inclusive fitness. This crucial evolutionary paradigm effectively explained the paradox of why animals may sometimes be observed to act altruistically, and it also indicated that to be able to discriminate in favor of their own kin, animals must first be able to recognize them.

Many fish have the sensory and cognitive ability to distinguish between kin and nonkin (Griffiths, 2003; Ward and Hart, 2003). Further, fine-scale recognition in fishes, that is, discriminating among conspecifics, is primarily based on chemical cues in many species. This provides the basis for a range of kin-selected behaviors amongst fishes, including reductions in aggression or competitiveness, providing active assistance to kin, including parental care, and preferentially social association with kin, all of which should increase the overall fitness of both the donor and its kin. The fitness benefits that accrue from kin-selected activities have important implications not only for the individuals themselves but also in the wider context of fisheries management because understanding the way in which fishes structure their interactions may inform decisions made by fishery managers and conservation biologists.

5.3 EXAMPLES OF KIN RECOGNITION AND KIN-BIASED BEHAVIOR ON THE BASIS OF CHEMICAL CUES IN FISHES

Kin recognition and kin-biased behavior can operate either within or between generations. In the latter case, many species of fish provide parental care to their offspring. Typically, this occurs at or near a central nest, where the young are guarded by their aggressively territorial parents. Most cichlids species provide a degree of care for their young; and in many cases, parental adults are able to recognize their offspring and to distinguish them from unrelated juveniles of the same size (McKaye and Barlow, 1976), primarily via the use of chemical cues (Kuhme, 1964a, 1964b; Myrberg, 1975). Male bluegill sunfish (*Lepomis macrochirus*) recognize and move toward chemical cues derived from their own offspring in preference to unrelated fry (Neff and Sherman, 2005). The authors ascribed the males' ability to recognize their offspring to self-referent phenotype matching, whereby the males compare their own chemical cues with those released by their offspring (Neff and Sherman, 2005). This enables the parent, or parents, to direct their parental care to the appropriate young, and also in some instances to add unrelated young to their own brood, presumably in order to dilute their predation risk (McKaye and McKaye, 1977; Wisenden and Keenleyside, 1992, 1994). The young of species such as convict cichlids are also able to recognize the chemical cues of kin and to orientate toward these cues if they get separated from them (Wisenden and Dye, 2009). It is not clear whether the young use cues arising from siblings or from their parents, or both, to achieve this. Fry also use chemical cues for recognition of their parents. Russock (1990) showed that young Mozambique tilapia (*Oreochromis mossambicus*) are able of using chemical cues to recognize their mother. In another cichlid species, the Midas fish (*Cichlasoma citrinellum*), the evidence is more equivocal. Noakes and Barlow (1973) found that fry used chemical cues to recognize their parents, whereas, by contrast, Barnett (1982) reported that the fry were unable to distinguish between their own mother and a female conspecific. The mechanism that underlies the recognition by offspring of their parent in cichlids may be based on a suite of cues. For example, Wisenden and Dye (2009) showed that convict cichlid fry could learn chemical cues associated with their "home" area, and that this ability was contingent upon those chemical cues being presented in conjunction with visual cues of their parents and sibs.

In some species of nesting fishes, such as cichlids and anabantids, juvenile fish remain at their natal site and help their parents raise subsequent generations of offspring (Wong and Balshine, 2011). By delaying their dispersal, sometimes for as much as a year (indeed the parents may forcibly expel lingering subadults), these "helpers at the nest" presumably gain greater fitness benefits than if they were to establish their own territories and reproduce themselves (Taborsky, 1984). Among such species, perhaps the best studied is the cichlid species, *Neolamprologus brichardi*. Nests of this species have been reported with up to four separate juvenile size classes, suggesting the simultaneous presence of four batches of young (Taborsky and Limberger, 1981). The ability to recognize kin in this species was suggested by the findings of a study by Hert (1985) that reported high levels of aggression by residents toward outsiders trying to join an established family. Le Vin, Mable, and Arnold (2010) found that in *N. brichardi*'s congener, *N. pulcher*, kin recognition was achieved primarily through the use of chemical cues. In this case, the stimulus fish were reared separately to the focal fish, suggesting the use of phenotype matching as the mechanism for recognition.

Once the most vulnerable stages of early life are complete and parental care ceases, members of a brood may continue to form social associations with their siblings as they leave the nest. Of course, social association with kin is also possible among members of species without parental care as long as members of kin cohorts occur in the same range and individuals are able to recognize their kin. Arguably, social groups formed of kin individuals could realize fitness benefits for group members, over and above the benefits enjoyed by members of social groups whose membership is not structured by relatedness (Krause and Ruxton, 2002).

In a study on guppies, Evans and Kelley (2008) found that pairs of full siblings spent more time in close proximity with one another than did pairs of half-siblings. This has important implications for shoal cohesion, which is an important factor in promoting the antipredator function of groups. Similarly, one could envisage the possibility that the antipredator benefits of group living could be amplified in social groups of related individuals through decreased aggression among group members, which potentially enables greater investment in vigilance behavior (Wisenden and Smith, 1998; Griffiths *et al.*, 2004). The performance of inherently risky yet valuable behaviors such as predator inspection might, in theory, be promoted in kin groups, again because of the potential for inclusive fitness benefits (Milinski *et al.*, 1997).

Despite this, however, evidence to date for the formation of kin social groups in free-ranging fishes is equivocal. Pouyaud *et al.* (1999) provided evidence of kin association in a juvenile mouthbrooding cichlid species, *Sarotherodon melanotheron*. In Eurasian perch (*Perca fluviatilis*), studies of the association patterns of juveniles show evidence of a social preference for kin (Gerlach *et al.*, 2001; Behrmann-Godel, Gerlach, and Eckmann, 2006). Sibling groups of free-ranging juvenile black perch (*Embiotoca jacksoni*) were reported by Sikkel and Fuller

(2010), who also noted that aggression was much greater between groups than within them. Buston *et al.* (2009) found that colonies of humbug damselfish, *Dascyllus aruanus*, often contain related juveniles. This is all the more surprising given that the larval dispersal stage of coral reef fishes that is presumed to scatter individuals from within broods. Set against these findings, however, are a number of papers that fail to find evidence of kin association among wild fishes (see Griffiths and Ward, 2011 for a review).

Given that a large and diverse range of species have been shown to demonstrate kin association preferences in the laboratory (Griffiths and Ward, 2011), this disconnect between laboratory observations and patterns of kin association in the field is troubling. This may be because of differences between the chemical ecology of the laboratory and field environments. Because chemical cues are the primary source of information on kinship, this will clearly have an effect on the efficiency of cue transmission. Laboratory experiments may sometimes provide the relevant chemical cues at an unnaturally high concentration and/or the presence of additional biologically generated chemical cues, whereas ubiquitous in the natural aquatic environment, are seldom present in the laboratory. In addition, laboratory-based studies are often carried out using batches of fish that for either their entire life, or for just part of it, were artificially constrained and forced to repeatedly interact, thereby developing a familiarity with their kin that is wholly unrepresentative of field conditions. Finally, assessments of kinship within shoals that are characterized as the mean relatedness of all group members may fail to capture subgroup association patterns. For example, shoals of fish may contain multiple subgroups, some of which may represent stable associations among kin individuals; hence, defining an appropriate null model requires considerable thought. In this instance, measurement of kin association may be achieved more effectively by use of social network approaches. Piyapong *et al.* (2011) examined association patterns of siblings using a network approach in free-ranging shoals of guppies. Although in low predation habitats, there was no evidence of kin association, in high predation areas pairs of siblings co-occurred significantly more than expected.

In salmonid species, shoaling with kin is most likely to occur during migration, because at other times the fish are often territorial and do not shoal. Again, the evidence for kin association during these times is equivocal. Olsen *et al.* (2004) investigated kin association patterns among individually tagged Atlantic salmon smolts (*Salmo salar*), finding that interindividual distances were much lower between kin individuals than between nonkin. Fraser, Duchesne, and Bernatchez (2005) reported subgroups of kin within shoals of adult and subadult brook char (*Salvelinus fontinalis*) and suggested that the associations between kin individuals could be stable and relatively long lasting. By contrast, Palm *et al.* (2008) found no evidence of kin associations in Atlantic salmon in the Baltic Sea that were surveyed using drift nets. Nonetheless, they did find a trend toward individuals from the same natal rivers to co-occur, which may indicate the possibility of intragroup kin associations.

Much work has been done on kin-biased behavior in salmonids at the juvenile territorial stages of the life cycle. At this point in their lives, many species aggressively defend small feeding territories, although often the mosaic of territories may be fairly dynamic, with territories sometimes overlapping, whereas dominant fish may range across several other territories in an effort to locate food (Armstrong, Huntingford, and Herbert, 1999; Martin-Smith and Armstrong, 2002). Griffiths and Armstrong (2002) found that sibling juvenile Atlantic salmon shared territorial resources for longer than unrelated fish and consumed more food in the presence of kin. Siblings are also less aggressive to one another than unrelated fish (Brown and Brown, 1993), which translates into faster growth (Brown, Brown, and Wilson, 1996; Greenberg *et al.*, 2002; Gerlach *et al.*, 2007). Odors likely play a role in these processes, but this is yet to be demonstrated.

In habitats where competition among individuals is intense, for example, where resources are scarce, it may actually be advantageous to avoid kin. Griffiths, Armstrong, and Metcalfe (2003) studied winter sheltering behavior in Atlantic salmon and found that fish avoided sharing the same shelter as kin, possibly because each additional individual that joins others occupying a shelter may impose costs on those already within. Further, if it is true that the more genetically similar two individuals are, the more similar their patterns of resource exploitation will be, then it may pay kin to disperse to avoid the cost of greater competition between relations on this basis. A study by Carlsson and Carlsson (2002) showed that relatedness correlated positively with interindividual distance in juvenile brown trout (see also Griffiths and Armstrong, 2001). The avoidance of kin is also crucially important to avoid selecting a closely related individual as a mate (Charlesworth and Charlesworth, 1987). Nonetheless, there is thus far little experimental evidence of kin avoidance in fish mate choice. Guppies, for example, seem to be indifferent to the relatedness of potential mates (Viken, Fleming, and Rosenqvist, 2006; Pitcher, Rodd, and Rowe, 2008; Guevara-Fiore, Rosenqvist, and Watt, 2010). This is despite the known ability of guppies to be able to discriminate kin from nonkin (Hain and Neff, 2007). Moreover, the cichlid, *Pelvicachromis taeniatus*, shows a preference for related individuals as mates (Thunken *et al.*, 2007), possibly because of the greater cooperation observed between related individuals in their parental care relative to unrelated individuals.

Another form of kin-biased behavior may be seen in the avoidance of kin cannibalism. While fish, particularly parental fish, will eat their offspring under some conditions (Manica, 2002), livebearers such as guppies and mollies preferentially avoid their own young and preferentially predate unrelated fry (Loekle, Madison, and Christian, 1982). Female sticklebacks will occasionally raid the nests of males and eat the eggs contained within; however, there is evidence to suggest that females are capable of recognizing and avoiding the nests in which they have spawned on the basis of chemical cue recognition (Fitzgerald and van Havre, 1987).

5.4 POTENTIAL MECHANISMS—TYPES OF CUE INVOLVED IN CHEMICALLY MEDIATED RECOGNITION OF KIN

A range of chemical cues may be employed in kin recognition, and these are typically derived from (or rather contained within) excretions produced by the donor. For example, Olsen (1987) and Moore, Ives, and Kell (1994) reported that the cues enabling conspecific and kin recognition in juvenile Atlantic salmon are contained within the urine of the fish. Excretory products such as urine will clearly contain a spectrum of different chemicals that contribute to a typical signature mixture (Wyatt, 2010). Some components of this mixture inform the receiver about its relatedness to the donor, whereas others convey information about the donor's physiological state and aspects of its social status including recent dietary history (Olsen, Grahn, and Lohm, 2003). Although these components may be used by the receiver in addition to the cues on relatedness, for example, in mate choice decisions, this section mainly describes the present state of our knowledge of chemical kin recognition cues and returns to the topic of state and context dependency of chemical cues later in this chapter.

Research into the mechanisms underlying recognition of kin by fishes has recently focused on the role of the major histocompatibility complex (MHC). The MHC is an area of the vertebrate genome that codes for glycoproteins involved in immunological recognition. As a by-product, the MHC also appears to affect the chemical cues that an individual produces (Brown and Eklund, 1994; Eizaguirre *et al.*, 2009). Several mechanisms have been postulated. In one of these fragments of MHC-related peptides are shed by cells and excreted from the body, and these peptides are now known to contribute to an individual's signature mixture, carrying information about its MHC genotype (Milinski *et al.*, 2005). In the context of kin recognition, MHC-mediated chemical cues are thought to be of particular value because the cues of kin, which share large parts of their genomes, are likely to be more similar than the cues of nonkin. Moreover, individuals will be able to reference their own chemical cues and use this as a template to compare the cues of another individual—the more similar the chemical cues of another fish are to its own, the more closely related that individual is likely to be. In an elegant series of experiments, Olsen, Grahn, and Lohm (2002) showed that the strength of the preference of juvenile Arctic char (*Salvelinus alpinus*) to associate with the chemical cues of siblings was mediated by the similarity of these siblings' MHC genotypes to their own. Further, they demonstrated no preference for associating with kin odors when offered a choice between the chemical cues of a sibling that had an MHC genotype that was different to that of the choosing fish versus a nonsibling that had an identical MHC genotype. Thus, Olsen, Grahn, and Lohm (2002) were able to demonstrate a primary (albeit not exclusive) role of MHC-derived cues in mediating kin recognition in these fish. Similarly, Rajakaruna *et al.* (2006) showed that while an MHC class II gene was important in kin recognition for juvenile Atlantic salmon (*S. salar*) and brook trout (*Salvelinus fontinalis*), the fish were able to use other chemical cues derived from other components of the genotype to recognize kin. As yet, however, there have been few studies examining the role of MHC-derived cues in promoting kin recognition in nonsalmonid species. One important exception to this is the research carried out on mate choice in fishes, particularly that on three-spined sticklebacks (*Gasterosteus aculeatus*) where chemical cues based on the MHC are used by females to avoid kin as potential mates and instead to optimize the diversity of their offsprings' MHC genotype by outbreeding (Reusch *et al.*, 2001; Aeschlimann *et al.*, 2003; Wegner *et al.*, 2003a; Wegner, Reusch, and Kalbe, 2003b; Forsberg *et al.*, 2007).

While MHC-derived cues offer a potentially widespread mechanism for kin recognition across fish species, the question of whether the recognition template is innate, based on phenotype matching, is actively learned through social experience, or is some combination of these remains to be fully resolved. The point nevertheless remains that experience is not needed with phenotype matching. Fish among all vertebrate animals have the greatest diversity of reproductive strategies: some produce huge numbers of eggs and scatter these to fend for themselves from the outset; others produce fewer eggs, instead of investing in diligent parental care. These reproductive strategies should theoretically play a role in shaping the value of kin recognition and the mechanisms by which it may be achieved. The offspring of the so-called broadcast spawners are likely to have far less opportunity to interact with their kin during early life in comparison to the young of species that lay eggs in nests and guard their brood. As a result, broadcast spawners may have to rely on innate kin-recognition mechanisms, whereas increasing levels of parental care should tend to predict the development of a learned recognition template based on early life familiarity with siblings.

Among salmonids, which given their huge economic importance are perhaps unsurprisingly the most studied group in this context, a number of studies have highlighted the importance of phenotype matching in kin recognition (Quinn and Busack, 1985; Olsen, 1989; Winberg and Olsen, 1992; Brown, Brown, and Crosbie, 1993; Olsen and Winberg, 1996). This mechanism that differs from imprinting allows for the recognition of kin based on a template that may be formed either during the egg stage or very soon after hatching, while the fry feed endogenously. Alternatively, phenotype matching may be performed based on self-referencing using some heuristic along the lines of "those who smell most like me may be kin" (Holmes and Sherman, 1982). Empirical support for this idea comes from a fascinating study by Thunken *et al.* (2009) who exposed cichlids, *P. taeniatus*, to a choice between their own chemical cues and those of a related individual. The fish demonstrated a strong preference for their own cues, which, if recognition is based on a self-referencing template, would obviously provide an exact match to the template. Among salmonids, the recognition template may be learned during an early life stage. Quinn and Busack (1985) reported that juvenile coho salmon, *Oncorhynchus kisutch*, showed a preference for water containing the chemical cues of unfamiliar kin over water containing cues from unfamiliar nonkin despite the broods having been separated approximately 1 h post fertilization. This suggests that they either learn the generalized smell of a sibling and can later use this to identify all siblings, or that they may instead self-reference their own smell against that of another individual. Using Arctic char, Winberg and Olsen (1992) similarly separated broods at the egg stage; however in this case, while fish reared with siblings showed the ability to recognize other, unfamiliar siblings, fish that had been reared in isolation did not manifest any preference for siblings. Taken together, these results suggest that although prior interactions with particular kin individuals may not be necessary an individual to be able subsequently to recognize them, some exposure to kin individuals is necessary to develop a general recognition template for kin in these species.

More recently, the zebrafish (*Danio rerio*) has been used to examine the phenomenon of "olfactory memory" and template formation. Harden *et al.* (2006) described patterns of increased gene expression in the developing olfactory system of embryos 48–72 h post fertilization in response to the presence of an artificial odorant. Although not directly linked to the recognition of kin, these results neatly demonstrate the sensitivity of embryos to key external chemosensory cues at an early stage of development, prior to hatching. Investigations into kin recognition in zebrafish have shown not only the ability to recognize related individuals based on phenotype matching but also that this ability is enhanced by social experience (Gerlach and Lysiak, 2006). The specificity of timing in the development of the recognition template for kin in zebrafish was reported in a ground-breaking study by Gerlach *et al.* (2008). In this study, exposure to kin on the sixth day following fertilization was required for the kin recognition template to develop. The study further excluded self-referent phenotype matching in this species, since isolated larvae did not imprint on their own cues. Even more remarkably, the larvae failed to imprint on nonkin at this time, which the authors suggest may be indicative of a genetic predisposition, and sensitivity to the chemical cues of kin. Although the zebrafish recognition template appears to be robust to mistakes in imprinting, a study on livebearing fish suggested that such templates have at least the potential for flexibility in other species. Warburton and Lees (1996) raised juvenile guppies with the young of a closely related species, the swordtail (*Xiphophorus hel-leri*). Subsequently, the guppies raised with heterospecifics displayed a preference for associating with swordtails in preference to their own species, which is the more typical pattern among guppies. Clearly, a fascinating next step would be to test the development of a preference for kin among guppies that have been raised from birth among nonkin.

The use of phenotype matching as a mechanism in the recognition of kin has also been documented in a range of other species; however, there appears to be no consistent role for familiarity in either supporting or enhancing this. In the cooperatively breeding cichlid, *N. pulcher*, juveniles appear to use phenotype matching for recognizing kin, but familiarity appears not to play any role (Le Vin, Mable, and Arnold, 2010). By contrast, (Hain and Neff, 2007) found that guppies (*Poecilia reticulata*) taken from a population at the Paria River in Trinidad used both phenotype matching and familiarity in kin recognition. When kin groups are reared in close proximity, the ability of fish to develop familiarity is manifested, and this may act to enhance patterns of social preference for kin. Both Courtenay *et al.* (1997) and Frommen, Luz, and Bakker (2007) determined that familiarity increased social preferences for kin in this way in juvenile coho salmon and three-spined sticklebacks, respectively.

Finally, there are examples in the literature of fish requiring visual cues, in addition to olfactory cues, to recognize kin. Arnold (2000) reported that Lake Eacham rainbowfish (*Melanotaenia eachamensis*) provided solely with chemical cues showed only weak kin recognition abilities, whereas individuals that were provided with both cues performed much better in discriminating kin. Similarly, Steck, Wedekind, and Milinski (1999) found that sticklebacks that were provided solely with chemical cues were unable to recognize kin, in contrast to other studies that demonstrate kin recognition abilities in this species when given both chemical and visual cues (van Havre and Fitzgerald, 1988; see also Mehlis, Bakker, and Frommen, 2008). There are a number of potential reasons why additional cues would strengthen a response including the ability to enhance transmission of the relatedness signal (e.g., Ward and Mehner, 2010), and it is possible that the presence of conspecific visual cues in shoaling species acts to calm an isolated test fish and promote accuracy in its decision-making. The primary importance of chemical cues in kin recognition between fishes appears to be well established; however, an experimental protocol that titrates visual cues against chemical cues could be used to examine this further. One could imagine a binary choice of, on the one side of a test arena, kin visual cues in conjunction with non kin chemical cues; whereas on the other side of the arena nonkin visual cues presented simultaneously with kin chemical cues sensu Ward, Axford, and Krause, (2002).

5.5   RECOGNITION OF UNRELATED CONSPECIFICS VIA CHEMICAL CUES

There now exists considerable evidence that fish are able to distinguish among unrelated conspecifics and are able to do so irrespective of the context and in the absence of any obvious behavioral or phenotypic cues. On the basis of this recognition, they respond more favorably to some conspecifics than to others; for example, they may manifest a social association preference for particular individuals, or they may be consistently less aggressive toward some than to others. This phenomenon is often referred to as "familiarity," and it plays a significant role in structuring the social organization of thousands of fish species.

In the wild, fish manifest a wide variety of patterns of social organization. Thousands of species of fish form into shoals or into dominance hierarchies. For example, Shaw (1978) estimated that around one-quarter of all fish species shoal throughout their lives, whereas over half of species shoal at some stage of their life history. All such patterns of social organization are predicated upon social recognition. The formation and maintenance both of shoals and of dominance hierarchies are mediated extensively by the recognition of chemical cues (Solomon, 1977; Hara, 1992; Rosenthal and Lobel, 2006). When chemical recognition fails, social organization may break down. This is demonstrated by the effect of the so-called "info-disrupting" aquatic contaminants that can block fishes' chemical recognition capabilities, leading to the break-up of shoals and the dissolution of dominance hierarchies (Fisher, Wong, and Rosenthal, 2006; Ward, Duff, and Currie, 2006; Ward *et al.*, 2008).

Shoaling fish are capable of discriminating between certain subsets of conspecifics in their local population, or even between individuals, and they use this ability to inform their choices of shoaling partners. There are now numerous examples in the literature of fish demonstrating an association preference for familiar individuals over unfamiliar ones (Griffiths, 2003; Ward and Hart, 2003). Functionally, this behavior is known to be adaptive as it generally enhances the advantages of shoaling behavior by reducing per capita risk of predation (Chivers, Brown, and Smith, 1995) and improving foraging efficiency (Ward and Hart, 2005, 2007).

Similarly, there are many examples of fish showing a reduction in aggression toward familiar conspecifics by comparison with unfamiliars. This may be in the context of territoriality whereby fish that hold territories are less aggressive to their familiar near neighbors than to conspecific interlopers, the so-called "dear enemy" effect (Jaeger, 1981). Or it may be in the context of dominance hierarchies, the maintenance of which are obviously dependent on the ability to recognize other individuals within the hierarchy. While the benefits that individual members of the hierarchy gain from this form of social organization may vary considerably, in general such hierarchies stabilize and reduce the frequency of aggressive interactions within the hierarchy. In addition to the energetic savings and the lowering of the risk of injury through fighting that this provides, membership of a hierarchy typically enables fish to spend a greater proportion of their time budget engaged in more profitable activities, increasing foraging activity and allowing more effective antipredator vigilance (Griffiths *et al.*, 2004).

5.6 EXAMPLES OF RECOGNITION OF FAMILIARS VIA CHEMICAL CUES

Dozens of studies have now documented an association preference for familiar individuals in fish species (Ward and Hart, 2003). The most frequently used laboratory protocol employed to test for the expression of familiarity is the binary choice test. Here, a single focal fish is introduced to an aquarium containing two stimulus shoals of conspecifics, one of which the focal fish has prior social experience of and the other that the focal fish has not previously (or recently) encountered. The stimulus shoals are constrained by barriers that prevent the focal fish from actually joining either shoal; but depending on the type of barrier, the focal fish can gather visual cues, chemical cues or both to recognize the stimulus fish. The focal fish can then decide which shoal to associate with; and in the overwhelming majority of cases, the focal fish manifests a social preference for the familiar shoal.

The majority of such studies as set up in such a way as to allow a choosing fish to access both visual and chemical cues arising from the stimulus shoal. There have been a smaller number of studies that provide the focal fish with cues pertaining to only one sensory modality, either visual or chemical. These are obviously of particular interest in the context of this chapter, because we can get an idea of the sensory basis of social familiarity preferences. Brown and Smith (1994) presented fathead minnows (*Pimephales promelas*) in a binary choice test with different combinations of sensory treatments. Intriguingly, they found that chemical cues, either in conjunction with visual cues or without, were necessary for the focal fish to express a preference for natural (i.e., familiar shoalmates). When provided with visual cues alone, focal fish showed no preference for either of the stimulus shoals. A similar test using sticklebacks provided an identical result to this (Ward, Hart, and Krause, 2004) although the preference for a familiar shoal was stronger when both visual and chemical cues were available than when only chemical cues were presented (see also van Havre and Fitzgerald, 1988). It should be pointed out, however, that this does not necessarily imply that visual cues were necessary for recognition. Instead, it seems possible that the additional sensory feedback provided by the visual cues merely informed the focal fish that it had located the shoal that it identified using its chemosenses. Of the tests that have provided visual cues only, there have been reported a shoaling preference for familiars in two species, the guppy (Warburton and Lees, 1996; Lachlan, Crooks, and Laland, 1998) and the bluegill sunfish (Dugatkin and Wilson, 1992).

Chemical cue preferences in fishes are also frequently examined using a fluvarium and provide a flowing water variation to the binary choice test described earlier. There are now numerous examples of fishes discriminating between unrelated conspecifics at a general level of recognition using this approach. For example, Courtenay *et al.* (1997) reported that juvenile coho salmon (*Oncorhynchus kisutch*) were able to distinguish between members of their own local population and members of a different population on the basis of chemical cues, preferring the cues of the more familiar local population. Recognition and preference for members of the same population may provide important addition cues in migration. Selset and Doving (1980) showed that adult char (*S. alpinus*) are attracted to water conditioned using smolts from their own population. Beyond the salmonids, Behrmann-Godel, Gerlach, and Eckmann (2006) found that perch (*P. fluviatilis*) from Lake Constance showed a preference for chemical cues derived from individuals of their own local population over those derived from members of a different but sympatric population. The focal fish in this case was determined as being unrelated to the fish that provided the stimuli. Marine species are under-represented in this field; however in a study on catfish, *Plotosus lineatus*, Matsumura, Matsunaga, and Fusetani (2004) revealed that focal individuals were strongly attracted to the chemical cues of their shoalmates. As the authors point out, it is possible that this may involve kin recognition as much as the recognition of unrelated familiars, which remains to be resolved.

The differences in the results and the conclusions drawn about the relative importance of particular sensory cues seem to suggest that social familiarity, or rather the level of familiarity, may vary across species. For some species, individual recognition may be necessary to enable strategic behavior, such as identifying good foraging partners, or those who cheat, or to remember previous mates. For most shoaling species, which might encounter hundreds or thousands of conspecifics over the course of each day, a general level of recognition may be sufficient to promote the expression of familiarity preferences. This would not necessarily allow identification of specific individuals, but would enable recognition of some more general identity, perhaps shared by conspecifics in a localized population. Ward *et al.* (2009) examined the mechanisms underpinning social recognition of familiars in guppies and in sticklebacks. They found that guppies were capable of individual recognition of conspecifics as well as being able to differentiate between groups of conspecifics based on some general cues. By contrast, while sticklebacks were also capable of differentiating between groups, there was no evidence that they were capable of individual recognition. As hinted earlier, this might reflect differences in the species' ecology. Guppies may be trapped in pools during the dry season and forced to interact repeatedly with the same individuals over an extended period. In addition, their promiscuous mating system may promote individual recognition—remembering previous mates potentially enables the fish to allocate their reproductive efforts toward novel partners. Sticklebacks used in this experiment were taken from lakes with large populations of this species where the high encounter rate means that identification of individuals may not usually be worthwhile, or even possible. Instead, a more general level of recognition of the local population allows individuals to navigate within their social environment, to participate in the benefits of familiarity and potentially to gather locally relevant information.

Individual recognition of unrelated individuals, as opposed to the recognition of some general group or subpopulation-level character, is also known to occur in many fish on the basis of chemical cues. Todd, Atema, and Bardach (1967) and Carr and Carr (1985) demonstrated the ability of yellow bullheads (*Ictalurus natalis*) and brown bullheads (*I. nebulosus*), respectively, to recognize particular individual conspecifics via chemical cues alone using classical conditioning techniques. Such individual recognition may form the basis for the stabilization of dominance relationships in these species. Reductions in aggression as a result of familiarity have been shown across a range of species (three-spine sticklebacks (Utne-Palm and Hart, 2000), juvenile brown trout (Hojesjo *et al.*, 1998), Arctic char (Seppa *et al.*, 2001), Atlantic salmon (O'Connor, Metcalfe, and Taylor, 2000). It seems likely that chemical cues play a major role in the recognition of familiar individuals in these cases. Goncalves-de-Freitas *et al.* (2008) reported that patterns of aggression between Nile tilapia were affected by level of waterflow. Those individuals in a water-renewing treatment showed greater aggression to conspecifics, presumably on the basis that chemical communication is diminished by water throughput. In a further study of Nile tilapia, Giaquinto and Volpato (1997) separated the sensory cues that passed between individuals. They found that the availability of chemical cues of conspecifics, and particularly the chemical cues of familiar conspecifics, acted to reduce observed levels of aggression. As the authors point out however, it is difficult to determine whether this pattern is the result of individual recognition, or as a result of recognition of some general cue relating to dominance, or a mix of both.

5.7 POTENTIAL MECHANISMS—TYPES OF CUE INVOLVED IN CHEMICAL RECOGNITION OF UNRELATED CONSPECIFICS

For some species, familiarity may be driven by specific individual recognition; whereas in others, the expression of familiarity may only require recognition of a more general group- or population-specific cue. The components of this cue may reflect differences in species' sensory ecology. Some species rely extensively on intraspecific chemical communication, others far less so. In addition, the physiological requirements of living in a freshwater or marine environment (or indeed in an estuarine one) have implications for the kinds of cue used by fishes.

Fishes may be viewed as "leaky bags" (Atema, 1996), constantly emanating a stream of chemicals though their gills, skin, urine, and feces. These chemicals can be used as cues to identify their producer. To date, a number of cue sources have been implicated in the recognition of familiar conspecifics. Courtenay *et al.* (1997) found that juvenile coho salmon were attracted to cues released by the feces of members of their own population. Chemical cues contained in the urine are known to be important in kin recognition (Moore, Ives, and Kell, 1994; Olsen *et al.*, 1998) and dominance signaling (Barata *et al.*, 2007, 2008); it seems likely that these are also used in discriminating familiar individuals. In addition, mucus from the skin of fish may also be the source of chemical cues. Matsumura, Matsunaga, and Fusetani (2004) work on marine striped catfish identified a phosphatidylcholine molecular species in the fish's mucus that is used in familiar group recognition. Doving, Nordeng, and Oakley (1974) suggested that chemicals contained in the mucus were involved in population recognition in Arctic char. Furthermore, an analysis of the mucus and intestinal contents of another salmonid, the Atlantic salmon, by Stabell, Selset, and Sletten (1982) found molecular evidence for between-group differences, which they suggested might form the basis of population-level recognition. A more detailed cross-comparison of the chemo-attraction to conspecific intestinal contents, urine, mucus, and amino acids in Atlantic salmon found that intestinal contents provide the strongest response in test subjects, but did not find any evidence of test group familiarity preferences (Fisknes and Doving, 1982; see also Zhang and Hara, 2009). Species-recognition processes are reviewed elsewhere in this book (Sorensen and Baker, 2015) and may reflect similar mechanisms.

The mix of chemicals yielded by any fish from some or all of these sources contributes to its chemical signature, which may be learned and used by others to identify it. Interestingly, however, since all of these chemical sources are effectively by-products of the animal's metabolism, they are not fixed, but can vary according to intrinsic factors (i.e., physiological changes) and extrinsic factors such as the animal's environment, or diet. This provides a strong contrast to chemical communication by pheromones, which by their nature are largely fixed and immutable. The effect of diet on a chemical signature can even be discerned by comparatively anosmic humans, as anyone who has eaten asparagus or garlic or a range of other odiferous foodstuffs will testify! Diet can have an equally major effect on fishes' chemical signatures. Bryant and Atema (1987) studied the effect of a change of diet on chemical communication in yellow bullhead catfish. Familiarity is based on chemical recognition in this species and is responsible for reducing levels of aggression between familiar individuals (Todd, Atema, and Bardach, 1967). The authors found that a change in diet precipitated a change in urine-borne amino acids and, ultimately, the chemical signature of the fish to the extent that they were no longer recognized as familiars by conspecifics.

Cues relating to the diet of members of social species are known to mediate their association preferences. Specifically, fish show an association preference for fish that have eaten the same diet as themselves, which suggests chemical self-referencing (Arctic char: Olsen, Grahn, and Lohm, 2003; three-spined sticklebacks: Ward, Hart, and Krause, 2004; Ward *et al.*, 2005). Further, Ward, Hart, and Krause (2004) and Ward *et al.* (2005) showed that fish are also more socially attracted to conspecifics that have occupied a similar habitat (in terms of water chemistry) than to fish from a different habitat. But while the fish used diet and habitat-derived chemical cues in their social association decisions, they did not use any cues relating to previous social experience. This suggests the possibility that although the mechanism behind familiarity is often assumed to be the learning of individual identities through previous social interactions with them, it may instead be based on the shared environmental and dietary experience of fish from the same area, expressed in the chemical cues that they release. Indeed, sticklebacks showed no preference for individuals that they had previously interacted with over those that they had not when recent habitat and dietary-derived cues were controlled for (Ward *et al.*, 2009). Ward, Webster, and Hart (2007) tested this mechanism in the field and showed that fish transplanted between habitats changed their social association preferences over time; although they initially preferred individuals from their original habitat, this changed over time such that they ultimately preferred individuals from their new habitat. The timeframe over which this shift occurs was examined by Webster *et al.* (2007) who reported that fish reversed their association preferences in this way over a period of approximately 2h. The exact biochemical basis of this phenomenon is as yet unknown.

5.8   THE FUTURE OF RESEARCH IN FISH CHEMICAL COMMUNICATION

Much work remains to be done before we can confidently claim that we fully understand the recognition mechanisms underlying the social organization of fishes. This is more than ever an extremely important field of research given the parlous state of many of the world's fisheries and the increasing reliance on aquaculture, both of which are typically based on fish species with strong patterns of social organization.

Technological advances in the past few years have allowed us a far greater insight into the process of social recognition via chemical cues. For many years, it was impossible for researchers confronted with a test animal that showed no response to chemical cues to distinguish between the possibilities that the individual had not detected the cues or that it had detected the cues but had decided not to act upon this. The recent development and improvements of techniques such as electro-olfactogram recording allow these hitherto troublesome alternatives to be separated and enable us to determine to a much greater extent whether fish responses are governed by sensory shortcomings or by strategic behavior. At present, it is, of course, relatively early days for this type of work; procedures are highly invasive and necessarily have to occur outside of any ecological context. Nonetheless, this may not always be the case. The tantalizing prospect of fish implanted with physiological sensors that relay information in real time is close at hand and such techniques may ultimately be adapted for use in physiological recording.

It is difficult to counter the argument that laboratory experiments in fish chemical ecology may sometimes give an exaggeratedly high impression of the response of fish to chemical cues because the water used in experiments is often artificially clean and is usually devoid of other organisms. Moreover, human contamination of aquatic habitats is widespread, and this also makes a contribution—a frequently deleterious contribution—to the chemical environment. This is known to have serious implications for fish communication in many cases, since the chemicals act as "infodisruptors" (Olsen and Hoglund, 1985; Lurling and Scheffer, 2007; Ward *et al.*, 2008; Tierney *et al.*, 2010; see also Chapter 10). Ultimately, this means that chemical communication channels in natural environment are subject to large amounts of chemical noise, derived both from sympatric organisms and from aquatic contamination. It is assumed that fish chemical communication has evolved to cope with at least some of this interference and that fish are likely to be highly sensitive to certain "infochemicals." In many laboratory protocols, however, these chemicals are often presented in the absence of the typical biochemical noise. The result is that we get an artificially high signal-to-noise ratio and a concomitant artificially high response, which has a limited ability to inform us about the fish's natural behaviour. It remains, therefore, a vital challenge to incorporate a far greater degree of environmental relevance into experiments.

5.9   SUMMARY

Chemical communication is fundamental to the ability of fish to recognize kin and familiar conspecifics. There is little evidence that pheromones play any major role in intraspecific recognition, however. Instead, the cues involved represent signature mixtures, which are either learned or compared with a self-referent template, for recognition to occur. The chemical cues that are used to achieve recognition of kin appear to have a different basis to those that are involved in the recognition of familiars. Where the former conveys information about the genotype of an individual, the latter conveys information about such things as physiological state, recent diet, and habitat characteristics. But while it may be possible to consider these cues separately from a functional perspective, it is important to note that all the chemical emanations from a fish contribute to its chemical signature, potentially allowing a receiver to assess a large amount of information and to use this to achieve recognition of either a general or an individual characteristic.

Despite the clear importance of fish social organization from both a conservation and fishery management standpoint, much work remains to be done if we are to claim that we have a good understanding of the mechanisms that support it. With particular reference to chemosensory communication and social recognition, there are very few data on marine species. More generally, while a picture is emerging of variation in social recognition between species, attempts remain to be made to understand the phylogeny and the evolutionary basis of this. Finally, it is absolutely crucial in the context of studying fish responses to chemical cues that a far greater emphasis is placed on natural relevance in experiments.

REFERENCES

Aeschlimann, P.B., Haberli, M.A., Reusch, T.B.H. *et al.* (2003) Female sticklebacks *Gasterosteus aculeatus* use self-reference to optimize MHC allele number during mate selection. *Behavioural Ecology and Sociobiology*, **54**, 119–126.

Armstrong, J.D., Huntingford, F.A., and Herbert, N.A. (1999) Individual space use strategies of wild juvenile Atlantic salmon. *Journal of Fish Biology*, **55**, 1201–1212.

Arnold, K.E. (2000) Kin recognition in rainbowfish (*Melanotaenia eachamensis*): sex, sibs and shoaling. *Behavioural Ecology and Sociobiology*, **48**, 385–391.

Atema, J. (1996) Eddy chemotaxis and odor landscapes: exploration of nature with animal sensors. *Biological Bulletin*, **191**, 129–138.

Barata, E.N., Hubbard, P.C., Almeida, O.G. *et al.* (2007) Male urine signals social rank in the Mozambique tilapia (*Oreochromis mossambicus*). *BMC Biology*, **5**, 54.

Barata, E.N., Fine, J.M., Hubbard, P.C. *et al.* (2008) A sterol-like odorant in the urine of Mozambique tilapia males likely signals social dominance to females. *Journal of Chemical Ecology*, **34**, 438–449.

Barnett, C. (1982) The chemosensory responses of young cichlid fish to parents and predators. *Animal Behaviour*, **30**, 35–42.

Behrmann-Godel, J., Gerlach, G., and Eckmann, R. (2006) Kin and population recognition in sympatric Lake Constance perch (*Perca fluviatilis* L.): can assortative shoaling drive population divergence? *Behavioural Ecology and Sociobiology*, **59**, 461–468.

Brown, J.L. and Eklund, A. (1994) Kin recognition and the major histocompatibility complex—an integrative review. *American Naturalist*, **143**, 435–461.

Brown, G.E. and Brown, J.A. (1993) Do kin always make better neighbors—the effects of territory quality. *Behavioural Ecology and Sociobiology*, **33**, 225–231.

Brown, G.E. and Smith, R.J.F. (1994) Fathead minnows use chemical cues to discriminate natural shoalmates from unfamiliar conspecifics. *Journal of Chemical Ecology*, **20**, 3051–3061.

Brown, G.E., Brown, J.A., and Crosbie, A.M. (1993) Phenotype matching in juvenile rainbow-trout. *Animal Behaviour*, **46**, 1223–1225.

Brown, G.E., Brown, J.A., and Wilson, W.R. (1996) The effects of kinship on the growth of juvenile Arctic char. *Journal of Fish Biology*, **48**, 313–320.

Bryant, B.P. and Atema, J. (1987) Diet manipulation affects social-behavior of catfish—importance of body odor. *Journal of Chemical Ecology*, **13**, 1645–1661.

Buston, P.M., Fauvelot, C., Wong, M.Y.L., and Planes, S. (2009) Genetic relatedness in groups of the humbug damselfish *Dascyllus aruanus*: small, similar-sized individuals may be close kin. *Molecular Ecology*, **18**, 4707–4715.

Carlsson, J. and Carlsson, J.E.L. (2002) Micro-scale distribution of brown trout: an opportunity for kin selection? *Ecology of Freshwater Fish*, **11**, 234–239.

Carr, M.G. and Carr, J.E. (1985) Individual recognition in the juvenile brown bullhead (*Ictalurus nebulosus*). *Copeia*, **1985**, 1060–1062.

Charlesworth, D. and Charlesworth, B. 1987. Inbreeding depression and its evolutionary consequences. *Annual Review of Ecology and Systematics*, **18**, 237–268.

Chivers, D.P., Brown, G.E., and Smith, R.J.F. (1995) Familiarity and shoal cohesion in fathead minnows (*Pimephales promelas*)—implications for antipredator behavior. *Canadian Journal of Zoology-Revue Canadienne De Zoologie*, **73**, 955–960.

Courtenay, S.C., Quinn, T.P., Dupuis, H.M.C. *et al.* (1997) Factors affecting the recognition of population-specific odours by juvenile coho salmon. *Journal of Fish Biology*, **50**, 1042–1060.

Doving, K.B., Nordeng, H., and Oakley, B. (1974) Single unit discrimination of fish odors released by char (*Salmo-alpinus* L.) populations. *Comparative Biochemistry and Physiology*, **47**, 1051–1063.

Dugatkin, L.A. and Wilson, D.S. (1992) The prerequisites for strategic behavior in bluegill sunfish, *Lepomis macrochirus*. *Animal Behaviour*, **44**, 223–230.

Eizaguirre, C., Yeates, S.E., Lenz, T.L. *et al.* (2009) MHC-based mate choice combines good genes and maintenance of MHC polymorphism. *Molecular Ecology*, **18**, 3316–3329.

Evans, J.P. and Kelley, J.L. (2008) Implications of multiple mating for offspring relatedness and shoaling behaviour in juvenile guppies. *Biology Letters*, **4**, 623–626.

Fisher, H.S., Wong, B.B.M., and Rosenthal, G.G. (2006) Alteration of the chemical environment disrupts communication in a freshwater fish. *Proceedings of the Royal Society B-Biological Sciences*, **273**, 1187–1193.

Fisknes, B. and Doving, K.B. (1982) Olfactory sensitivity to group-specific substances in atlantic salmon (*Salmo salar* L.). *Journal of Chemical Ecology*, **8**, 1083–1092.

Fitzgerald, G.J. and Vanhavre, N. (1987) The adaptive significance of cannibalism in sticklebacks (gasterosteidae, pisces). *Behavioral Ecology and Sociobiology*, **20**, 125–128.

Forsberg, L.A., Dannewitz, J., Petersson, E., and Grahn, M. (2007) Influence of genetic dissimilarity in the reproductive success and mate choice of brown trout - females fishing for optimal MHC dissimilarity. *Journal of Evolutionary Biology*, **20**, 1859–1869.

Fraser, D.J., Duchesne, P., and Bernatchez, L. (2005) Migratory char schools exhibit population and kin associations beyond juvenile stages. *Molecular Ecology*, **14**, 3133–3146.

Frommen, J.G., Luz, C., and Bakker, T.C.M. (2007) Kin discrimination in sticklebacks is mediated by social learning rather than innate recognition. *Ethology*, **113**, 276–282.

Gerlach, G. and Lysiak, N. (2006) Kin recognition and inbreeding avoidance in zebrafish, Danio *rerio*, is based on phenotype matching. *Animal Behaviour*, **71**, 1371–1377.

Gerlach, G., Schardt, U., Eckmann, R., and Meyer, A. (2001) Kin-structured subpopulations in Eurasian perch (*Perca fluviatilis* L.). *Heredity*, **86**, 213–221.

Gerlach, G., Hodgins-Davis, A., Macdonald, B., and Hannah, R.C. (2007) Benefits of kin association: related and familiar zebrafish larvae (*Danio rerio*) show improved growth. *Behavioral Ecology and Sociobiology*, **61**, 1765–1770.

Gerlach, G., Hodgins-Davis, A., Avolio, C., and Schunter, C. (2008) Kin recognition in zebrafish: a 24-hour window for olfactory imprinting. *Proceedings of the Royal Society B-Biological Sciences*, **275**, 2165–2170.

Giaquinto, P.C. and Volpato, G.L. (1997) Chemical communication, aggression, and conspecific recognition in the fish Nile tilapia. *Physiology and Behavior*, **62**, 1333–1338.

Goncalves-de-Freitas, E., Teresa, F.B., Gomes, F.S., and Giaquinto, P.C. (2008) Effect of water renewal on dominance hierarchy of juvenile Nile tilapia. *Applied Animal Behaviour Science*, **112**, 187–195.

Greenberg, L.A., Hernnas, B., Bronmark, D. *et al.* (2002) Effects of kinship on growth and movements of brown trout in field enclosures. *Ecology of Freshwater Fish*, **11**, 251–259.

Griffiths, S.W. (2003) Learned recognition of conspecifics by fishes. *Fish and Fisheries*, **4**, 256–268.

Griffiths, S.W. and Armstrong, J.D. (2001) The benefits of genetic diversity outweigh those of kin association in a territorial animal. *Proceedings of the Royal Society of London Series B-Biological Sciences*, **268**, 1293–1296.

Griffiths, S.W. and Armstrong, J.D. (2002) Kin-biased territory overlap and food sharing among Atlantic salmon juveniles. *Journal of Animal Ecology*, **71**, 480–486.

Griffiths, S.W. and Ward, A.J.W. (2011) Social recognition of conspecifics, in *Fish Learning & Behaviour* (eds C. Brown, K.N. Laland, and J. Krause), Chapman & Hall, London.

Griffiths, S.W., Armstrong, J.D., and Metcalfe, N.B. (2003) The cost of aggregation: juvenile salmon avoid sharing winter refuges with siblings. *Behavioral Ecology*, **14**, 602–606.

Griffiths, S.W., Brockmark, S., Hojesjo, J., and Johnsson, J.I. (2004) Coping with divided attention: the advantage of familiarity. *Proceedings of the Royal Society of London Series B-Biological Sciences*, **271**, 695–699.

Guevara-Fiore, P., Rosenqvist, G., and Watt, P.J. (2010) Inbreeding level does not induce female discrimination between sibs and unrelated males in guppies. *Behavioral Ecology and Sociobiology*, **64**, 1601–1607.

Hain, T.J.A. and Neff, B.D. (2007) Multiple paternity and kin recognition mechanisms in a guppy population. *Molecular Ecology*, **16**, 3938–3946.

Hamilton, W.D. (1963) Evolution of altruistic behavior. *American Naturalist*, **97**, 354–356.

Hamilton, W.D. (1964a) Genetical evolution of social behaviour 2. *Journal of Theoretical Biology*, **7**, 17–52.

Hamilton, W.D. (1964b) Genetical evolution of social behaviour I. *Journal of Theoretical Biology*, **7**, 1–16.

Hara, T.J. (ed.) (1992) *Fish Chemoreception*, Chapman & Hall, London.

Harden, M.V., Newton, L.A., Lloyd, R.C., and Whitlock, K.E. (2006) Olfactory imprinting is correlated with changes in gene expression in the olfactory epithelia of the zebrafish. *Journal of Neurobiology*, **66**, 1452–1466.

Hert, E. (1985) Individual recognition of helpers by the breeders in the cichlid fish lamprologus-brichardi (Poll, 1974). *Zeitschrift Fur Tierpsychologie-Journal of Comparative Ethology*, **68**, 313–325.

Hojesjo, J., Johnsson, J.I., Petersson, E., and Jarvi, T. (1998) The importance of being familiar: individual recognition and social behavior in sea trout (*Salmo trutta*). *Behavioral Ecology*, **9**, 445–451.

Holmes, W.G. and Sherman, P.W. (1982) The ontogeny of kin recognition in 2 species of ground-squirrels. *American Zoologist*, **22**, 491–517.

Jaeger, R.G. (1981) Dear enemy recognition and the costs of aggression between salamanders. *American Naturalist*, **117**, 962–974.

Keenleyside, M.H.A. (1955) Some aspects of the schooling behaviour of fish. *Behaviour*, **8**, 83–248.

Krause, J. and Ruxton, G.D. (2002) *Living in Groups*, Oxford University Press, Oxford.

Kuhme, W. (1964a) Eine chemisch ausgelöste Brutpflegereaktion bei cichliden (Pisces). *Naturwissenschaften*, **51**, 20–21.

Kuhme, W. (1964b) Eine chemisch ausgelöste Schwarmreaktion bei jungen Cichliden (Pisces). *Naturwissenschaften*, **51**, 120–121.

Lachlan, R.F., Crooks, L., and Laland, K.N. (1998) Who follows whom? Shoaling preferences and social learning of foraging information in guppies. *Animal Behaviour*, **56**, 181–190.

Le Vin, A.L., Mable, B.K., and Arnold, K.E. (2010) Kin recognition via phenotype matching in a cooperatively breeding cichlid, *Neolamprologus pulcher*. *Animal Behaviour*, **79**, 1109–1114.

Loekle, D.M., Madison, D.M., and Christian, J.J. (1982) Time dependency and kin recognition of cannibalistic behavior among poeciliid fishes. *Behavioral and Neural Biology*, **35**, 315–318.

Lurling, M. and Scheffer, M. (2007) Info-disruption: pollution and the transfer of chemical information between organisms. *Trends in Ecology & Evolution*, **22**, 374–379.

Manica, A. (2002) Filial cannibalism in teleost fish. *Biological Reviews*, **77**, 261–277.

Martin-Smith, K.M. and Armstrong, J.D. (2002) Growth rates of wild stream-dwelling Atlantic salmon correlate with activity and sex but not dominance. *Journal of Animal Ecology*, **71**, 413–423.

Mateo, J.M. (2004) Recognition systems and biological organization: the perception component of social recognition. *Annales Zoologici Fennici*, **41**, 729–745.

Matsumura, K., Matsunaga, S., and Fusetani, N. (2004) Possible involvement of phosphatidylcholine in school recognition in the catfish, *Plotosus lineatus*. *Zoological Science*, **21**, 257–264.

Mckaye, K.R. and Barlow, G.W. (1976) Chemical recognition of young by midas cichlid, cichlasoma-citrinellum. *Copeia*, **1976**, 276–282.

Mckaye, K.R. and Mckaye, N.M. (1977) Communal care and kidnapping of young by parental cichlids. *Evolution*, **31**, 674–681.

Mehlis, M., Bakker, T.C.M., and Frommen, J.G. (2008) Smells like sib spirit: kin recognition in three-spined sticklebacks (*Gasterosteus aculeatus*) is mediated by olfactory cues. *Animal Cognition*, **11**, 643–650.

Milinski, M., Luthi, J.H., Eggler, R., and Parker, G.A. (1997) Cooperation under predation risk: experiments on costs and benefits. *Proceedings of the Royal Society of London Series B-Biological Sciences*, **264**, 831–837.

Milinski, M., Griffiths, S., Wegner, K.M. *et al.* (2005) Mate choice decisions of stickleback females predictably modified by MHC peptide ligands. *Proceedings of the National Academy of Sciences of the United States of America*, **102**, 4414–4418.

Moore, A., Ives, M.J., and Kell, L.T. (1994) The role of urine in sibling recognition in Atlantic salmon *Salmo salar* parr *Proceedings of the Royal Society of London Series B-Biological Sciences*, **255**, 173–180.

Myrberg, A.A. (1975) The role of chemical and visual stimuli in the preferential discrimination of young by the cichlid fish, *Cichlasoma nigrofasciatum*. *Zeitschrift far Tierpsychologie*, **37**, 274–297.

Neff, B.D. and Sherman, P.W. (2005) In vitro fertilization reveals offspring recognition via self-referencing in a fish with paternal care and cuckoldry. *Ethology*, **111**, 425–438.

Noakes, D.L.G. and Barlow, G.W. (1973) Ontogeny of parent-contacting in young cichlasoma-citrinellum (Pisces, Cichlidae). *Behaviour*, **46**, 221–255.

O'Connor, K.I., Metcalfe, N.B., and Taylor, A.C. (2000) Familiarity influences body darkening in territorial disputes between juvenile salmon. *Animal Behaviour*, **59**, 1095–1101.

Olsen, K.H. (1987) Chemoattraction of juvenile arctic charr *Salvelinus-alpinus* (L.) to water scented by conspecific intestinal content and urine. *Comparative Biochemistry and Physiology A-Physiology*, **87**, 641–643.

Olsen, K.H. (1989) Sibling recognition in juvenile arctic charr, *Salvelinus-alpinus* (L.). *Journal of Fish Biology*, **34**, 571–581.

Olsen, K.H. and Hoglund, L.B. (1985) Reduction by a surfactant of olfactory mediated attraction between juveniles of arctic charr, *Salvelinus-alpinus* (L.). *Aquatic Toxicology*, **6**, 57–69.

Olsen, K.H. and Winberg, S. (1996) Learning and sibling odor preference in juvenile arctic char, *Salvelinus alpinus* (L.). *Journal of Chemical Ecology*, **22**, 773–786.

Olsen, K.H., Grahn, M., Lohm, J., and Langefors, A. (1998) MHC and kin discrimination in juvenile Arctic charr, *Salvelinus alpinus* (L.). *Animal Behaviour*, **56**, 319–327.

Olsen, K.H., Grahn, M., and Lohm, J. (2002) Influence of MHC on sibling discrimination in Arctic char, *Salvelinus alpinus* (L.). *Journal of Chemical Ecology*, **28**, 783–795.

Olsen, K.H., Grahn, M., and Lohm, J. (2003) The influence of dominance and diet on individual odours in MHC identical juvenile Arctic char siblings. *Journal of Fish Biology*, **63**, 855–862.

Olsen, K.H., Petersson, E., Ragnarsson, B. *et al.* (2004) Downstream migration in Atlantic salmon (*Salmo salar*) smolt sibling groups. *Canadian Journal of Fisheries and Aquatic Sciences*, **61**, 328–331.

Palm, S., Dannewitz, J., Jarvi, T. *et al.* (2008) No indications of Atlantic salmon (*Salmo salar*) shoaling with kin in the Baltic Sea. *Canadian Journal of Fisheries and Aquatic Sciences*, **65**, 1738–1748.

Pitcher, T.E., Rodd, F.H., and Rowe, L. (2008) Female choice and the relatedness of mates in the guppy (*Poecilia reticulata*). *Genetica*, **134**, 137–146.

Piyapong, C., Butlin, R.K., Faria, J.J. *et al.* (2011) Kin assortment in juvenile shoals in wild guppy populations. *Heredity*, **106**, 749–756.

Pouyaud, L., Desmarais, E., Chenuil, A. *et al.* (1999) Kin cohesiveness and possible inbreeding in the mouthbrooding tilapia *Sarotherodon melanotheron* (Pisces: Cichlidae). *Molecular Ecology*, **8**, 803–812.

Quinn, T.P. and Busack, C.A. (1985) Chemosensory recognition of siblings in juvenile coho salmon (*Oncorhynchus kisutch*). *Animal Behaviour*, **33**, 51–56.

Rajakaruna, R.S., Brown, J.A., Kaukinen, K.H., and Miller, K.M. (2006) Major histocompatibility complex and kin discrimination in Atlantic salmon and brook trout. *Molecular Ecology*, **15**, 4569–4575.

Reusch, T.B.H., Haberli, M.A., Aeschlimann, P.B., and Milinski, M. (2001) Female sticklebacks count alleles in a strategy of sexual selection explaining MHC polymorphism. *Nature*, **414**, 300–302.

Rosenthal, G.G. and Lobel, P.S. (2006) Communication, in *Behaviour and Physiology of Fish* (eds K.A. Sloman, R.W. Wilson, and S. Balshine), Academic Press, London.

Russock, H.I. (1990) The effect of natural chemical stimuli on the preferential behavior of oreochromis mossambicus (pisces, cichlidae) fry to maternal models. *Behaviour*, **115**, 315–326.

Selset, R. and Doving, K.B. (1980) Behavior of mature anadromous char (*Salmo-alpinus* L.) towards odorants produced by smolts of their own population. *Acta Physiologica Scandinavica*, **108**, 113–122.

Seppa, T., Laurila, A., Peuhkuri, N. *et al.* (2001) Early familiarity has fitness consequences for Arctic char (*Salvelinus alpinus*) juveniles. *Canadian Journal of Fisheries and Aquatic Sciences*, **58**, 1380–1385.

Shaw, E. (1978) Schooling fishes. *American Scientist*, **66**, 166–175.

Sikkel, P.C. and Fuller, C.A. (2010) Shoaling preference and evidence for maintenance of sibling groups by juvenile black perch *Embiotoca jacksoni*. *Journal of Fish Biology*, **76**, 1671–1681.

Solomon, D.J. (1977) Review of chemical communication in freshwater fish. *Journal of Fish Biology*, **11**, 363–376.

Sorensen, P.W. and Baker, C. (2015) Species-specific pheromones and their roles in shoaling, migration, and reproduction: a critical review and synthesis, in *Fish Pheromones and Related Cues* (eds Peter W. Sorensen and Brian D. Wisenden), John Wiley & Sons, Inc., Hoboken.

Stabell, O.B., Selset, R., and Sletten, K. (1982) A comparative chemical study on population-specific odorants from atlantic salmon. *Journal of Chemical Ecology*, **8**, 201–217.

Steck, N., Wedekind, C., and Milinski, M. (1999) No sibling odor preference in juvenile three-spined sticklebacks. *Behavioral Ecology*, **10**, 493–497.

Taborsky, M. 1984. Broodcare helpers in the cichlid fish lamprologus-brichardi—their costs and benefits. *Animal Behaviour*, **32**, 1236–1252.

Taborsky, M. and Limberger, D. (1981) Helpers in fish. *Behavioral Ecology and Sociobiology*, **8**, 143–145.

Thunken, T., Bakker, T.C.M., Baldauf, S.A., and Kullmann, H. (2007) Direct familiarity does not alter mating preference for sisters in male *Pelvicachromis taeniatus* (Cichlidae). *Ethology*, **113**, 1107–1112.

Thunken, T., Waltschyk, N., Bakker, T.C.M., and Kullmann, H. (2009) Olfactory self-recognition in a cichlid fish. *Animal Cognition*, **12**, 717–724.

Tierney, K.B., Baldwin, D.H., Hara, T.J. *et al.* (2010) Olfactory toxicity in fishes. *Aquatic Toxicology*, **96**, 2–26.

Todd, J.H., Atema, J., and Bardach, J.E. (1967) Chemical communication in social behaviour of a fish—Yellow Bullhead (*Ictalurus natalis*). *Science*, **158**, 672–673.

Utne-Palm, A.C. and Hart, P.J.B. (2000) The effects of familiarity on competitive interactions between three-spined sticklebacks. *Oikos*, **91**, 225–232.

van Havre, N. and Fitzgerald, G.J. (1988) Shoaling and kin recognition in the three-spined stickleback (*Gasterosteus aculeatus* L.). *Biology of Behaviour*, **13**, 190–201.

Viken, A., Fleming, I.A., and Rosenqvist, G. (2006) Premating avoidance of inbreeding absent in female guppies (*Poecilia reticulata*). *Ethology*, **112**, 716–723.

Warburton, K. and Lees, N. (1996) Species discrimination in guppies: learned responses to visual cues. *Animal Behaviour*, **52**, 371–378.

Ward, A.J.W. and Hart, P.J.B. (2003) The effects of kin and familiarity on interactions between fish. *Fish and Fisheries*, **4**, 348–358.

Ward, A.J.W. and Hart, P.J.B. (2005) Foraging benefits of shoaling with familiars may be exploited by outsiders. *Animal Behaviour*, **69**, 329–335.

Ward, A.J.W. and Mehner, T. (2010) Multimodal mixed messages: the use of multiple cues allows greater accuracy in social recognition and predator detection decisions in the mosquitofish, *Gambusia holbrooki*. *Behavioral Ecology*, **21**, 1315–1320.

Ward, A.J.W., Axford, S., and Krause, J. (2002) Mixed-species shoaling in fish: the sensory mechanisms and costs of shoal choice. *Behavioural Ecology and Sociobiology*, **52**, 182–187.

Ward, A.J.W., Hart, P.J.B., and Krause, J. (2004) The effects of habitat- and diet-based cues on association preferences in three-spined sticklebacks. *Behavioral Ecology*, **15**, 925–929.

Ward, A.J.W., Holbrook, R.I., Krause, J., and Hart, P.J.B. (2005) Social recognition in sticklebacks: the role of direct experience and habitat cues. *Behavioural Ecology and Sociobiology*, **57**, 575–583.

Ward, A.J.W., Duff, A.J., and Currie, S. (2006) The effects of the endocrine disrupter 4-nonylphenol on the behaviour of juvenile rainbow trout (*Oncorhynchus mykiss*). *Canadian Journal of Fisheries and Aquatic Sciences*, **63**, 377–382.

Ward, A.J.W., Webster, M.M., and Hart, P.J.B. (2007) Social recognition in wild fish populations. *Proceedings of the Royal Society B-Biological Sciences*, **274**, 1071–1077.

Ward, A.J.W., Duff, A.J., Horsfall, J.S., and Currie, S. (2008) Scents and scents-ability: pollution disrupts chemical social recognition and shoaling in fish. *Proceedings of the Royal Society B-Biological Sciences*, **275**, 101–105.

Ward, A.J.W., Webster, M.M., Magurran, A.E. *et al.* (2009) Species and population differences in social recognition between fishes: a role for ecology? *Behavioral Ecology*, **20**, 511–516.

Webster, M.M. and Hart, P.J.B. (2007) Prior association reduces kleptoparasitic prey competition in shoals of three-spined sticklebacks. *Animal Behaviour*, **74**, 253–258.

Webster, M.M., Goldsmith, J., Ward, A.J.W., and Hart, P.J.B. (2007) Habitat-specific chemical cues influence association preferences and shoal cohesion in fish. *Behavioural Ecology and Sociobiology*, **62**, 273–280.

Wegner, K.M., Kalbe, M., Kurtz, J. *et al.* (2003a) Parasite selection for immunogenetic optimality. *Science*, **301**, 1343–1343.

Wegner, K.M., Reusch, T.B.H., and Kalbe, M. (2003b) Multiple parasites are driving major histocompatibility complex polymorphism in the wild. *Journal of Evolutionary Biology*, **16**, 224–232.

Winberg, S. and Olsen, K.H. (1992) The influence of rearing conditions on the sibling odor preference of juvenile arctic charr, *Salvelinus-alpinus* L. *Animal Behaviour*, **44**, 157–164.

Wisenden, B. and Dye, T. (2009) Young convict cichlids use visual information to update olfactory homing cues. *Behavioural Ecology and Sociobiology*, **63**, 443–449.

Wisenden, B.D. and Keenleyside, M.H.A. (1992) Intraspecific brood adoption in convict cichlids—a mutual benefit. *Behavioral Ecology and Sociobiology*, **31**, 263–269.

Wisenden, B.D. and Keenleyside, M.H.A. (1994) The dilution effect and differential predation following brood adoption in free-ranging convict cichlids (Cichlasoma-Nigrofasciatum). *Ethology*, **96**, 203–212.

Wisenden, B.D. and Smith, R.J.F. (1998) A re-evaluation of the effect of shoalmate familiarity on the proliferation of alarm substance cells in ostariophysan fishes. *Journal of Fish Biology*, **53**, 841–846.

Wisenden, B.D. and Stacey, N.E. (2005) Fish semiochemicals and the evolution of communication networks, in *Animal Communication Networks* (ed. P.K. McGregor), Cambridge University Press, Cambridge.

Wong, M. and Balshine, S. (2011) The evolution of cooperative breeding in the African cichlid fish, Neolamprologus pulcher. *Biological Reviews*, **86**, 511–530.

Wrede, W.L. (1932) Versuche uber den Artduft der Ellritze. *Zeitschrift far vergleichende Physiologie*, **17**, 510–519.

Wyatt, T.D. (2010) Pheromones and signature mixtures: defining species-wide signals and variable cues for identity in both invertebrates and vertebrates. *Journal of Comparative Physiology A-Neuroethology Sensory Neural and Behavioral Physiology*, **196**, 685–700.

Zhang, C. and Hara, T.J. (2009) Lake char (*Salvelinus namaycush*) olfactory neurons are highly sensitive and specific to bile acids. *Journal of Comparative Physiology A-Neuroethology Sensory Neural and Behavioral Physiology*, **195**, 203–215.

Chapter 6
Chemical Cues That Indicate Risk of Predation

Brian D. Wisenden

Minnesota State University Moorhead, Moorhead, Minnesota, USA

6.1 INTRODUCTION

Chemical cues released passively as a by-product of ecological interactions provide a rich and reliable source of public information to guide behavioral decision-making. Predation is an omnipresent and unforgiving agent of selection that is reliably correlated with context-specific chemical cues. Consequently, natural selection has led fish to evolve finely tuned mechanisms to first detect chemical information about predation risk and then execute adaptive behavioral responses to minimize exposure to these risks. The regular interval of reviews of this topic (Smith, 1992; Chivers and Smith, 1998; Wisenden, 2003; Wisenden and Stacey, 2005; Wisenden and Chivers, 2006; Ferrari, Wisenden, and Chivers, 2010; Chivers, Brown, and Ferrari, 2013) reflects the pace and volume of new research on how fish and other aquatic animals use chemical information to ameliorate predation risk.

In this chapter, the principal sources of chemical information about predation risk and the large and active literature on how injury-released chemical alarm cues from conspecifics mediate these interactions are briefly reviewed. Alarm cues are chemicals released by cells and tissues of prey damaged in the process of a predator attack. Because alarm cues are released only in the context of predation, they reliably inform nearby receivers of the presence of an actively foraging predator. Thus, there is strong selection on receivers to elaborate mechanisms to detect and discern the quality of this information and to respond accordingly. I focus on these interactions because conspecific chemical alarm cues are the class of compounds and mixtures that come closest to the concept of pheromone. From there, what little is known about the chemistry of these cues and the anatomical and physiological mechanisms that detect them are summarized.

6.2 CHEMICAL INFORMATION ABOUT PREDATION RISK

In addition to damage-released chemical alarm cues, predator odor and disturbance cues are chemical cues that relay information about predation risk, and provide information about position in the predation sequence (Fig. 6.1A). The predation sequence parses predation events into discrete steps, each of which produces stage-specific chemical cues. Predator odor is the most obvious chemical by which prey detect the presence of predation risk. From the perspective of prey, predator odor is a

Fish Pheromones and Related Cues, First Edition. Edited by Peter W. Sorensen and Brian D. Wisenden.
© 2015 John Wiley & Sons, Inc. Published 2015 by John Wiley & Sons, Inc.

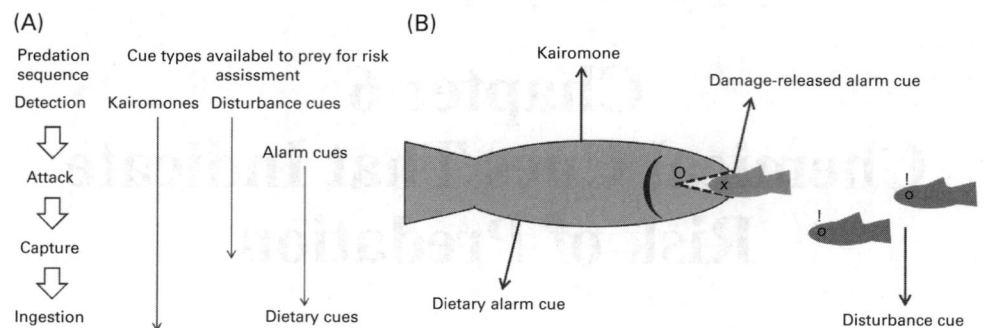

Figure 6.1. Sources of chemical information about predation risk come from the odor of the predator (kairomone), disturbance cues released by startled prey, damage-released alarm cues from predation in progress, and postingestion cues released from the predator's diet. (A) Ecological context of when each cue type becomes available in the predation sequence; (B) source of each cue type.

kairomone, that is, a chemical that is released by one species that can provide benefits to heterospecific receivers without providing benefits to the sender. A second type of chemical cue used by prey for detecting risk are chemicals released by disturbed but uninjured prey (herein termed "disturbance cues"). Both of these cues are used at the detection stage of the predation sequence. Kairomones and disturbance cues cause prey to increase frequency of vigilance behaviors and prepare for evasive maneuvers. Chemical alarm cues are released from damaged conspecific tissues during the attack and capture stages and indicate clear and present danger from an actively foraging predator. These cues invoke intense antipredator behavioral responses. A type of cue related to an alarm cue is a dietary cue, which are chemicals of ingested prey released from the digestive tract of predators (Fig. 6.1). Thus, kairomones comprise multiple chemical constituents; the background species-specific signature odor of the predator with additional components from the predator's diet.

6.2.1 Cues Associated with Pre-Attack Detection of Predation Risk

At the detection stage of predation, chemicals passively released from predators (kairomones) or disturbance cues released by startled or distressed (but otherwise uninjured) prey can serve to inform prey about the presence and nature of predation risk. Detecting and learning to recognize kairomones confer demonstrable survival benefits in encounters with predators (Berejikian *et al.*, 1999; Mirza and Chivers, 2000, 2002; Gazdewich and Chivers, 2002). Adaptive use of fish kairomones is also well studied in zooplankton (Tollrian and Harvell, 1999) and amphibia (e.g., Schoeppner and Relyea, 2009). In fishes, some salmonid species have innate recognition of odors of their predators (Berejikian, Tezak, and LaRae, 2003; Vilhunen and Hirvonen, 2003; Hawkins, Magurran, and Armstrong, 2007; Scheurer *et al.*, 2007; see Section 3.3); however, the majority (dozens of studies on minnows, characins, salmonids, gobies, cichlids, centrarchids, percids— see Chivers and Smith, 1998; Ferrari, Wisenden, and Chivers, 2010 for recent reviews) do not have innate recognition of predator odors and must acquire recognition of predator odor through associative learning.

Fish learn to associate correlates of predation events, such as predator odor, when those novel stimuli are paired with the release of chemical alarm cues. Consequently, a single pairing of novel neutral stimuli (e.g., odor of an unfamiliar predator) with damage-released chemical alarm cues (e.g., skin extract of a conspecific) is sufficient to impart near-permanent association of danger with the novel predator odor. Thereafter, encounters with the novel predator odor invoke the same suite of antipredator behaviors as the behavioral response to skin extract. Göz (1941) and Magurran (1989), both working with European minnows (*Phoxinus phoxinus*), were the first to record this form of learning. Suboski (1990) coined the label "releaser-induced recognition learning" to describe the critical role that damage-released alarm cues have in triggering the formation of association between predation risk and a novel stimulus. Because predation is such a common occurrence, particularly during early life stages, surviving fishes have ample opportunity to acquire recognition of novel odors associated with predators in their area (Chivers and Smith, 1994a, 1995a; Pollock *et al.*, 2003). Fish can also learn to associate danger with habitat-specific chemical signatures (Chivers and Smith, 1995b), visual appearance of the predators (Chivers and Smith, 1994b), or the sound of predators (Wisenden *et al.*, 2008). The selective advantage of this learning paradigm is that fish quickly arm themselves with the ability to detect early indicators of predation risk through multiple sensory modalities. From the viewpoint of natural selection, early detection of predator presence allows prey the opportunity to execute antipredator responses and completely evade detection by the predator.

Another route by which prey detect the presence of a predator is by chemical stimuli released from the digestive system of the predator (Chivers and Mirza, 2001) (Fig. 6.1). This mechanism allows prey to recognize novel fish as dangerous even if they have no previous experience with that species of predator (Mathis and Smith, 1993a) and use these dietary cues to associate that predator with predation risk in future encounters (Mathis and Smith, 1993b; Brown and Godin, 1999; Mirza and Chivers, 2003). Notable exceptions to this general rule are two studies that found no evidence that prey can detect dietary cues. Feminella and Hawkins (1994) showed that tadpoles of tailed frogs (*Ascaphus truei*) respond to dietary alarm cues from predatory giant salamanders (*Dicamptodon* spp.), cutthroat trout (*Salmo clarkii*) and brook trout (*Salvelinus fontinalis*) but not to the dietary alarm cues of shorthead sculpin (*Cottus confusus*). Similarly, Maniak, Lossing, and Sorensen (2000) found no evidence that Eurasian ruffe (*Gymnocephalus cernuus*) can detect conspecific dietary cues that had passed through the gut of northern pike (*Esox lucius*). Thus, it appears that some predators, such as shorthead sculpin, have the ability to mask or breakdown recognizable components of alarm cues, or perhaps that some fish, such as ruffe, do not have the ability to detect alarm cues after they have been processed in the gut of a predator. In a comparison of digestion efficiency between two predators, the mudminnows (an esociform, *Umbra limi*) and bluegill sunfish (a centrarchid, *Lepomis macrochirus*), fathead minnows (*Pimephales promelas*) were equally able to detect dietary alarm cues suggesting that either the more derived predator (sunfish) were no better at masking dietary cues than ancestral ones or that minnows have evolved the ability to detect dietary cues from multiple predator types (Sutrisno *et al.*, 2014).

6.2.2   Cues Associated with Post-Attack Detection of Predation Risk

When a predator succeeds in contacting prey, its teeth, mandibles, bills, raptorial forelimbs, and so on, cause damage to prey tissue and result in the passive release of highly context-specific compounds. These chemical cues are reliably correlated with an actively foraging predator and, therefore indicate imminent predation risk to prey. This general response is widespread among aquatic organisms from ciliates to amphibia because of the prevalence of predation and the ability of water to transmit semiochemicals (Chivers and Smith, 1998; Wisenden, 2003; Ferrari, Wisenden, and Chivers, 2010).

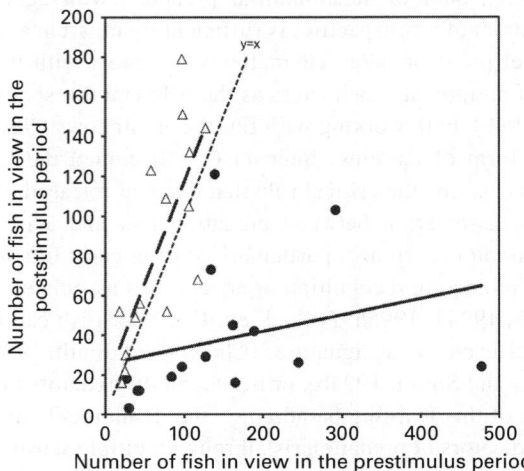

Figure 6.2. Number of wild fish in view of a submerged video camera before and after the release of water (open triangles, dashed line) or skin extract (solid circles, solid line) into the littoral zone of a boreal lake in Minnesota, USA. Skin extract was derived from redbelly dace (*Phoxinus eos*). There was no change in area use following release of water but a significant decrease in area use following release of skin extract. Data taken from Wisenden and Barbour (2005).

6.3 USE OF CHEMICAL INFORMATION ABOUT PREDATION RISK

Responses of natural populations of unconstrained free-living freshwater littoral fishes (fathead minnows *P. promelas*, redbelly dace *P. eos*, brook stickleback *Culaea inconstans*, finescale dace *Chrosomus neogaeus*) show clear avoidance of areas where conspecific skin extract has been released (Fig. 6.2) (Wisenden *et al.*, 2004b; Wisenden and Barbour, 2005; Friesen and Chivers, 2006; Wisenden *et al.*, 2010).

Antipredator responses are not limited to changes in behavior. Crucian carp and goldfish have the capacity to increase the ratio of body depth:length in response to damage-released chemical alarm cues (Brönmark and Miner, 1992; Stabell and Lwin, 1997; Chivers, Zhao, and Ferrari, 2007a). Greater body depth reduces the risk of predation by gape-limited predators. Greater body depth may also permit greater maneuverability in structured littoral zones where prey take refuge when they co-occur with predators. Life history traits can also be affected by damage-released chemical alarm cues associated with predation risk, including earlier time of egg hatch (Mirza, Chivers, and Godin, 2001) and an earlier time to sexual maturity in guppies (Dzikowski *et al.*, 2004).

6.3.1 Active Space and Active Time of Chemical Alarm Cues

To understand how chemical alarm cues function in the ecology of fishes, information is needed about active space (i.e., volume of water chemically labeled as dangerous following a predation event) and active time (i.e., duration of time that an area remains chemically labeled). Dilution series data from the lab (Lawrence and Smith, 1989) estimated that 1 cm^2 of fathead minnow (*P. promelas*) skin activates 58 000 L, a volume equivalent to a sphere with a radius of 2.4 m, or a cylinder 1 m tall with a radius of 4.3 m. Field-based estimates of active space generally agree with this estimate. When skin extract of fathead minnows or northern redbelly dace was released within a meter of shore at depths of about 1 m, active space over a 2 h interval was between 2 and 8 m

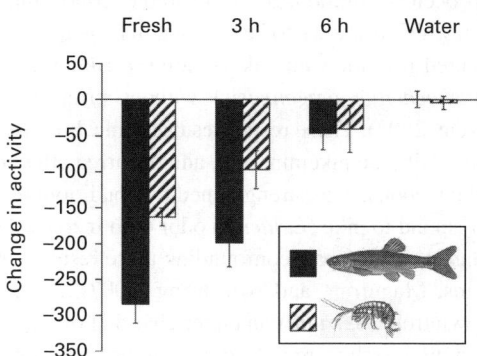

Figure 6.3. Active time of damage-released chemical alarm cues for fathead minnows (solid fill) and amphipod crustaceans (hatched fill). Minnow skin extract and whole-body squashes of amphipods were maintained at 18°C for 0, 3, or 6h before being frozen, and compared against a blank water control. Data taken from Wisenden *et al.* (2009). (*See insert for color representation of the figure.*)

(Wisenden, 2008). Remarkably, this remained true for skin extract derived from either fathead minnows or redbelly dace and was consistently the case for responses by conspecifics and hetero-specifics, including brook stickleback (Wisenden, 2008). These data suggest a universal ecological safe radius of 2–8 m around any predation event.

Active time of biological activity of chemical alarm cues is determined by biochemical degradation of the chemical constituents present in raw skin extract and advection (mass water movement). Behavioral responses by fathead minnows to fresh skin extract was the same as responses to skin extract that had been aged for 3h at 18°C but significantly stronger than responses to skin extract that had been aged for 6h at 18°C (Wisenden *et al.*, 2009). Cue aged 6h was not significantly different from blank water controls, indicating that active time was between 3 and 6h. To test if this active time was due to a breakdown of one or more specific constituents of minnow skin, or something more general for all aquatic taxa, Wisenden *et al.* (2009) repeated the experiment with gammarid amphipods that presumably involve different (and as yet unidentified) chemical molecules and olfactory receptors than used by minnows. Active time of amphipod chemical alarm cues was also between 3 and 6h, respectively. Thus, minnows and amphipods have similar ecological periods of predation risk despite different cue chemistries and physiological mechanisms for detecting them (Fig. 6.3). By contrast, Ferrari, Messier, and Chivers (2008a) showed that responses of wood frog tadpoles (*Rana sylvatica*) to conspecific chemical alarm cues ceased in less than 2h.

Taken together, estimates of active time and active space suggest that behavioral responses to semiochemicals is dependent on more than the detection of stimulus X by receptor Y. This is perhaps because behavioral responses to predation risk play out in an ecological theater mediated by sensory input from multiple modalities, shaped and augmented by highly efficient associative learning and tempered by experience with temporal variation in predator hunting behavior.

6.3.2 Innate Recognition of Predator Odors

Fish differ from many invertebrates and amphibian larvae by not relying on genetically based olfactory templates for predator recognition. There is no obvious reason for why fish rely almost entirely on learning for recognition of predators. Learning has advantages over fixed genetic templates in being flexible to variation in predator identity over temporal and spatial scales, and over ontogenetic shifts in habitat and vulnerability to gape-limited predators. (see 6.3.3) The few examples of innate genetic recognition of predators all occur in salmonids. For example, coho salmon (*Oncorhynchus kisutch*) respond innately to L-serine, which is known to be released in large quantities by bears (Rehnberg and Schrek, 1986). Hatchery-reared juvenile Chinook salmon (*O. tshawytscha*) respond to odor cues of northern pikeminnow (*Ptychocheilus oregonensis*) without prior experience with the predator (Berejikian, Tezak, and LaRae, 2003). These responses occur in Chinook salmon derived from populations allopatric or sympatric with the pikeminnows, and occurred after controlling for predator diet. However, the intensity of the response was strengthened by conditioning training (described below). Atlantic salmon (*S. salar*) respond to pike (*E. lucius*) odor with increased rates of ventilation, but the pike had been fed dead Atlantic salmon parr confounding these results with conspecific dietary cues of the test species (Hawkins, Magurran, and Armstrong, 2007). Arctic char (*Salvelinus alpinus*) respond to odor cues of brown trout (*S. trutta*) on either char diet or a nonchar diet, but responded to odor of pikeperch only when pikeperch were fed a diet of arctic char (Vilhunen and Hirvonen, 2003). Steelhead trout (*O. mykiss*) from a population that had been separated from exposure to their predator Dolly Varden (*Salvelinus malmo*) for 80 years still retained an innate response to Dolly Varden odor (controlled for dietary cues) that was equal in intensity to responses of sympatric populations of steelhead trout (Scheurer *et al.*, 2007).

Even though fathead minnows have been a convenient model organism for studies of acquired recognition of predator odor (Brown, 2003), they too show some evidence of innate bias. When fathead minnows were conditioned with skin extract to fear either the visual presentation of a northern pike or a goldfish, they learned to fear each one but retained recognition of pike for at least 40 days, whereas fear of goldfish was not retained (Chivers and Smith, 1994b).

6.3.3 Learned Recognition of Novel Predators

This topic has been reviewed in detail elsewhere (Brown, 2003; Ferrari *et al.*, 2007; Chivers, Brown, and Ferrari, 2013), and remains an active area of new research because learning plays a major role in mediating behavioral responses to chemical information. In the parlance of experimental psychology, conspecific alarm cues contained in skin extract act as an unconditioned stimulus (US), that is, a stimulus that evokes an innate response without prior experience. When skin extract (US) is paired with a neutral novel stimulus (the conditioned stimulus, CS), fish associate the CS with predation risk. This form of associative learning occurs after a single simultaneous pairing of US and CS. Göz (1941) was the first to observe this form of learning, but it was Suboski (1990) who suggested that alarm cues act as a "releaser" for this special case of associative learning. Suboski (1990) showed that zebrafish could be conditioned to fear morpholine (a nonbiological odorant) when it was paired with conspecific skin extract. Hall and Suboski (1995) did a similar study in which they conditioned zebrafish to fear an arbitrary visual stimulus (flashing red light).

Fathead minnows, zebrafish, brook stickleback, convict cichlids, walleye, gobies, various salmonids—virtually every fish species ever tested—associate novel stimuli with predation risk if it is correlated with release of alarm cue (Ferrari, Wisenden, and Chivers, 2010). Learned recognition can be a superior strategy over a hard-wired genetically based recognition template because fishes face a range of predators that shift dramatically with ontogeny, and vary widely over short- and long-term scales in time and space. Although fathead minnows and northern pike have presumably co-existed for many thousands of years, inexperienced minnows have no recognition of pike odor. This suggests that an innate recognition template does not confer selective advantage over learned recognition; otherwise, minnows would have evolved an innate template for pike long ago. As briefly summarized in Section 2.1, lab experiments on a range of species (cyprinids, salmonids, gobiids, gasterosteids, and percids) demonstrate that predator-naïve fish use associative learning to recognize predators by odor, sight, or sound. Additional laboratory work has shown that behavioral responses to predator odor is nuanced in subtle and sophisticated ways commensurate with the degree of risk indicated by the cue (Chivers, Brown, and Ferrari, 2013). For example, minnows, char, perch, and catfish distinguish age classes solely on the basis of the odors they produce. For example, small brook char (4.3 cm standard length, SL) and large brook char (7.6 cm SL) respond most intensely to skin extract of size-matched conspecifics than they did to skin extract of the opposite size class (Mirza and Chivers, 2002a). Fathead minnows respond more intensively to the odor of small pike (239 mm SL) than they do to the odor of large pike (645 mm SL), perhaps because small-size classes of pike represent a greater threat to minnows (Kusch, Mirza, and Chivers, 2004). Pintado (*Pseudoplatystoma coruscans*), a carnivorous and cannibalistic catfish, respond with feeding behaviors to body odor of conspecifics smaller than themselves but not to those larger than themselves (Giaquinto and Volpato, 2005). Minnows and stickleback can also learn to associate risk to a novel odor when the novel odor is paired with observation of an overt antipredator response of nearby conspecific or heterospecific to the introduced kairomone without any alarm cues present (Mathis, Chivers, and Smith, 1996).

By pairing novel kairomones with graded concentrations of conspecific alarm cues, Ferrari *et al.* (2005) showed that fathead minnows learned graded responses proportional to the perceived level of threat. If given multiple conditioning sessions using differing concentrations of alarm cue, minnows give precedence to the kairomone-alarm cue pairing with the strongest indicator of risk (Ferrari and Chivers, 2006). Moreover, minnows that associate risk with the kairomone of one predator species can extend or generalize an association of risk with other phylogenetically similar predators (Ferrari *et al.*, 2007; Ferrari, Messier, and Chivers, 2008b). Predator-conditioning training is also a potential tool for fisheries management whereby predator-naïve hatchery-reared fish are trained to recognize common predators before they are stocked into natural water bodies (Brown and Smith, 1998; Berejikian *et al.*, 1999; Chivers and Mirza, 2000; Mirza and Chivers, 2002b; Wisenden *et al.*, 2004a).

Although the recognition of risk confers enormous fitness benefits, the ease with which associations occur presents a liability in that fish may easily form fearful associations with irrelevant stimuli. Fathead minnows use motion to focus learning on salient stimuli only. If skin extract is released when there are two objects present, and one object bobs vertically while the other remains stationary, then minnows associate risk with the one in motion and ignore the other (Wisenden and Harter, 2001).

6.4 CHEMISTRY OF CHEMICAL CUES PREDATION RISK IN FISHES

In contrast to other classes of semiochemicals used by fish reported in this book, the chemistry of cues used for mediating predator–prey interactions is conspicuously and woefully understudied. The reasons for why greater progress has been made in understanding the chemistry of semiochemicals associated with reproduction may be because stimulus–response interactions in reproductive interactions are less variable, less dependent on learning, and therefore produce cleaner data than those associated with predator–prey interactions.

Testing for active chemical constituents in predator–prey interactions is complicated in that cues may be released in multiple contexts (detection, postattack, postingestion), from multiple sources (conspecifics, predator, ecologically similar or phylogenetically similar heterospecifics) and the enormous role that learning plays in shaping behavioral responses to these cues. Much work remains to be done on parsing individual components that trigger antipredator responses. Most of the detailed work on cue chemistry has focused on the *Schreckstoff* skin-derived alarm cue derived from skin extract of fishes in the superorder Ostariophysi.

6.4.1 Disturbance Cues

Startled (but not injured) crayfish (Hazlett, 1985, 1990) and tadpoles (Kiesecker *et al.*, 1999) release cues that increase predator vigilance in nearby conspecifics. These initial observations have been extended to fishes for percids (Wisenden, Chivers, and Smith, 1995), salmonids (Mirza and Chivers,

2002b; Vavrek *et al.*, 2008), and cichlids (Vavrek *et al.*, 2008). The principal focus of study for characterizing the active ingredient of disturbance cues has been urinary ammonia although it is possible that detectable levels of circulating corticosteroids and other endogenous correlates of stress are released by stressed individuals (Lebedeva, Vosilene, and Golovkina, 1994). Thus far, there is no evidence that startled prey regulate the timing or rate of urine release, or that components in urine other than ammonia play a role in the alarming quality of disturbance cues.

6.4.2 Skin Extract

Alarm responses are mediated by ciliated olfactory sensory neurons in fathead minnows and yellow perch (Dew *et al.*, 2014). The activity of neurons on the posterior surface of the medial region of the olfactory bulb in crucian carp, *Carassius carassius*, is proportional to the concentration of conspecific skin extract used as a test stimulus (Hamdani and Døving, 2003). Other test stimuli that occur in skin extract, such as amino acids, nucleotides and taurolithocholic acid had no effect on these sensory neurons. Thus, these olfactory neurons appear to be specifically tuned to components in conspecific extract. Moreover, these neurons also respond, albeit to a lesser degree, to skin extract of heterospecific common carp, tench, and bream (Lastein, Hamdani, and Døving, 2008). These sensory neurons project onto the medial bundle of the medial olfactory tract (mMOT). Overt antipredator behavioral responses to conspecific skin extract do not occur when the mMOT has been experimentally severed (Fig. 6.4) (Hamdani *et al.*, 2000).

The chemistry of the active ingredients in skin extract has been studied mainly in crucian carp and fathead minnows. Early work on chemically characterizing these cues began in the 1940s (see Døving and Lastein (2009) for an excellent review that includes the early papers published in German by R. Hüttel and colleagues and by W. Pfeiffer *et al.*). These initial investigations narrowed the chemical composition of alarm substance to purines or pterins, compounds with small molecular weights ($\ll 500\,Da$). Subsequent fractioning experiments placed the active components in skin extracts of minnows to be 1100–$1400\,Da$ or more (Lebedeva, Malyukina, and Kasumyan, 1975; Kasumyan and Ponomarev, 1987). Relatively recent lab and field tests with hypoxanthine-3 *N*-oxide (small molecular weight) show that this molecule, and similar ones with a nitrogen-oxide side group,

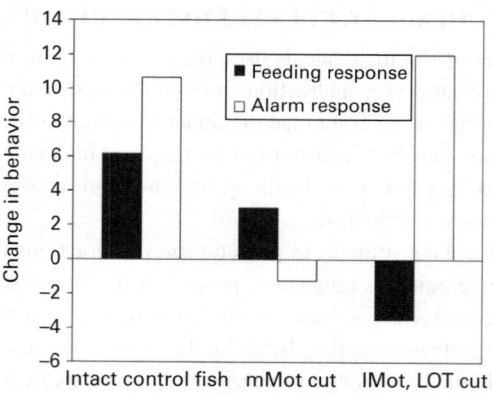

Figure 6.4. Behavioral responses of crucian carp in response to alanine (food odor) or conspecific skin extract (alarm odor) for control fish, fish that had had the medial bundle of the medial olfactory tract (mMOT) severed, or fish that had had the lateral bundle of the medial olfactory tract (lMOT) and the lateral olfactory tract (LOT) severed. Sensory processing of chemical cues associated with predation risk is curtailed only when the mMOT is cut, whereas feeding responses are reduced when the lMOT and LOT are cut. Data taken from Hamdani *et al.* (2000).

is effective in eliciting overt alarm reactions in fathead minnows (Pfeiffer *et al.*, 1985; Brown *et al.*, 2000, 2003; Brown, Adrian, and Shih, 2001). Another recent laboratory study showed that precipitating protein from a solution of minnow skin extract deactivated its biological activity (Wisenden *et al.*, 2009) suggesting a role for carrier proteins for solubility or perhaps even as a time keeper of biodegradation that receivers can use to assess time elapsed since cue release. Most recently, Mathuru *et al.* (2012) worked with fractionated skin extract of zebrafish and found biological activity in two fractions with molecular weights of about 1 and 30 kDa, respectively. The low-molecular-weight (LMW) fraction induced darting, whereas the high-molecular-weight (HMW) fraction induced slow swimming. Subsequent tests on the HMW indicated the presence of long polymers, possibly polysaccharides, ultimately resolving to glycosaminoglycans (GAGs). Further testing led to the GAG chondroitin sulfate from sharks (GAGS have not yet been studied in ostariophysians). Mathuru *et al.* 2012 went further and demonstrated the ability of chondroitin sulfate to stimulate the mediodorsal glomerular region of the olfactory bulbs of zebrafish, consistent with electrophysiological recordings of crucian carp exposed to skin extract (Hamdani and Døving, 2003). By contrast, hypoxanthine-3 N-oxide showed no stimulation of the mediodorsal region of the olfactory bulb. The juxtaposition of behavioral testing and physiological recordings by Mathuru *et al.* (2012) make the chondroitin hypothesis the strongest data to date on the chemical nature of alarm cues in any fish. Nevertheless, Mathuru *et al.* (2012) specifically point out that GAGS cannot explain the entire activity of skin and postulate that alarm cues may be mixtures. Moreover, Mathuru *et al.* (2012) cannot account for species specificity of alarm cues. Cross-species reactions of fish to heterospecific alarm cues decline in intensity with phylogenetic distance (Schütz, 1956), and therefore alarm cues likely comprise a mixture of compounds that confer both alarm and species specificity.

6.5 COMMUNICATION: ARE PREDATOR–PREY CHEMICALS "PHEROMONES"?

Historically, all of these forms of chemical information have been considered as pheromones (Smith, 1992; Chivers and Smith, 1998). However, the integration of communication theory into this field has brought greater scrutiny over the usage of terms. In this book, we adopt the definition of pheromone proposed by Wyatt (2010) that a pheromone is a compound or mixture of compounds that is species-wide, evokes an innate response in conspecific receivers and is an evolved signal (maintained by selection on both sender and receiver). Disturbance and alarm cues are more than species wide (as cross-species reactions are well documented for each), evoke innate responses but also learned ones too, and evidence to date (reviewed later) does not indicate an evolved signal, at least not in the context of alarm signaling (Wisenden and Stacey, 2005; Wisenden and Chivers, 2006; Chivers, Brown, and Ferrari, 2007b; Ferrari, Wisenden, and Chivers, 2010; Bradbury and Vehrencamp, 2011). Significantly, disturbance and alarm cues differ from chemicals released in the context of reproduction in that there is no plausible candidate mechanism by which an alarm "signal" could ever potentially evolve because senders to do not accrue fitness benefits from alerting conspecifics and heterospecifics to the presence of danger. Therefore, whatever their chemistry turns out to be, these compounds should best be referred to as chemical cues rather than chemical signals, and therefore, not "true" pheromones as defined by Wyatt (2010).

An evolved chemical *signal*, as per currently accepted communication theory (Bradbury and Vehrencamp, 2011), requires reciprocal selection between sender and receiver in that (i) receivers accrue fitness benefit from evolving receptors and adaptive behavioral responses to chemical information sent by the sender and (ii) receiver responses to the signal that contribute to the fitness of the sender and thus shape the evolution of the signal. Senders consequently evolve specialization for the production and release of chemical signals (see chapter 7 on Communication by Wisenden, 2015).

Table 6.1. Hypotheses Summarized by Smith (1992) for Mechanisms by Which Fitness Benefits can Accrue to the Sender of an Alarm Signal in Any Sensory Modality.

Hypothesis	Action	Comments
Information receiver is self		
Sensory assessment	Ambush avoidance Visual ranging	This not likely to occur in the olfactory modality because the speed of dissemination of chemicals in water is very slow.
Information receiver is other prey		
Reduce predator success in area or on prey species	Reduced success of future attacks	It is difficult to imagine a scenario where chemical cues permanently alter future predator success.
Initiate group crypsis (reduction in movement)	Sender benefits from decreased chance of detection	Senders could benefit from disturbance cues because urine is under partial voluntary control but not from injury-released alarm cues.
Initiate tighter shoal cohesion	Sender benefits from decreased probability of capture.	Senders could benefit from disturbance cues because urine is under partial voluntary control but not from injury-released alarm cues.
Manipulate other prey to flee and distract predator	Sender benefits from decreased probability of capture.	Senders could benefit from disturbance cues because urine is under partial voluntary control but not from injury-released alarm cues.
Save "dear enemy" neighbors	Sender benefits from retaining helpful neighbors.	This is not a likely scenario because very few fish species have territorial neighbors that become familiar and less antagonistic.
Save mate	Sender benefits from avoiding cost of lost reproductive success or time and risk finding a new mate.	This is not a likely scenario because very few fish mate monogamously.
Save commensals	Sender benefits from retaining helpful individuals.	This is not a likely scenario because very few fish species have interspecific commensal interactions.

Information receiver is a predator

Pursuit invitation	Sender benefits by inducing attack at a time when it is prepared to evade the attack.	Senders could benefit from disturbance cues because urine is under partial voluntary control but not from injury-released alarm cues.
Pursuit deterrence	Sender benefits by alerting the predator that the element of surprise is lost; therefore, attack will not be successful.	Senders could benefit from disturbance cues because urine is under partial voluntary control but not from injury-released alarm cues.
Confuse or surprise predator	Sender has an opportunity to escape.	This may work in systems such as cephalopods where a chemical under voluntary release disables the visual and olfactory system of their predator. No such analog is known for fish.
Aposematism	Sender deters attack by signaling unpalatability.	This has never been tested but may occur in toxic species such as members of the tetraodontidae.
Injure predator	Sender uses chemical weaponry to disable or injure predator.	This has never been tested but may occur in toxic species such as members of the tetraodontidae.
Attract secondary predator	Sender can escape while predators fight each other.	Mathis *et al.* (1995); Wisenden and Thiel (2002) showed that predators are attracted to skin extract of minnows. In staged predator–predator encounters, Chivers *et al.* (1996) showed that minnows escaped in 5 out of 13 trials when a second pike was released after the first pike had seized it, compared with 0 out of 13 trials when a second predator was not released.

Information receiver are relatives

Save offspring	Sender accrues direct fitness benefits by genes it shares with its offspring.	Because adult fish are hundreds of times the size of their offspring, adults and offspring do not share predators. No known examples of chemical defense of eggs or young are known.
Save kin	Sender accrues indirect fitness benefits by genes it shares with its kin	This is an unlikely scenario because there is no evidence for genetic structure in fish shoals.

Chemical cues passively released during a predation event provide useful information to receivers. This puts selection on receivers in that those that are best able to detect and respond adaptively to this information are more likely than others to evade predation long enough to reproduce. Thus, natural selection steeply promotes the sensory biology and neural wiring to maximize the effective use of this information. Evidence for sender-side selection is lacking. Injury-released chemical cues was once thought to reach signal status in the fishes of the superorder Ostariophysi (minnows, characins, catfishes, suckers, etc.) because these fishes possess specialized club cells in the epidermis (Smith, 1992). Because Ostariophysans appeared to be unique in showing both behavioral alarm reactions to conspecific skin extract and possession of epidermal club cells, it followed logically that club cells were a component of an alarm signaling system (Pfeiffer, 1977). The mechanism(s) by which fitness benefits might accrue to the alarm signal sender and justify the metabolic cost of maintaining club cells (Wisenden and Smith, 1997, 1998) attracted the attention of evolutionary ecologists (Smith, 1992; Williams, 1992). Smith (1992) listed 16 hypotheses for potential mechanisms that could produce selfish fitness benefits to senders of injury-released alarm signals (Table 6.1). Only one of these 16 hypotheses—the attraction of secondary predators—has received empirical support. An alternative hypothesis for epidermal club cell function is that these cells perform an immune function against pathogens that attack through the skin. Neither of these hypotheses relies on an alarm signaling function for the elaboration and maintenance of epidermal club cells over evolutionary time.

6.5.1   Attraction of Secondary Predators Hypothesis

This hypothesis maintains that the chemicals contained within club cells of fathead minnows are an attractant to predators (Mathis, Chivers, and Smith, 1995; Wisenden and Thiel, 2002). When a predator detects a predation event in progress, they have an opportunity to pirate a meal away from the other predator, or perhaps even an opportunity to eat the other predator while it is distracted with its meal. These predator–predator interactions can provide the prey opportunity for escape (Chivers, Brown, and Smith, 1996). The more that prey invest in club cells, and the higher the concentration of detectable chemical compounds in these cells, the more likely they are to attract a secondary predator and escape. While these data provide one scenario for the maintenance of epidermal club cells, they do so outside of the context of any alarm signaling function.

6.5.2   Immune Function Hypothesis

Club cells of the Ostariophysi, and analogous club cells in species in the percidae, may play a role in the immune system because the epidermis is the first line of defense against tissue damage caused by ultraviolet radiation, pathogens and parasites, and secondary bacterial infection following abrasion or injury of any kind (Smith, 1992). This hypothesis stimulated simultaneous experimentation in several labs that all cooperatively combined multiple data sets into a single publication to provide multiple and independent lines of evidence showing that epidermal club cell density increases in response to ultraviolet radiation, pathogenic attack by *Saprolegnia* spores and epidermal infection by trematode cercariae (Chivers *et al.* 2007b; Halbgewachs *et al.*, 2009). Interestingly, club cell proliferation does not occur when fathead minnows are attacked by a species of trematode that uses fathead minnows as its intermediate host, thus specialized to evade the immune system of fathead minnows (James, Wisenden, and Goater, 2009). The immune system hypothesis is thus broadly supported. In both the secondary-predator attraction hypothesis and the immune function hypothesis, the fitness benefits that led to the origin and maintenance of club cells is entirely independent from their subsequent role as a component of chemical alarm cues. Receiver-side selection in response to passively released public information is the most plausible and parsimonious explanation for the evolution of behavioral responses to chemical alarm cues (Wisenden, 2000; Wisenden and Stacey, 2005; Wisenden and Chivers, 2006; Ferrari, Wisenden, and Chivers, 2010; Wyatt, 2010).

6.5.3 Epitaph for the Moniker "Alarm Substance" Cells

Pfeiffer (1977) assayed a range of fish species for alarm reactions to conspecific skin extract and concluded that only Ostariophysan species gave strong reactions. Notably, a defining character of the Ostariophysan fishes is the presence of epidermal club cells, which led to the hypothesis that these cells were the source of "*Schreckstoff*," literally alarm substance, and hence these cells became known as alarm substance cells (Smith, 1992). Since the time of Pfeiffer (1977), many studies have demonstrated strong alarm reactions by nonostariophysan fish (e.g., stickleback, centrarchids, cichlids, salmonids, esociformes, poecilliids, gobies, and walleye) to chemical cues of injured conspecifics (Chivers and Smith, 1998, Ferrari, Wisenden, and Chivers, 2010). These other species (other than percids) lack epidermal club cells. Thus, a causal link between epidermal club cells and an alarm cue is tenuous (Wisenden and Stacey, 2005; Wisenden and Chivers, 2006; Ferrari, Wisenden, and Chivers, 2010). Moreover, larval fathead minnows do not possess club cells; and yet, their skin extract elicits alarm reactions from conspecifics (Carreau *et al.*, 2008). Finally, nonfish aquatic taxa, all of which lack epidermal club cells, show antipredator responses to chemical cues released by injured conspecifics (Chivers and Smith, 1998; Wisenden, 2003; Ferrari, Wisenden, and Chivers, 2010). Behavioral responses to alarm cues can be explained entirely by receiver-side selection. We can, therefore, lay to rest the moniker "alarm substance cells" for epidermal club cells of the Ostariophysi, and analogous cells in the percidae (Smith, 1992; Chivers *et al.*, 2007b). To continue to use this term is a misrepresentation of the function and evolutionary origin of these cells.

6.6 CONCLUSION AND DIRECTIONS FOR FUTURE RESEARCH

Predation risk is assessed chemically through passively released by-products of the predation sequence; predator odor (kairomones), disturbance cues, injury-released alarm cues, and postingestion dietary cues. Adaptive behavioral responses to this public information arose through receiver-side selection to minimize risk of predation. Although there remain several potential mechanisms to be tested (Smith, 1992), no evidence marshaled to date suggests anything other than the parsimonious explanation that chemically mediated predator–prey interactions are anything more than correlational by-products of ecological interactions between predator and prey. These are not pheromones *sensu* Wyatt (2010).

Every review of this literature since Smith (1992) has called for the urgent need for chemical characterization of the active molecules in these cues and the mapping of the receptors that detect them. Research on crucian carp (receptors), fathead minnows, and zebrafish (cue chemistry) have made important contributions in this direction but much more needs to be done.

Future work will increasingly explore the role of information networks on the evolution of receptors, cognition and behavioral responses. The integration of the physiological mechanisms of detection with the ecological function and behavioral responses is a rich cross-section that few investigators have crossed (Mathuru *et al.*, 2012). A perennial deficit is the dearth of field-based studies.

Finally, chemical ecologists are well positioned to help monitor and manage the effects of anthropogenic pollutants and oceanic acidification resulting from increasing CO_2 levels associated with global climate change (Olsen, 2015; Dew *et al.*, 2014; Ferrari *et al.*, 2011a,b; Lürling, 2012).

REFERENCES

Berejikian, B.A., Smith, R.J.F., Tezak, E.P. *et al.* (1999) Chemical alarm signals and complex hatchery rearing habitats affect antipredator behavior and survival of chinook salmon (*Oncorhynchus tshawytscha*) juveniles. *Canadian Journal of Fisheries and Aquatic Sciences*, **56**, 830–838.

Berejikian, B.A., Tezak, E.P., and LaRae, A.L. (2003) Innate and enhanced predator recognition in hatchery-reared chinook salmon. *Environmental Biology of Fishes*, **67**, 241–251.

Bradbury, J.W. and Vehrencamp, S.L. (2011) *Principles of Animal Communication*, 2nd edn. Sinauer, Sunderland, MA.

Brönmark, C. and Miner, J.G. (1992) Predator-induced phenotypical change in body morphology in crucian carp. *Science*, **258**, 1348–1350.

Brown, G.E. (2003) Learning about danger: chemical alarm cues and local risk assessment in prey fishes. *Fish and Fisheries*, **4**, 227–234.

Brown, G.E. and Godin, J-G.J. (1999) Who dares, learns: chemical inspection behaviour and acquired predator recognition in a characin fish. *Animal Behaviour*, **57**, 475–481.

Brown, G.E. and Smith, R.J.F. (1998) Acquired predator recognition in juvenile rainbow trout (*Oncorhynchus mykiss*): conditioning hatchery-reared fish to recognize chemical cues of a predator. *Canadian Journal of Fisheries and Aquatic Sciences*, **55**, 611–617.

Brown, G.E., Adrian, J.C. Jr., Smyth, E. *et al.* (2000) Ostariophysan alarm pheromones: laboratory and field tests of the functional significance of nitrogen-oxides. *Journal of Chemical Ecology*, **26**, 139–154.

Brown, G.E., Adrian, J.C. Jr., and Shih, M. (2001) Behavioural responses of fathead minnows (*Pimephales promelas*) to hypoxanthine-3-N-oxide at varying concentrations. *Journal of Fish Biology*, **58**, 1465–1470.

Brown, G.E., Adrian, J.C. Jr., Naderi, N.T. *et al.* (2003) Nitrogen-oxides elicit antipredator responses in juvenile channel catfish, but not convict cichlids or rainbow trout: conservation of the Ostariophysan alarm pheromone. *Journal of Chemical Ecology*, **29**, 1781–1796.

Carreau-Green, N.D., Mirza, R.S., Martinez, M.L., and Pyle, G.G. (2008) The ontogeny of chemically mediated antipredator responses of fathead minnows *Pimephales promelas*. *Journal of Fish Biology*, **73**, 2390–2401.

Chivers, D.P. and Mirza, R.S. (2001) Predator diet cues and the assessment of predation risk by aquatic vertebrates: a review and prospectus, in *Chemical Signals in Vertebrates*, vol. **9** (eds. A. Marchlewska-Koj, J.J. Lepri, and D. Müller-Schwarze). Plenum Press, New York, pp. 277–284.

Chivers, D.P. and Smith, R.J.F. (1994a) The role of experience and chemical alarm signalling in predator recognition by fathead minnows, *Pimephales promelas*. *Journal of Fish Biology*, **44**, 273–285.

Chivers, D.P. and Smith, R.J.F. (1994b) Fathead minnows, *Pimephales promelas*, acquire predator recognition when alarm substance is associated with the sight of unfamiliar fish. *Animal Behaviour*, **48**, 597–605.

Chivers, D.P. and Smith, R.J.F. (1995a) Free-living fathead minnows rapidly learn to recognize pike as predators. *Journal of Fish Biology*, **46**, 949–954.

Chivers, D.P. and Smith, R.J.F. (1995b) Fathead minnows (*Pimephales promelas*) learn to recognize chemical stimuli from high-risk habitats by the presence of alarm substance. *Behavioral Ecology*, **6**, 155–158.

Chivers, D.P. and Smith, R.J.F. (1998) Chemical alarm signalling in aquatic predator-prey systems: a review and prospectus. *Écoscience*, **5**, 338–352.

Chivers, D.P., Brown, G.E., and Smith, R.J.F. (1996) The evolution of chemical alarm signals: attracting predators benefits alarm signal senders. *American Naturalist*, **148**, 649–659.

Chivers, D.P., Zhao, X., and Ferrari, M.C.O. (2007a) Linking morphological and behavioural defences: prey fish detect the morphology of conspecifics in the odour signature of their predators. *Ethology*, **113**, 733–739.

Chivers, D.P., Wisenden, B.D., Hindman, C.J. *et al.* (2007b) Epidermal 'alarm substance' cells of fishes maintained by non-alarm functions: possible defence against pathogens, parasites and UVB radiation. *Proceedings of the Royal Society Series B*, **274**, 2611–2619.

Chivers, D.P., Brown, G.E., and Ferrari, M.C.O. (2013) The sophistication of predator odor recognition by minnows. *Chemical Signals in Vertebrates* XII (eds M.L. East and M. Denhard), Springer Verlag, Germany, pp. 247–257.

Dew, W.A., Azizishirazi, A., and Pyle, G.G. (2014) Contaminant-specific targeting of olfactory sensory neuron classes: Connecting neuron class impairment with behavioural deficits. *Chemosphere*, **112**, 519–525.

Døving, K.B. and Lastein, S. (2009) The alarm reaction in fishes – odorants, modulations of responses, neural pathways. *Annals of the New York Academy of Science*, **1170**, 413–423.

Dzikowski, R., Hulata, G., Harpaz, S., and Karplus, I. (2004) Inducible reproductive plasticity of the guppy *Poecilia reticulata* in response to predation cues. *Journal of Experimental Zoology Part A: Comparative Experimental Biology*, **301A**, 776–782.

Feminella, J.W. and Hawkins, C.P. (1994) Tailed frog tadpoles differentially alter their feeding behavior in response to non-visual cues from four predators. *North American Benthological Society*, **13**, 310–320.

Ferrari, M.C.O. and Chivers, D.P. (2006) Learning threat-sensitive predator avoidance: how do fathead minnows incorporate conflicting information? *Animal Behaviour*, **71**, 19–26.

Ferrari, M.C.O., Trowell, J.J., Brown, G.E., and Chivers, D.P. (2005) The role of leaning in the development of threat-sensitive predator avoidance in fathead minnows. *Animal Behaviour*, **70**, 777–784.

Ferrari, M.C.O., Gonzalo, A., Messier, F., and Chivers, D.P. (2007) Generalization of learned predator recognition: an experimental test and framework for future studies. *Proceedings of the Royal Society of London Series B*, **274**, 1853–1859.

Ferrari, M.C.O., Messier, F., and Chivers, D.P. (2008a) Degradation of chemical cues under natural conditions: risk assessment by larval frogs. *Chemoecology*, **17**, 263–266.

Ferrari, M.C.O., Messier, F., and Chivers, D.P. (2008b) Can prey exhibit threat-sensitive generalization of predator recognition? Extending the predator recognition continuum hypothesis. *Proceedings of the Royal Society of London Series B*, **275**, 1811–1816.

Ferrari, M.C.O., Wisenden B.D., and Chivers, D.P. (2010) Chemical ecology of predator-prey interactions in aquatic ecosystems: a review and prospectus. *Canadian Journal of Zoology*, **88**, 698–724.

Ferrari, M.C.O., Dixson, D.L., Munday, P.L. *et al.* (2011a) Intrageneric variation in anti-predator responses of coral reef fishes to ocean acidification: implications of projecting climate change on marine communities. *Global Change Biology*, **17**, 2980–2986.

Ferrari, M.C.O., McCormick, M.I., Munday, P.L. *et al.* (2011b) Putting prey and predator into the CO2 equation - qualitative and quantitative effects of ocean acidification on predator-prey interactions. *Ecology Letters*, **14**, 1143–1148.

Friesen, R.G. and Chivers, D.P. (2006) Underwater video reveals strong avoidance of chemical alarm cues by prey fishes. *Ethology*, **112**, 339–345.

Gazdewich, K.J. and Chivers, D.P. (2002) Acquired predator recognition by fathead minnows: influence of habitat characteristics on survival. *Journal of Chemical Ecology*, **28**, 439–445.

Giaquinto, P.C. and Volpato, G.L. (2005) Chemical cues related to conspecific size in pintado catfish, *Pseudoplatystoma coruscans*. *Acta Ethologia*, **8**, 65–69.

Göz, H. (1941) Über den Art-und Individualgeruch bei Fishen. *Zeitschrift für vergleichende Physiologie*, **29**, 1–45.

Halbgewachs, C.F., Marchant, T.A., Kusch, R.C., and Chivers, D.P. (2009) Epidermal club cells and the innate immune system of minnows. *Biological Journal of the Linnaean Society*, **98**, 891–897.

Hall, D. and Suboski, M.D. (1985) Visual and olfactory stimuli in learned release of alarm reactions by zebra danio fish (Brachydaniorerio). *Neurobiology of Learning and Memory*, **63**, 229–240.

Hamdani, E.-H. and Døving, K.B. (2003) Sensitivity and selectivity of neurons in the medial region of the olfactory bulb to skin extract from conspecifics in crucian carp, *Carassius carasius*. *Chemical Senses*, **28**, 181–189.

Hamdani, E.-H., Stabell, O.B., Alexander, G., and Døving, K.B. (2000) Alarm reaction in the crucian carp is mediated by the medial bundle of the medial olfactory tract. *Chemical Senses*, **25**, 103–109.

Hawkins, L.A., Magurran, A.E., and Armstrong, J.D. (2007) Innate abilities to distinguish between predator species and cue concentration in Atlantic salmon. *Animal Behaviour*, **73**, 1051–1057.

Hazlett, B.A. (1985) Disturbance pheromones in the crayfish *Orconectes virilis*. *Journal of Chemical Ecology*, **11**, 1695–1711.

Hazlett, B.A. (1990) Source and nature of disturbance-chemical system in crayfish. *Journal of Chemical Ecology*, **16**, 2263–2275.

James, C.T., Wisenden, B.D., and Goater, C.P. (2009) Epidermal club cells do not protect fathead minnows against trematode cercariae: a test of the anti-parasite hypothesis. *Journal of the Linnean Society*, **98**, 884–890.

Kasumyan, A.O. and Ponomarev, V.Y. (1987) Biochemical features of alarm pheromone in fish of the order Cypriniformes. *Journal of Evolutionary Biochemical Physiology*, **23**, 20–24.

Kiesecker, J.M., Chivers, D.P., Marco, A. *et al.* (1999) Identification of a disturbance signal in larval red-legged frogs, *Rana aurora*. *Animal Behaviour*, **57**, 1295–1300.

Kusch, R.C., Mirza, R.S., and Chivers, D.P. (2004). Making sense of predator scents: investigating the sophistication of predator assessment abilities of fathead minnows. *Behavioral Ecology and Sociobiology*, **55**, 551–555.

Lastein, S., Hamdani, E.-H., and Døving, K.B. (2008) Single unit responses to skin odorants from conspecifics and heterospecifics in the olfactory bulb of crucian carp Carassius carassius. *The Journal of Experimental Biology*, **211**, 3529–3535.

Lawrence, B.J. and Smith, R.J.F. (1989) Behavioural response of solitary fathead minnows, *Pimephales promelas*, to alarm substance. *Journal of Chemical Ecology*, **15**, 209–219.

Lebedeva, N.Y, Malyukina, G.A., and Kasumyan, A.O. (1975) The natural repellent in the skin of cyprinids. *Journal of Ichthyology*, **15**, 472–480.

Lebedeva, N.Y, Vosilene, M.Z.Y., and Golovkina, T.V. (1994) Aspects of stress in rainbow trout, *Salmo gairdneri*, release of chemical alarm signals. *Journal of Ichthyology*, **33**, 66–74.

Lürling, M. (2012) Infodisruption: pollutants interfering with the natural chemical information conveyance in aquatic systems, in *Chemical Ecology of Aquatic Systems* (eds. C. Brönmark, and L.-A. Hansson). Oxford University Press, New York, pp. 250–271.

Magurran, A.E. (1989) Acquired recognition of predator odour in the European minnow (Phoxinusphoxinus). *Ethology*, **82**, 216–223.

Maniak, P.J., Lossing, R.D., and Sorensen, P.W. (2000) Injured Eurasion ruffe, (*Gymnocephalus cernuus*), release an alarm pheromone that could be used to control their dispersal. *Journal of Great Lakes Research*, **26**, 183–195.

Mathis, A. and Smith, R.J.F. (1993a) Chemical labelling of northern pike (*Esox lucius*) by the alarm pheromone of fathead minnows (*Pimephales promelas*). *Journal of Chemical Ecology*, **19**, 1967–1979.

Mathis, A. and Smith, R.J.F. (1993b) Fathead minnows, *Pimephales promelas*, learn to recognize northern pike, *Esox lucius*, as predators on the basis of chemical stimuli from minnows in the pike's diet. *Animal Behaviour*, **46**, 645–656.

Mathis, A., Chivers, D.P., and Smith, R.J.F. (1995) Chemical alarm signals: predator-deterrents or predator attractants? *American Naturalist*, **146**, 994–1005.

Mathis, A., Chivers, D.P., and Smith, R.J.F. (1996) Cultural transmission of predator recognition in fishes: intraspecific and interspecific learning. *Animal Behaviour*, **51**, 185–201.

Mathuru, A.J., Kibat C., Cheong, W.F. *et al.* (2012) Chondroitin fragments are odorants that trigger fear behavior in fish. *Current Biology*, **22**, 538–544.

Mirza, R.S. and Chivers, D.P. (2000) Predator-recognition training enhances survival of brook trout: Evidence from laboratory and field-enclosure studies. *Canadian Journal of Zoology*, **78**, 2198–2208.

Mirza, R.S. and Chivers, D.P. (2002a). Brook charr (*Salvelinus fontinalis*) can differentiate chemical alarm cues produced by different size classes of conspecifics. *Journal of Chemical Ecology*, **28**, 555–564.

Mirza, R.S. and Chivers, D.P. (2002b) Behavioural responses to conspecific disturbance chemicals enhance survival of juvenile brook charr, *Salvelinus fontinalis*, during encounters with predators. *Behaviour*, **139**, 1099–1110.

Mirza, R.S. and Chivers, D.P. (2003) Fathead minnows learn to recognize heterospecific alarm cues they detect in the diet of a known predator. *Behaviour*, **140**, 1359–1370.

Mirza, R.S., Chivers, D.P., and Godin, J.-G.J. (2001) Brook charr alevins alter timing of nest emergence in response to chemical cues from a fish predator. *Journal of Chemical Ecology*, **27**, 1775–1785.

Olsén, H. (2015) Effects of pollutants on olfactory detection and responses to chemical cues including pheromones in fish, in *Fish Pheromones and Related Cues* (eds Peter W. Sorensen and Brian D. Wisenden), John Wiley & Sons, Inc., Hoboken.

Pfeiffer, W. (1977) Distribution of fright reaction and alarm substance cells in fishes. *Copeia*, **1977**, 653–665.

Pfeiffer, W., Riegelbauer, G., Meier, G., and Scheibler, B. (1985) Effect of hypoxanthine-3(N)-oxide and hypoxanthine-1(N)-oxide on central nervous excitation of the black tetra *Gymnocorymbus ternetzi* (Characidae, Ostariophysi, Pisces) indicated by dorsal light response. *Journal of Chemical Ecology*, **11**, 507–523.

Pollock, M.S., Chivers, D.P., Mirza, R.S., and Wisenden, B.D. (2003) Fathead minnows learn to recognize chemical alarm cues of introduced brook stickleback. *Environmental Biology of Fishes*, **66**, 313–319.

Rehnberg, B.G. and Schreck, C.B. (1986) The olfactory L-serine receptor in coho salmon: biochemical specificity and behavioral response. *Journal of Comparative Physiology A: Neuroethology, Sensory, Neural and Behavioral Physiology*, **159**, 61–67.

Scheurer, J.A., Berejikian, B.A., Thrower, F.P. *et al.* (2007) Innate predator recognition and fright response in related populations of *Oncorhynchus mykiss* under different predation pressure. *Journal of Fish Biology*, **70**, 1057–1069.

Schoeppner, N.M. and Relyea, R.A. (2009) Interpreting the smells of predation: how alarm cues and kairomones induce prey defences. *Functional Ecology*, **23**, 1114–1121.

Schütz, E. (1956) VergleichendeUntersuchungenuber die SchreckreaktionbeiFischen und deren Verbreitung. *Zeitschrift furvergleichendePhysiologie*, **38**, 84–135.

Smith, R.J.F. (1992) Alarm signals in fishes. *Reviews in Fish Biology and Fisheries*, **2**, 33–63.

Stabell, O.B. and Lwin, M.S. (1997) Predator-induced phenotypic changes in crucian carp are caused by chemical signals from conspecifics. *Environmental Biology of Fishes*, **49**, 145–149.

Suboski, M.D. (1990) Releaser-induced recognition learning. *Psychological Review*, **97**, 271–284.

Sutrisno, R., Schotte, P.M., Schultz, S.K., and Wisenden, B.D. (2014) Chemical arms race between predator and prey: a test of predator digestive counter-measures against chemical labeling by dietary cues of prey. *Journal of Freshwater Ecology*, **29**, 17–23.

Tollrian, R. and Harvell, D. (1999) The ecology and evolution of inducible defenses. Princeton University Press, Princeton, NJ.

Vavrek, M.A., Elvidge, C.K., Decarie, R. *et al.* (2008) Disturbance cues in freshwater prey fishes: do juvenile convict cichlids and rainbow trout respond to ammonium as an 'early warning' signal? *Chemoecology*, **18**, 255–261.

Vilhunen, S. and Hirvonen, H. (2003) Innate antipredator responses of Arctic charr (*Salvelinus alpinus*) depend on predator species and their diet. *Behavioral Ecology and Sociobiology*, **55**, 1–10.

Von Frisch, K. (1941) Über einen Schreckstoff der Fischaut und seine biologische Bedeutung. *Zeitschrift fur Vergleichende Physiololgie*, **29**, 46–145.

Williams, G.C. (1992) *Natural Selection: Domains, Levels and Challenges*. Oxford University Press, Oxford.

Wisenden, B.D. (2000) Olfactory assessment of predation risk. *Philosophical Transactions of the Royal Society*, **355**, 1205–1208.

Wisenden, B.D. (2003) Chemically mediated strategies to counter predation, in *Sensory Processing in Aquatic Environments* (eds. S.P. Collin and N.J. Marshall). Springer, New York, pp. 236–251.

Wisenden, B.D. (2008) Active space of chemical alarm cue in natural fish populations. *Behaviour*, **145**, 391–407.

Wisenden, B.D. (2015) The cue-signal continuum: A hypothesized evolutionary trajectory for chemical communication in fishes, in *Fish Pheromones and Related Cues* (eds Peter W. Sorensen and Brian D. Wisenden), John Wiley & Sons, Inc., Hoboken.

Wisenden, B.D. and Barbour, K.A. (2005) Antipredator responses to skin extract of redbelly dace by free-ranging populations of redbelly dace and fathead minnows. *Environmental Biology of Fishes*, **72**, 227–233.

Wisenden, B.D. and Chivers, D.P. (2006) The role of public chemical information in antipredator behaviour, in *Fish Communication* (eds. F. Ladich, S.P. Collins, P. Moller, B.G. Kapoor). Science Publisher, Enfield, NH, pp. 259–278.

Wisenden, B.D. and Harter, K.R. (2001) Motion, not shape, facilitates association of predation risk with novel objects by fathead minnows (*Pimephales promelas*). *Ethology*, **107**, 357–364.

Wisenden, B.D. and Smith, R.J.F. (1997) The effect of physical condition and shoal-mate familiarity on proliferation of alarm substance cells in the epidermis of fathead minnows. *Journal of Fish Biology*, **50**, 799–808.

Wisenden, B.D. and Smith, R.J.F. (1998) A re-evaluation of the effect of shoalmate familiarity on the proliferation of alarm substance cells in fathead minnows. *Journal of Fish Biology*, **53**, 841–846.

Wisenden, B.D. and Stacey, N.E. (2005) Fish semiochemicals and the evolution of communication networks, in *Communication Networks* (ed. P.K. McGregor). Cambridge University Press, Cambridge, UK, pp. 540–567.

Wisenden, B.D. and Thiel, T.A. (2002) Field verification of predator attraction to minnow alarm substance. *Journal of Chemical Ecology*, **28**, 433–438.

Wisenden, B.D., Chivers, D.P., and Smith, R.J.F. (1995) Early warning in the predation sequence: a disturbance pheromone in Iowa darters (*Etheostoma exile*). *Journal of Chemical Ecology*, **21**, 1469–1480.

Wisenden, B.D., Klitzke, J., Nelson, R. *et al.* (2004a) Predator-recognition training of hatchery-reared walleye and a field test of a training method using yellow perch. *Canadian Journal of Fisheries and Aquatic Sciences*, **62**, 2144–2150.

Wisenden, B.D., Vollbrecht, K.A., and Brown, J.L. (2004b) Is there a fish alarm cue? Affirming evidence from a wild study. *Animal Behaviour*, **67**, 59–67.

Wisenden, B.D., Pogatshnik, J., Gibson, D. *et al.* (2008) Sound the alarm: Learned association of predation risk with novel auditory stimuli by fathead minnows (*Pimephales promelas*) and glowlight tetras (*Hemigrammus erythrozonus*) after single simultaneous pairings with conspecific chemical alarm cues. *Environmental Biology of Fishes*, **81**, 141–147.

Wisenden, B.D., Rugg, M.L., Korpi, N.L., and Fuselier, L.C. (2009) Estimates of active time of chemical alarm cues in a cyprinid fish and an amphipod crustacean. *Behaviour*, **146**, 1423–1442.

Wisenden, B.D., Binstock, C.L., Knoll, K.E. *et al.* (2010) Risk-sensitive information gathering by cyprinids following release of chemical alarm cues. *Animal Behaviour*, **79**, 1101–1107.

Wyatt, T.D. (2010) Pheromones and signature mixtures: defining species-wide signals and variable cues for identity in both invertebrates and vertebrates. *Journal of Comparative Physiology A*, **196**, 685–700.

Figure 4.1. Reproductive male (dark fish on left) and female (right) round goby on a rocky bottom of the St. Lawrence River at Morrisburg, Ontario. Photograph by Lonny Howard, September 5, 2010.

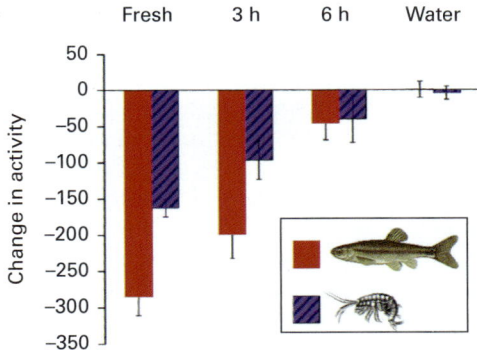

Figure 6.3. Active time of damage-released chemical alarm cues for fathead minnows (solid fill) and amphipod crustaceans (hatched fill). Minnow skin extract and whole-body squashes of amphipods were maintained at 18 °C for 0, 3, or 6 h before being frozen, and compared against a blank water control. Data taken from Wisenden *et al.* (2009).

Fish Pheromones and Related Cues, First Edition. Edited by Peter W. Sorensen and Brian D. Wisenden.
© 2015 John Wiley & Sons, Inc. Published 2015 by John Wiley & Sons, Inc.

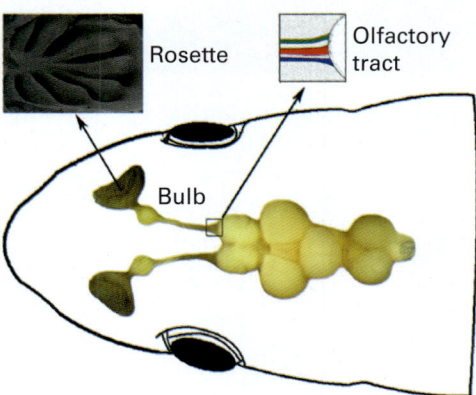

Figure 8.1. The brain of crucian carp illustrating the olfactory system. The dorsal view of the brain of a crucian carp in an outline of the head showing the olfactory rosette, the short olfactory nerves, and long olfactory tracts running from the olfactory bulbs to the telencephalon. The olfactory tracts divide as they enter the telencephalon. The lateral olfactory tract (LOT) is in green, the lateral part of the medial lateral olfactory tract (lMOT) is in red and the medial part of the medial olfactory tract (mMOT) is in blue. El Hassan Hamdani, Kjell B. Døving. The functional organization of the fish olfactory system. Progress in Neurobiology, vol. 82, issue 2. Copyright © Elsevier, 2007.

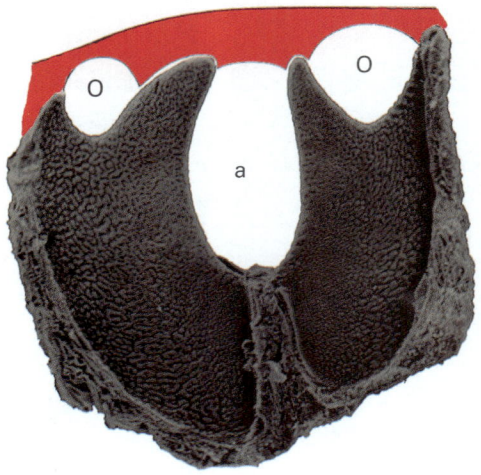

Figure 8.2.   Scanning electron micrograph (SEM) of a pair of lamellae from the olfactory organ of the cod, *Gadus morhua.* The dorsal lamella to the right is slightly smaller than the ventral one. The "top" that closes down on the organ has been drawn in red to illustrate how this organ directs water from the central atrium (a) to the peripheral outlet (o). The atrium opens to the anterior naris so that water is transported from the central region and out the posterior naris via the corridors between the lamellae. (K.B. Døving, unpublished).

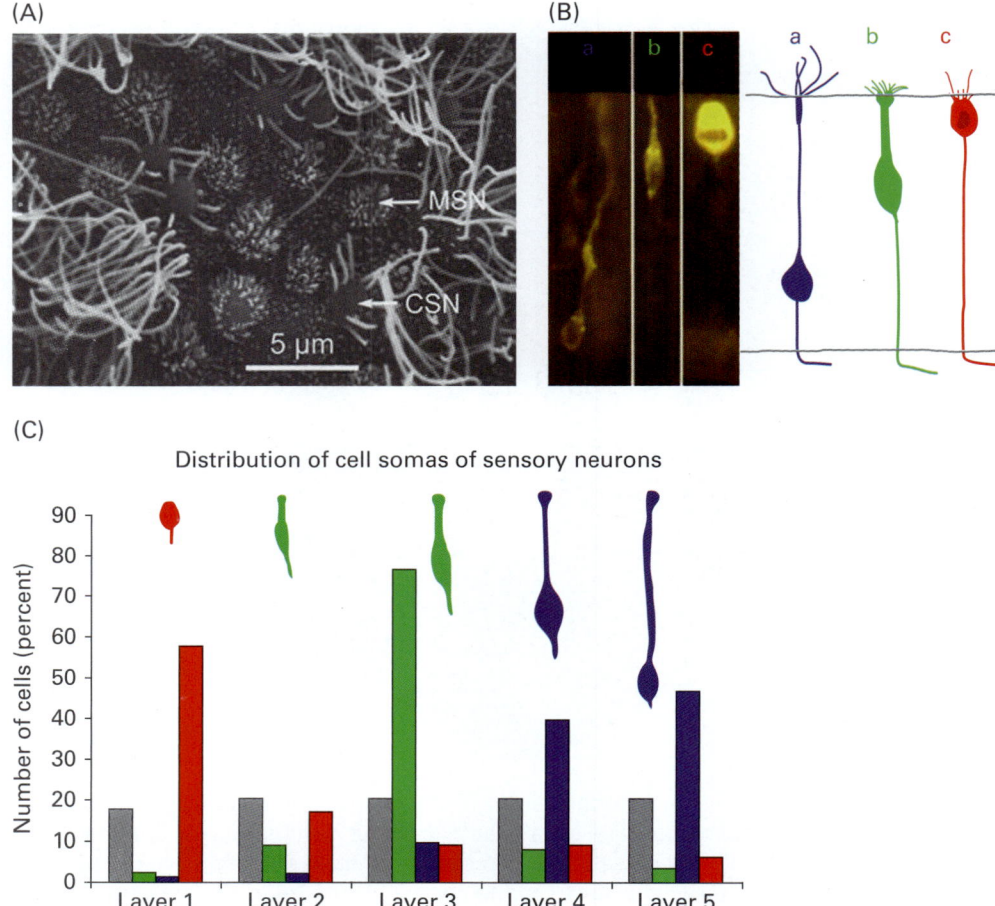

Figure 8.4. The olfactory epithelium and its sensory neurons. (A) Two types of OSNs, Ciliated (CSN), and microvillar (MSN) are seen in this SEM of the olfactory lamellae of an arctic char (*Salvelinus alpinus*). Courtesy of G. Thommesen. (B) Light microscope pictures made by DiI tracing techniques, left and cartoons right: (a) CSN; (b) MSN; (c) crypt cells. (C) Different morphologies of sensory neurons are located within different depths in the olfactory epithelium of crucian carp as revealed by retrograde staining CSNs (blue) are located near the basal lamina, MSNs (green) in the middle and crypt cells (red) at the upper most layer. The histogram demonstrates the percentage of different neurons that connect to the different parts of the olfactory tract. Red fibers connect to IMOT, green to LOT, and yellow to mMOT. The gray bars indicate distribution of cell somas for all cells in the epithelium.

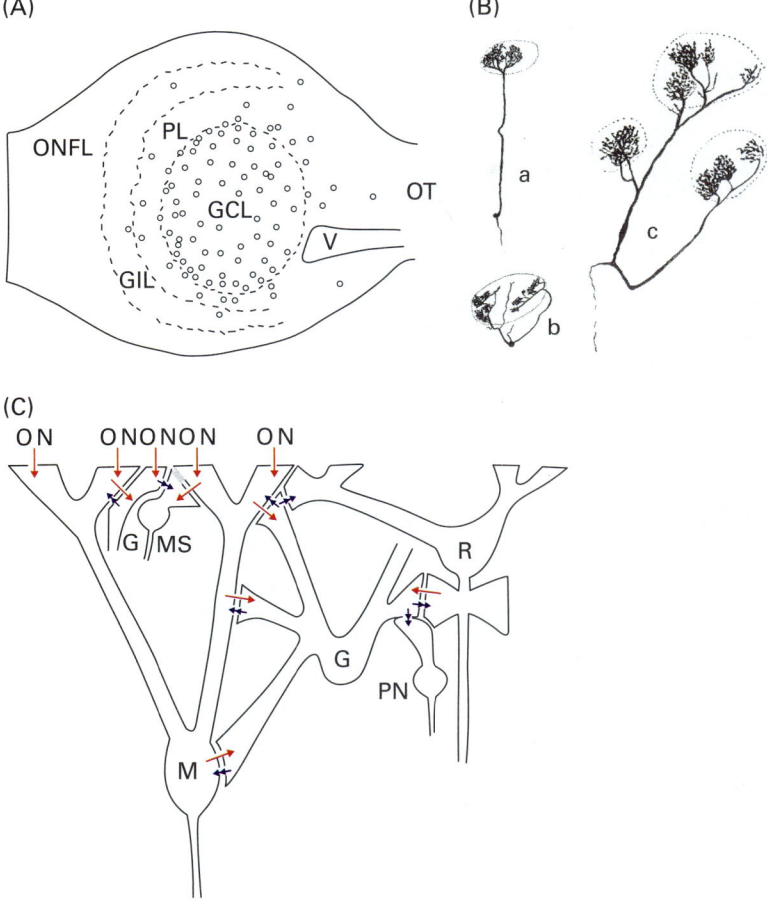

Figure 8.6. The fish olfactory bulb and its neurons. (A) Drawing of a sagittal section of the olfactory bulb of the Mediterranean barbel *Brissus meridionalis* demonstrating the layers. The abbreviations used in this figure are GCL, granule cell layer; GIL, glomerular layer; ONFL, olfactory nerve fiber layer; OT olfactory tract; PL, plexiform layer; V, ventricle. Alonso, J.R., Lara, J., Covenas, R., Aijon, J. (1988) Two types of mitral cells in the teleostean olfactory bulb. Neuroscience research communication 3, 113–118. (B) Different types of secondary neurons in the European smelt, *Osmerus eperalanus* (a) secondary neuron with a single dendrite to glomerulus. (b) secondary neuron with several dendrites to one glomerulus. (c) secondary neuron with dendrites to several glomeruli. Holmgren, N. (1920) Zur Anatomie und Histologie des Vorder- und Zwischenhirns der Knochenfische. Acta zoologica (Stockholm). Stockholm 1, 137–315. Copyright ©The Royal Swedish Academy of Sciences, 1920. (C) The putative arrangement of the synaptic connections of the bulbar neurons in the fish. The abbreviations used in this figure are G, granule cells; M(S), mitral cells/mixed-synapse cell; N, nest; ON, olfactory nerve terminal; PN, perinest cell; R, ruffed cell. The asymmetric synapses (excitatory) have been colored red and the symmetrical synapses (inhibitory) are colored blue. Kosaka, T., Hama, K. (1982) Synaptic organization in the teleost olfactory bulb. J Physiol (Paris) 78, 707–719.

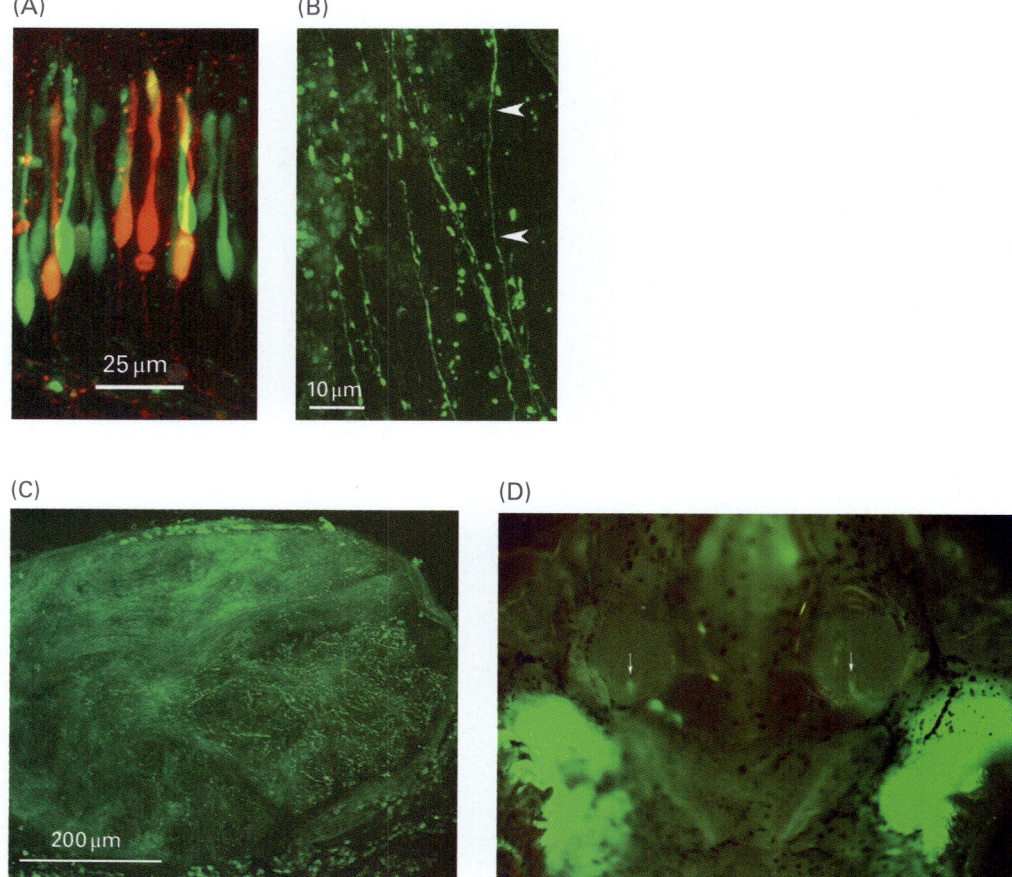

Figure 8.7. Olfactory sensory neurons visualized by endocytosis. (A) Exposing the olfactory organ of crucian carp with a bile salt taurolithocholate 10 nm together with dextran conjugated with Alexa 488 10 μm (green), thereafter a mixture of the alarm agonist hypoxanthine-3-*N*-oxide and dextran conjugated with Alexa 647 10 μm (red) caused staining of two sets of CSNs with long dendrites. The photo is a z-projection of 35 confocal images, covering a depth of 35 μm. Note the axons leaving the CSNs. Courtesy of Kenth-Arne Hansson (2011). (B) Exposing the olfactory organ of crucian carp to the bile salt taurolithocholate caused the dye to stain a subset of axons. In the olfactory nerve, the stained axons are seen to run in parallel with the long axis of the nerve. The stain is frequently seen as spots in the nerve, but regions with smooth staining are also observed (arrows). (C) Transverse section at the interface nerve and bulb. As the axons approach the interface at the anterior portion of the olfactory bulb, the axons are seen to run normal to the long axis of the nerve and aggregate in the dorsomedial portion (arrows). From Døving *et al.* (2011). (D) The CSNs responding to the bile salt taurolithocholate have been stained by ligand-selective endocytosis and surviving for 2 days after the dye has been transported to the olfactory bulb. The medial projection is seen as symmetrical points (arrows) in the olfactory bulb. From Døving *et al.* (2011).

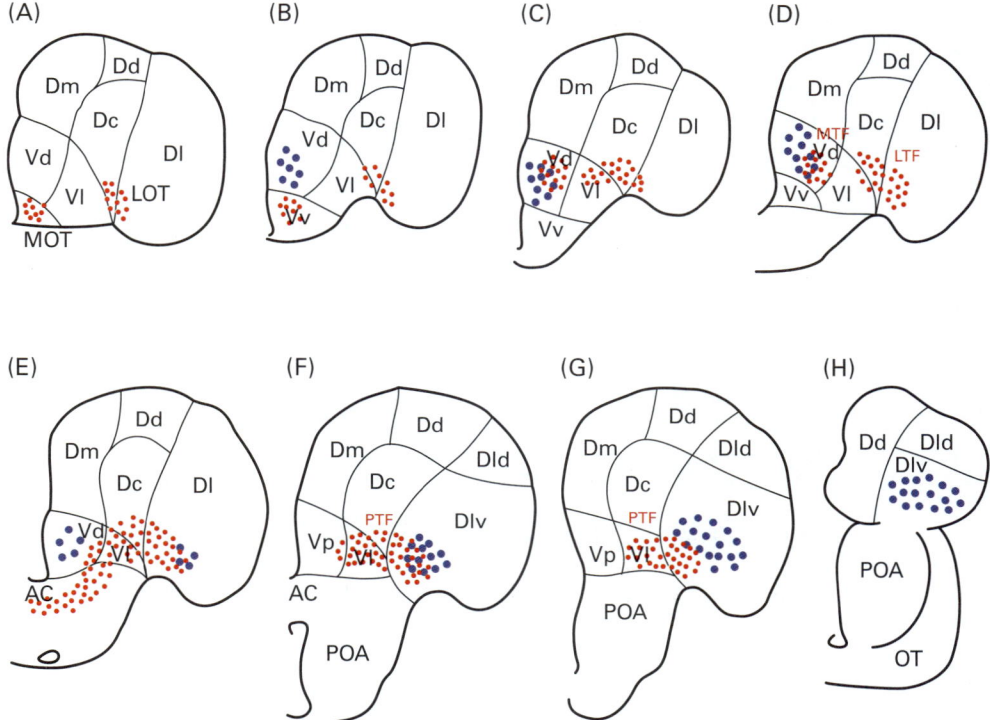

Figure 8.10. Telencephalic regions in goldfish connecting to the olfactory bulb. (A)–(H) are line drawings of frontal sections through the telencephalic hemisphere, illustrating the distribution of labeled neurons (blue dots), following primuline injections of the olfactory bulb. The sections are 150 μm apart. A, shows the area about 750 μm from the rostral end. The red dots indicate the terminal regions of degenerating axons 10 days after transection of the olfactory tract on the ipsilateral side. This figure was published in Brain Research, vol. 185, issue 2, Yoshitaka Oka, The origin of the centrifugal fibers to the olfactory bulb in the goldfish Carassius auratus. An experimental study using the fluorescent dye primuline as a retrograde tracer, pp. 215–225. Copyright © Elsevier, 1980. The abbreviations used in this figure are AC, anterior commissure; Dc, central part of area dorsalis; Dd, dorsal part of area dorsalis; Dld and Dlv, dorsal and ventral part of the lateral area dorsalis; Dm, medial part of area dorsalis; Dp, posterior part of area dorsalis; LOT, lateral olfactory tract; LTF, lateral terminal field; MTF, medial terminal field; OT, optic tract; POA, preoptic area; PTF, posterior terminal field; Vd, dorsal part of area ventralis; Vl, lateral part of area ventralis; Vp, posterior part of area ventralis; and Vv, ventral part of area ventralis.

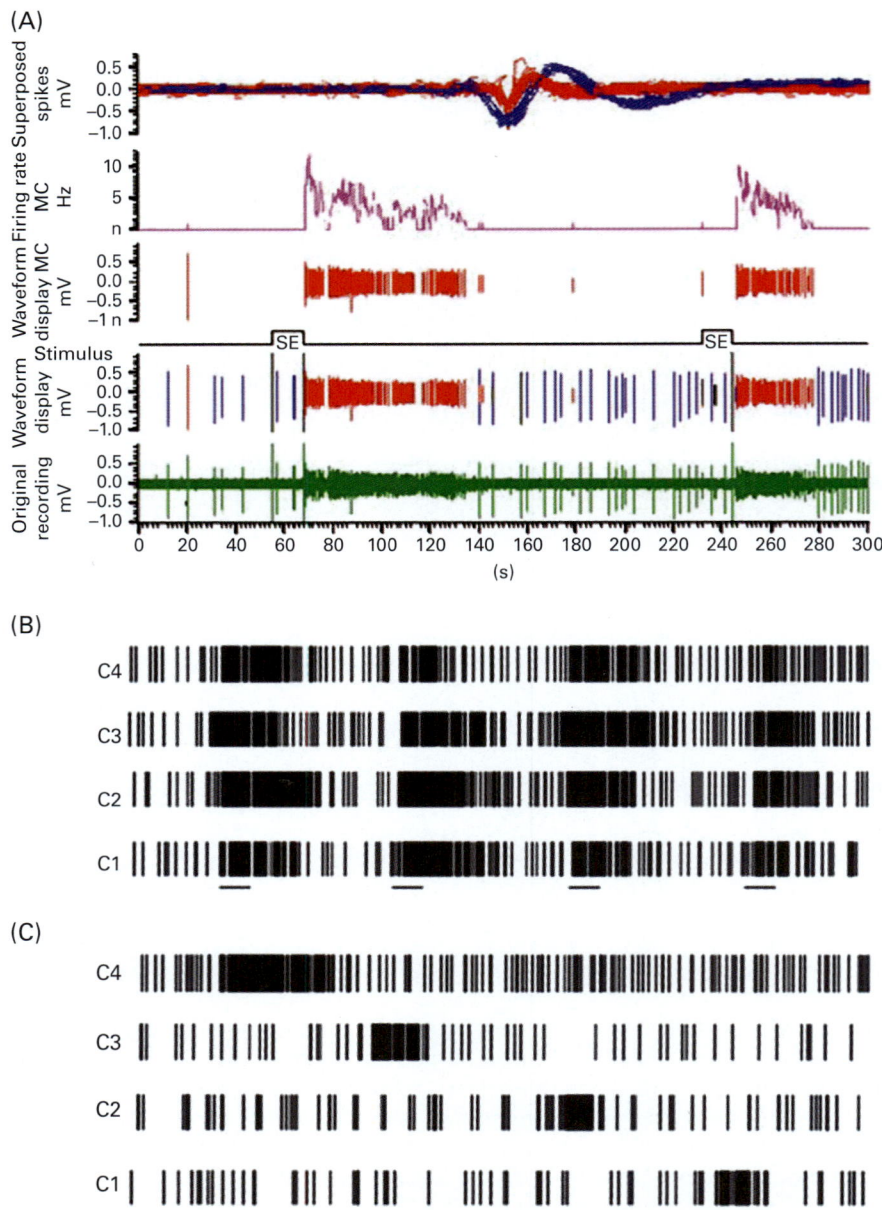

Figure 8.16. Distinction between bulbar neurons and gender differences in recorded responses. (A) Extracellular recordings from single units in the olfactory bulb of crucian carp. Analysis of the original recording (green) revealed two types of units. The type I unit (red) was activated by the stimulation with skin extract (SE), whereas the type II unit (blue) was inhibited during the stimulation period. The overlays shown in the upper tracing revealed the differences in appearance of the two types. El Hassan Hamdani, Kjell B. Døving. Sensitivity and Selectivity of Neurons in the Medial Region of the Olfactory Bulb to Skin Extract from Conspecifics in Crucian Carp, Carassius carassius. Chemical Senses, vol. 28, issue 3. Copyright © 2003, Oxford University Press. (B) and (C) The recordings from the single units in the olfactory bulb in crucian carp revealed dramatic differences between genders in the responses to sex pheromones. The units of the female olfactory bulb (B) did not show any discriminatory capacity. By contrast, the units in the male responded specifically to each of the four sex pheromones (A). Stine Lastein, El Hassan Hamdani, Kjell B. Døving. Gender Distinction in Neural Discrimination of Sex Pheromones in the Olfactory Bulb of Crucian Carp, Carassius carassius. Chemical Senses, vol. 31, issue 1. Copyright © 2006, Oxford University Press.

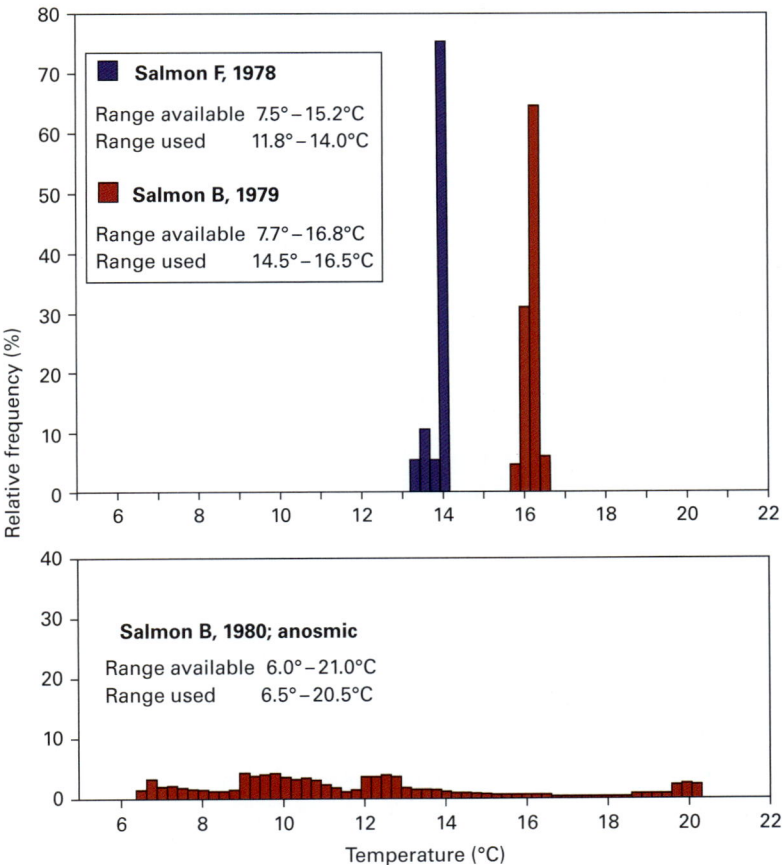

Figure 8.20. Swimming depth preferences of intact and anosmic salmon. The salmon with intact olfactory sense in 1978 and 1979 preferred a restricted layer of water with a particular temperature. The anosmic salmon in 1980 did not prefer a particular layer. Westerberg, H. (Goeteborg Univ. (Sweden), Oceanografiska Inst.) (1982) Ultrasonic tracking of Atlantic salmon (Salmo salar L.), 2: Swimming depth and temperature stratification. Report – Institute of Freshwater Research, Drottningholm 60: 102–120.

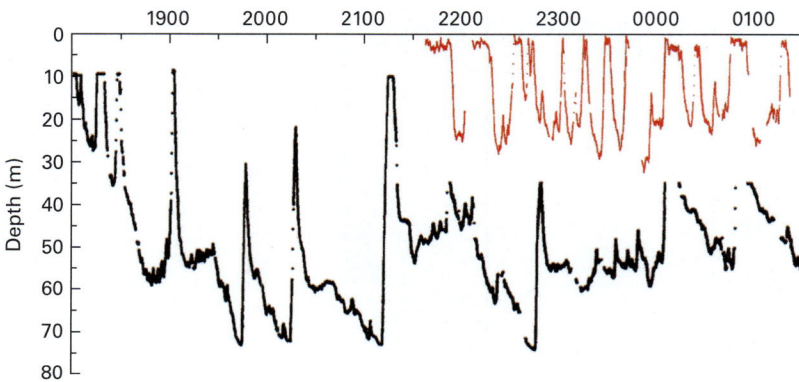

Figure 8.21. Swimming depth profiles of intact and anosmic salmon. The red trace indicates the depth profile of a salmon with an intact olfactory organ in a fjord system. Note that the salmon swims at specific depths, which are seldom below the sill depth of the fjord. However, the anosmic salmon (in black) indicates an aberrant behaviour, following the bottom contour. Kjell B. Døving, Håkan Westerberg, Peter B. Johnsen. Role of Olfaction in the Behavioral and Neuronal Responses of Atlantic Salmon, Salmo salar, to Hydrographic Stratification. Canadian Journal of Fisheries and Aquatic Sciences, vol. 42, issue 10. Copyright (c) 1985, NRC Research Press.

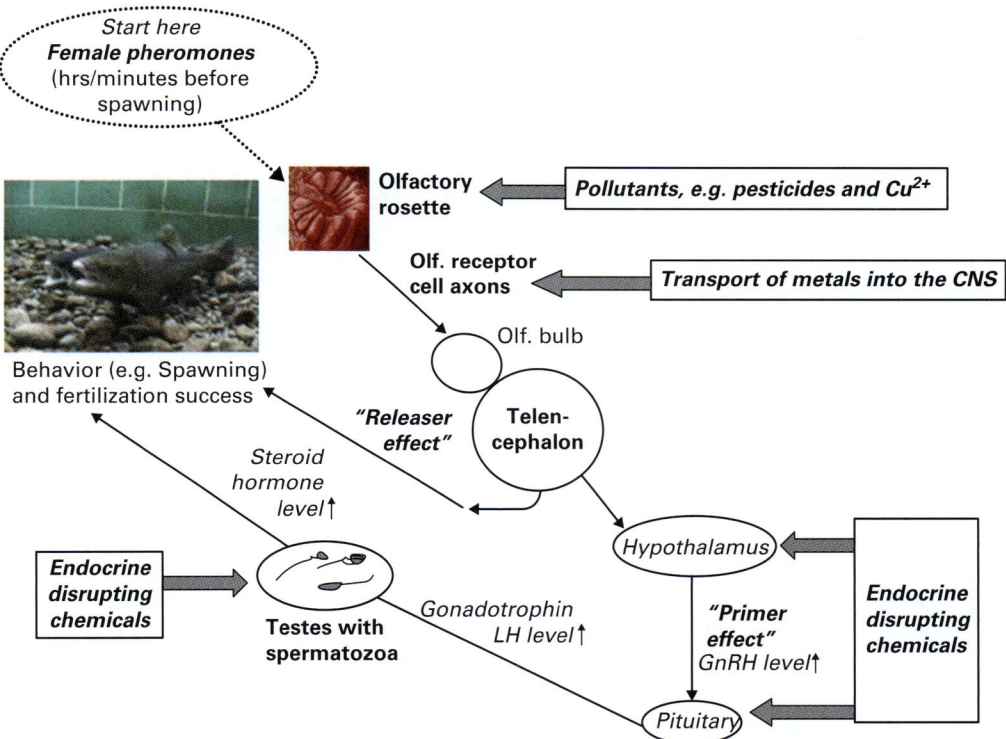

Figure 10.2. Possible negative effects of pollutants on various parts of chemical communication, from detection of pheromones to the resulting behavior and physiological responses. T. Breithaupt & M. Thiel. Chemical Communication in Crustaceans. Copyright 2010, Springer Business and Science Media.

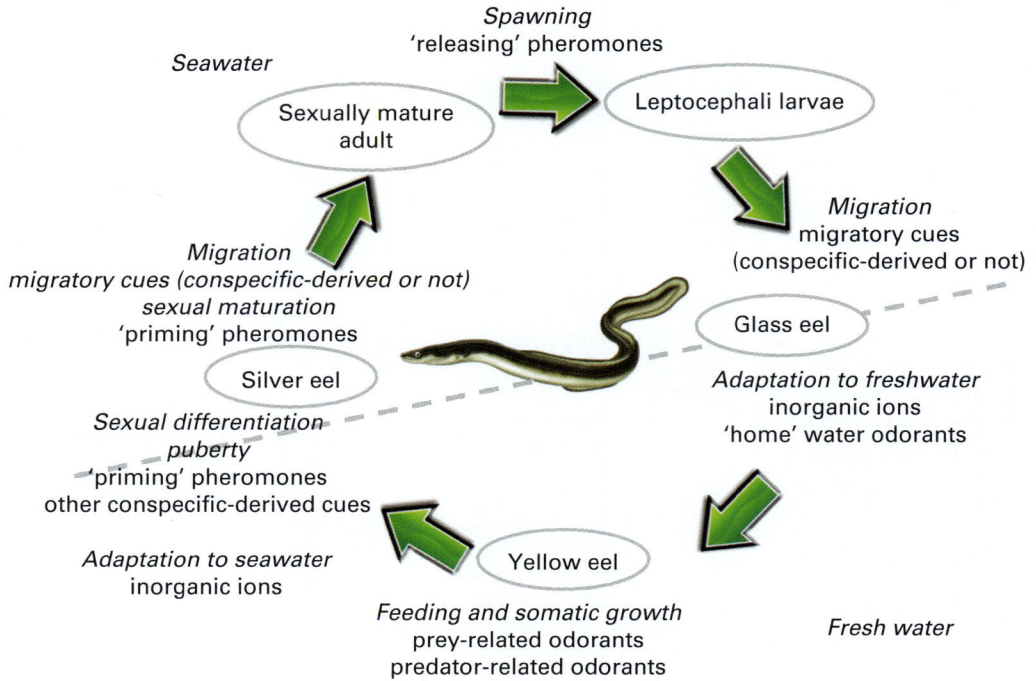

Figure 11.2. The life cycle of the freshwater eel. In addition to the scientific and economic importance of Anguillid eels, their complex life cycle allows "compartmentalization" of many facets of their biology, including olfaction and chemical communication. Somatic growth occurs largely in freshwater ("yellow eel" stage) where olfaction is important for prey detection and predator avoidance; chemical communication may be limited to the alarm response and/or conspecifics recognition. Under the influence of yet-unknown cues, the "silvering" process occurs; it involves the down-river migration to estuaries; environmental factors, such as changes in salinity and organics, may be important guides during this migration, but do conspecific-derived odorants play a role? During the long oceanic migration, eels mature under the influence of high pressure and the swimming process. But are eels also induced to sexual maturity by conspecific-derived odorants— pheromones—and/or is their migration guided in any part by odorant or "pheromone" trails? Eels do not eat during this migration, but their olfactory system remains functional. What are the identities and roles of reproductive pheromones in spawning (never directly observed under natural conditions)? Do the resultant larvae use the same or different chemical cues to guide their migration to the freshwater habitats of their parent?

Chapter 7
The Cue–Signal Continuum: A Hypothesized Evolutionary Trajectory for Chemical Communication in Fishes

Brian D. Wisenden

Minnesota State University Moorhead, Moorhead, Minnesota, USA

7.1 INTRODUCTION

Chemical information is ubiquitous and can be a powerful releaser of behavioral responses (Sorensen and Stacey, 1999; Wisenden and Chivers, 2006; Wyatt, 2010). Chemical information in the environment comes from abiotic and biotic sources. Abiotic cues allow fish to navigate to favorable habitat along gradients in pH, dissolved oxygen, or salinity. Chemical cues of biotic origin are called semiochemicals, after the Greek word *semion* for sign. Semiochemicals provide important information to receivers about every aspect of physiological and ecological function, including spatial and temporal variation in food availability, the diet choice of others, the identity of individual group members on the basis of shared phylogeny, familiarity or kinship, the presence and activity of predators, risk of parasitism, the dominance status of sexual rivals, and the location and reproductive readiness of potential mates and cues that play a role in parent–offspring interactions (Bradbury and Vehrencamp, 2011). The enormous fitness benefits of this information create selection pressure on the sensory biology of fish to favor elaboration of receptors to detect this information and cognitive processes to generate adaptive behavioral responses to this information. Wherever there is production of information by some individuals and exploitation of that information by other individuals, there is also potential for co-evolutionary processes to shape both the sensory biology of receivers to better detect these chemicals and for senders to specialize in production and release of semiochemicals (Stacey and Sorensen, 1991, Wisenden and Stacey, 2005, Wisenden and Chivers, 2006; Stacey, 2015).

Odors, chemical cues and signals, signature mixtures, pheromones and "substances" are terms used to describe the enormous diversity of semiochemicals that fish use to inform their behavioral decision-making. The absence of universally accepted standard nomenclature has created confusion because these terms do not necessarily reflect the evolutionary underpinning involved in the

Fish Pheromones and Related Cues, First Edition. Edited by Peter W. Sorensen and Brian D. Wisenden.
© 2015 John Wiley & Sons, Inc. Published 2015 by John Wiley & Sons, Inc.

production, detection, and processing of this information. To fully understand their ecological function, one needs an understanding of the evolutionary selection behind the origin and maintenance of semiochemical detection and production.

Historically, there have been two approaches to the study of fish semiochemicals: (1) by physiologists who study various aspects of the physiology and behavior of sex pheromones (e.g., Stacey and Sorensen, 2009) and (2) by behavioral ecologists who study how chemical cues mediate predator-prey interactions (e.g. Kelley, 2008; Ferrari, Wisenden, and Chivers, 2010). These parallel communities of researchers have developed largely independent literatures, each with its own lexicon for semiochemicals. The best way for the study of semiochemicals in fishes to move forward is for there to be integration of these literatures and with that an acceptance of a common vocabulary. Therefore, this chapter aims at organizing known semiochemicals along a hypothesized evolutionary trajectory from ancestral to derived. This framework will hopefully highlight opportunities for future research to fill in gaps in collective knowledge and to test predictions of this hypothesis.

7.2   A HYPOTHESIS FOR THE EVOLUTION OF SEMIOCHEMICALS

Orientation → Contextualization → Specialization of production → Specialization of release

A plausible sequence for the evolution of use and production of several broadly defined classes of semiochemicals (Table 7.1) is proposed here. The first and simplest form of behavioral response to external chemical information is *orientation*. The simplest form of orientation occurs in response to chemicals of abiotic origin, such dissolved oxygen, geological mineral cues and salinity, and so on, which provide important information about the environment. Chemo-orientation is used in the settlement of larvae of marine fishes, orientation cues for long-distance migratory behavior, and for microhabitat use. These are not semiochemicals *per se* because semiochemicals are only those chemicals that are released by living organisms.

Behavioral responsiveness to semiochemicals is a significant transition because it creates an opportunity for selection to act on the chemical nature of the cue itself. The simplest forms of semiochemicals provide information about the presence of conspecifics (competitors, prospective mates) and heterospecifics (prey, predators, competitors). The second stage of evolution of semiochemicals is *contextualization*. Semiochemicals become contextualized when they are reliably released during specific ecological or physiological interactions. Context allows selection to shape behavioral responses in sophisticated ways. Contextualized semiochemicals are still cues (i.e., not yet signals) because responses to semiochemicals are driven entirely by selection on receivers. Receiver response to a chemical cue does not contribute to fitness of the sender (Fig. 7.1). The first step toward elevating semiochemicals from cues to signals and, thus, true chemical communication, is the elaboration of tissues *specialized for production* of messenger chemicals, even if at first these chemicals are released for benefits other than manipulating receivers (Wisenden and Stacey, 2005). The ultimate category of semiochemicals is when there is the elaboration of tissues and behavioral mechanisms to *produce and control release of specialized* semiochemicals. Examples from fishes can be described for each of these stages in the evolutionary trajectory from chemical cues released passively as public information to chemical signals that are part of a co-evolved communication system (Sorensen and Stacey, 1999; Wisenden and Stacey, 2005).

7.3   ORIENTATION TO ENVIRONMENTAL SEMIOCHEMICALS

The simplest, and presumably most ancestral, form of chemical cues that reveal information about habitat suitability are chemicals that are released from the decay of organic matter and the geochemical makeup of the watershed, spatial and temporal variation in dissolved oxygen, and exposure to salinity (e.g., Leggett, 1977; Kramer, 1987; Odling-Smee and Braithwaite, 2003; Maes, Stevens, and Breine, 2007). An example of this phenomenon is diadromous fish that spend part of their life in

Table 7.1. Summary of a hypothesis for the evolution of semiochemicals used by fishes on the basis of the presence (1) or absence (0) of various selection pressures on sender (S) and/or receiver (R). The Cue/signal transition to communication occurs when senders show evidence of specialization for cue release.

| Semiochemical | Biotic | Contextual | Specialized issues | | Selection on |
			Production	Release	Sender/ receiver
Environmental cues					
Environmental parameters (pH, salinity, etc.)	0	0	0	0	R
Migratory cues (geochemistry)	0	0	0	0	R
Semiochemical cues					
Food odors (amino acids)	1	0	0	0	R
Odor of prey, predator, and conspecifics	1	0	0	0	R
Injury-released alarm cues	1	1	0	0	R
Recognition of familiarity and kinship (signature mixtures)	1	1	0	0	R
Sex steroids	1	1	1	0	R
Semiochemical signals (pheromones)					
Bile acids	1	1	1	0/1	R/S+R
Conspecific cues about social hierarchy	1	1	1	0/1	R/S+R
Sex pheromones	1	1	1	1	S+R

freshwater and part of their life in salt water. Anadromous salmonids mature in salt water and orient to chemical cues of their natal streams during the migration back to freshwater to spawn (Scholz *et al.*, 1976), which include organics and odors from plant, animals, and amino acids from biofilms (see Sorensen and Baker, 2015). Some catadromous fishes such as bull sharks spawn in salt water and orient to low salinity gradients to find freshwater to grow to adulthood (Heupel and Simpfendorfer, 2008). Eels returning to freshwater orient to organic compounds released by biofilm organisms (Sorensen, 1986, see Sorensen and Baker, 2015). The evolution of this kind of orientation results from selection on receivers to elaborate receptors and neural wiring to detect and respond adaptively to environmental chemicals of value to the fitness of the receiver (Table 7.1; Fig. 7.1).

7.4 PASSIVELY RELEASED BIOTIC SEMIOCHEMICALS WITHOUT ECOLOGICAL CONTEXT

The sum of biochemical processes occurring within aquatic animals generates a chemical profile, that is, a characteristic bouquet or mosaic of compounds that can uniquely identify the species (Sorensen and Baker, 2015). These chemical profiles can also provide information about the internal state of the animal such as hunger state, recent diet, and habitat use (Csanyi, 1985; Licht, 1989; Brown and Godin, 1999; Webster *et al.*, 2007; Ferrari, Wisenden, and Chivers, 2010). Molecules are often passively released from the mucus, gills, urine, and feces of fishes as a result of diffusion gradients and natural

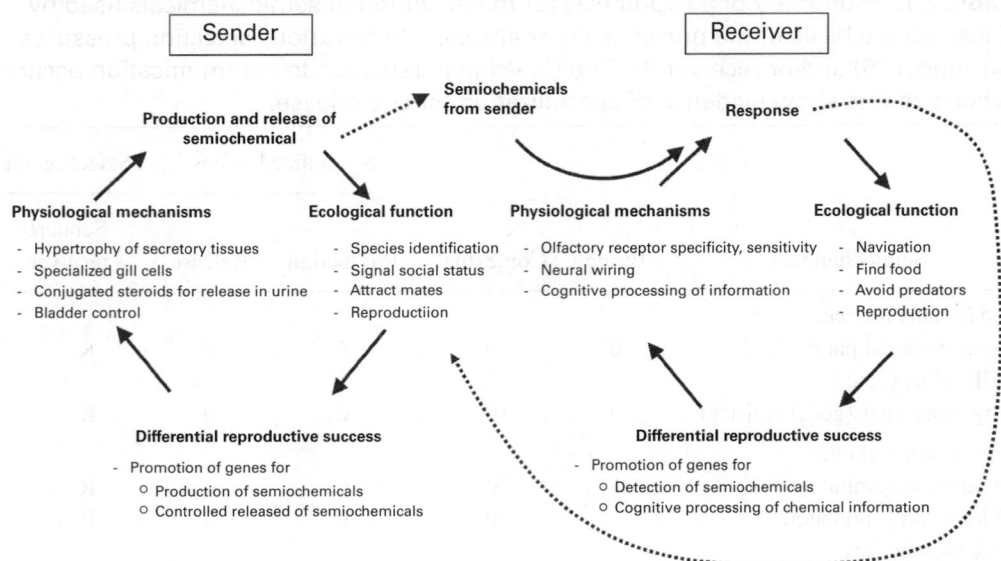

Figure 7.1. The evolution of signal production and release in senders and behavioral responses to semiochemicals in receivers. In receivers, genes that give rise to the proximate mechanisms of behavior (anatomical and physiological aspects of receptors, neural wiring of cognitive processes—the capacity and inclination to behave) are promoted when an individual responds adaptively to information contained in a semiochemical released into the environment. The semiochemical provides information about the context and timing for when a particular behavioral response will be most effective. When a response to semiochemicals is executed at the correct time and in the correct place, it is effective in achieving the ecological function (migration, foraging, predator avoidance, reproduction). Responses to semiochemicals evolve when individuals that are best at exploiting chemical information achieve greater reproductive success than individuals that do not respond as adaptively to semiochemicals. Genes for the production and contextual release of semiochemical signals by senders are promoted when the response of receivers confers a fitness benefit to senders. In a co-evolved communication system, natural selection promotes specialization for production and controlled release of the chemical signal (pheromone) to make the signal more detectable to receivers whose responses benefit the sender.

metabolic clearance. Detection of these molecules in the chemical profile reveals presence/absence information about specific organisms of fitness value to predators (odor of prey), to prey (odor of predators; Ferrari, Wisenden, and Chivers, 2010), species identification mating and migration (e.g. Scholz *et al.*, 1976; Fine and Sorensen, 2005; Sorensen *et al.*, 2005; Keefer *et al.*, 2006; Sorensen and Hoye, 2007), and settling site selection by larval coral reef fishes (Sweatman, 1988). Because these semiochemicals provide useful information, there is selection on receivers to detect and respond adaptively to these semiochemicals. However, because senders do not benefit from the response of receivers, there is no opportunity for selection to act on production of cues or control of their release (Fig. 7.1; Table 7.1).

7.5 PASSIVE, CONTEXTUAL RELEASE OF SEMIOCHEMICALS

When release of semiochemicals occurs only during a limited range of ecological contexts, then there will be opportunity for selection on receivers to tailor behavioral responses to those contexts. For example, fish respond to semiochemicals released from damaged conspecifics (alarm cues) that are released when a predator damages prey tissues in the process of attack and consumption of prey (Wisenden and Chivers, 2006; Kelley, 2008; Ferrari, Wisenden, and Chivers, 2010; Wisenden, 2015). Injury-released chemical cues are typically released only in the context of predation, and therefore reliably indicate the presence of predation risk. Responses to alarm cues are innate. Lab-reared minnows show full response intensity to conspecific extract (e.g., Wisenden *et al.*, 2010). Fishes are also adept at detecting injury-released cues from injured conspecifics in the diet of unfamiliar predators (Mathis and Smith, 1993; Wisenden, 2015). Although the chemical nature of alarm cues is not yet fully characterized for any species (Mathuru *et al.*, 2012), there is no evidence linking the source of alarm cues to a specialized structure for synthesis and release of an alarm cue. Even the much-discussed epidermal club cells of Ostariophysi and Percid fishes (Smith, 1992) are not directly linked to alarm cues (Chivers *et al.*, 2007; Carreau-Green *et al.*, 2008; Ferrari, Wisenden, and Chivers, 2010).

Another example of a contextual semiochemicals is those released from familiar individuals (Ward, Axford, and Krause, 2002, 2003; Ward and Hart, 2003; Ward, 2015). These cues are contextual because they provide information in addition to the simple presence/absence of a conspecific. The ability to discriminate familiar individuals leads to differential behavioral responses to these groups (Dugatkin and Wilson, 1993; Ward and Hart, 2003; Ward *et al.*, 2005; Webster *et al.*, 2007). This class of semiochemical is considered a signature mixture by Wyatt (2010) because responses result from learning either during a sensitive period early in development or based upon self-matching (e.g., Ward and Hart, 2003; Gerlach and Lysiak, 2006; Mehlis, Bakker, and Frommen, 2008; Le Vin, Mable, and Arnold, 2010; Green, Mirza, and Pyle, 2011; Ward, 2015).

Consistent with the types of semiochemicals discussed thus far in this chapter, information gathering of passively released contextual semiochemicals imposes selection pressure on receivers to detect and respond adaptively to chemical information, but it does not exert any selection pressure on senders to produce or regulate the release of semiochemicals.

7.6 PASSIVE, CONTEXTUAL RELEASE FROM TISSUES SPECIALIZED FOR THE PRODUCTION OF SEMIOCHEMICALS

Chemical compounds that regulate internal physiological processes for cell–cell communication are produced and secreted by specialized tissues and detected by specialized receptors on target tissues within the same individual. Having specialized tissues to produce a chemical compound(s) is the first step in interindividual chemical communication.

Bile is produced by specialized tissues in the liver, and many of the acids in bile are potent odorants for which there is low-threshold olfactory sensitivity in many species, including bile of conspecifics and heterospecifics (cross reactions among phylogenetically disparate species of European eel, *Anguilla anguilla*; goldfish, *Carassius auratus*; and Mozambique tilapia, *Oreochromis mossambicus*; Huertas *et al.*, 2010). Adult sea lamprey (*Petromyzon marinus*) use derivatives of larval bile acids as a migratory cue to find locations that contain larvae as an indicator of suitable nursery habitat for spawning (Sorensen *et al.*, 2005; Sorensen and Hoye, 2007; Fine and Sorensen, 2010).

In fishes, hormones produced by specialized tissues in the gonads regulate reproductive maturation. These hormones subsequently leak into the external environment where they reliably inform nearby conspecifics of the reproductive status of that individual (Sorensen and Scott, 1994; Sorensen and Stacey, 1999). Studies on phylogenetically disparate fish taxa suggest that hormonally based semiochemicals are more the rule than the exception (see Stacey, 2015). For example, male tilapia, *O. mossambicus*, distinguish between preovulatory and postovulatory females on the basis of chemical cues (Miranda *et al.*, 2005). Redfin shiners (*Notropis umbratilis*) spawn in the nests of green sunfish (*Lepomis cyanellus*). Spawning aggregations, territorial behavior, and courtship by redfin shiners can be induced by chemical cues associated with gamete release (milt, ovarian fluids) of the host sunfish even when these cues are experimentally released over substratum unsuitable for spawning (Hunter and Hasler, 1965). In this case, redfin shiners exploit publicly released sexual cues of a heterospecific to cue the context and timing of their own reproductive behaviors (Fig. 7.1). The absence of species specificity in these examples underscores how conspecific exploitation of these same semiochemicals could lead to specialization to control the quantity and context of release, and thus the evolution of pheromonal systems of communication.

Although there is once again selection on receivers to detect and respond adaptively to chemical information, there is only incipient selection on senders to elaborate mechanisms of synthesis and control of semiochemical release. Although the semiochemicals discussed thus far are passively released by-products of endogenous physiological processes, they set the stage for the evolution of specialized production and release of these cues because of the opportunity for receiver responses to confer fitness benefits to senders. Control of release by senders is necessary if exploitation of semiochemical cues is to become a component of a co-evolved communication system where receivers benefit from information in the semiochemical (and elaborate receptors and cognitive processes to respond adaptively) and senders benefit from manipulating receivers (and elaborate mechanisms to control production and release of the semiochemical signal).

7.7 ACTIVE, VOLUNTARILY RELEASED CHEMICAL COMPOUNDS FROM TISSUES SPECIALIZED FOR PRODUCTION, STORAGE, AND RELEASE OF SEMIOCHEMICALS

There are several clear examples of hyperdeveloped (beyond the immediate needs of intercellular communication) tissues for the synthesis and release of semiochemicals. Male sea lamprey possess specialized glandular gill cells to facilitate the release of the bile acid-derived sex pheromone 7α, 12α, 24-trihydroxy-5α-cholan-3-one-24-sulfate (also known as 3keto-petromyzonol sulfate, or 3ketoPZS) (Li *et al.*, 2002). Immunocytochemical experiments showed that 3ketoPZS is located in cells in the interlamellar region of the gills of prespermiating males and moves to the gill lamellae when males enter the spermiating phase of reproductive readiness (Siefkes *et al.*, 2003).

Seasonal fluctuations in the size of the seminal vesicles in *Clarias* catfish coincide with the release of compounds that attract females (Resink *et al.*, 1989). In gobies, the mesorchial region of the testes associated with the mesenteries, is hypertrophied and laden with Leydig cells that actively secrete conjugated steroids (a variety of 5β-reduced steroids, chiefly 17-oxo-5β-androstan-3α-yl, also known as etiocholanolone glucoronide, or ETIO-g) that attract females (Colombo, Bekvedere, and Pilati, 1977; Colombo *et al.*, 1980; Murphy, Stacey, and Corkum, 2001; Arbuckle *et al.*, 2005). Interestingly, the black goby has both territorial and sneaker male mating tactics, but only the territorial males possess a hypertrophied mesorchial gland (Locatello, Mazzoldi, and Rasotto, 2002). The absence of a mesorchial gland in sneaker males suggests selection for chemical crypsis (i.e., secondary loss of pheromone-producing tissue).

Fish that do not show obvious tissue specialization for production and release of sex pheromones may nevertheless have an intersexual cascade of pheromonal communication that governs gonadal maturation and spawning behavior that ensures simultaneous gamete release. For example, the release of 17, 20β-dihydroxy-4-pregn-3-one (17, 20βP), sulfated (17, 20βP-20S) and androstenone by pre-ovulatory female goldfish stimulate male reproductive behaviors and sperm production (Sorensen and Stacey, 1999). While 17, 20βP is released via the gills, 17,20βP-20S is released in pulses of urine, giving females voluntary control over its release (Sorensen *et al.*, 1995). Ovulation is accompanied by a 100-fold increase in prostaglandin $F_{2\alpha}$ ($PGF_{2\alpha}$) (Sorensen *et al.*, unpublished results). $PGF_{2\alpha}$ and a related metabolite 15-keto-$PGF_{2\alpha}$ are released primarily with pulses of urine, which increase in frequency when in the presence of a male, especially at the time of entering vegetation where oviposition occurs (Appelt and Sorensen, 2007). In other species, such as the Mossambique tilapia, *O. mossambicus*, and swordtail (*Xiphophorus birchmanni*), it is the males that increase the rate of urine release when in the presence of a pre-ovulatory female (Almeida *et al.*, 2005; Rosenthal *et al.*, 2011). Interestingly, anosmic male tilapia did not increase rate of urination in the presence of pre-ovulatory females indicating that the urination of males is based solely in response to pheromones released by pre-ovulatory females (Miranda *et al.*, 2005). Males in another cichlid, *Astatotilapia burtoni*, showed a similar urinary response to visual stimuli alone, with low rates of urination when presented with mouth-brooding (nonreceptive) females but increasing urination rate when presented with non-mouth-brooding females (Maruska and Fernald, 2012). Urine containing sex pheromones are released by *O. mossambicus* in contexts other than reproduction. Socially dominant males increase concentrations of urinary sex steroids (Oliveira, Almada, and Canario, 1996) and frequency of urine pulse release (Barata *et al.*, 2007) in male–male social interactions.

Taken together, these studies provide evidence from disparate phylogenies for selection on context and timing of release of behaviorally active chemical compounds by both sexes. By commonly accepted communication theory (e.g., Bradbury and Vehrencamp, 2011), these chemical cues should be considered signals in a co-evolved communication system, and thus, qualify as true pheromones.

7.8   DISCUSSION: EVOLUTION OF SEMIOCHEMICALS FROM CUES TO SIGNALS

This book on fish pheromones covers a range of semiochemicals released by conspecifics that can be arranged across the evolutionary spectrum from cues to signals, showing an incremental transition from basic information gathering to the elaboration of specialized receptors and mechanisms for production and release of pheromones. The term "pheromone" was first coined by Karlson and Lüscher (1959), as a form of chemical communication between conspecifics. Karlson and Lüscher (1959) defined pheromone using examples from insects (social Hymenoptera, Isoptera, Lepidoptera), sexual attractants in Crustacea, alarm pheromone [which we are now careful to call alarm *cues* because senders do not benefit from the response of receivers (Ferrari, Wisenden, and Chivers, 2010; Wisenden, 2015)] in minnows and territorial marking substances of carnivorous mammals. Wyatt (2009, 2010, 2013) reviewed 50 years of research on pheromones across many taxa and updated the definition: "Pheromones are molecules that are evolved signals which elicit a specific reaction, for example, a stereotyped behavior and/or a developmental process in a conspecific." (Wyatt, 2010). Clearly, fish produce and respond to semiochemicals that qualify as pheromones as per Wyatt's definition of the term. However, the majority of semiochemicals used by fishes are not pheromones by this definition. Most fish "pheromonal" systems are in fact precommunication processes in which receivers exploit chemical forms of public information while providing no selection on senders to specialize in production or release of semiochemicals. Future work on fish communication should emphasize the demonstration of evidences of specialization for detection, production, and context-specific active release of semiochemicals. To do that, we must first acquire much more information about the chemical nature of the compounds, often comprising mixtures of compounds, which fish use to inform their behavioral decision-making for migration, foraging, predator avoidance, and reproduction.

REFERENCES

Almeida, O.G., Miranda, A., Frade, P. *et al.* (2005) Urine as a social signal in the Mozambique tilapia (*Oreochromis mossambicus*). *Chemical Senses*, **30 S1,** i309–i310.

Appelt, C.W. and Sorensen, P.W. (2007) Female goldfish signal spawning readiness by altering when and where they release a urinary pheromone. *Animal Behaviour*, **74**, 1329–1338.

Arbuckle, W.J., Bélanger, A.J., Corkum, L.D. *et al.* (2005) In vitro biosynthesis of novel 5β-reduced steroids by the testis of the round goby. *Neogobius melanostomus. General and Comparative Endocrinology*, **140**, 1–13.

Barata, E.N., Hubbard, P.C., Almeida, O.G. *et al.* (2007) Male urine signals social rank in the Mozambique tilapia (*Oreochromis mossambicus*). *BMC Biology*, **5**, 54.

Bradbury, J.W. and Vehrencamp, S.L. (2011) *Principles of Animal Communication, 2nd Edn,* Sinauer, Sunderland, MA.

Brown, G.E. and Godin, J.-G. (1999) Who dares, learns: chemical inspection behaviour and acquired predator recognition in a characin fish. *Animal Behaviour*, **57**, 475–481.

Carreau-Green, N.D., Mirza, R.S., Martinez, M.L., and Pyle, G.G. (2008) The ontogeny of chemically mediated antipredator responses of fathead minnows *Pimephales promelas. Journal of Fish Biology*, **73**, 2390–2401.

Chivers, D.P., Wisenden, B.D., Hindman, C.J. *et al.* (2007) Epidermal "alarm substance" cells of fishes maintained by non-alarm functions: possible defence against pathogens, parasites and UVB radiation. *Proceedings of the Royal Society Series B*, **274**, 2611–2619.

Colombo, L., Bekvedere, P.C., and Pilati, A. (1977) Biosynthesis of free and conjugated 5β-reduced androgens by the testis of the black goby, *Gobius jozo* L. *Bollettino di Zoologia*, **44**, 131–134.

Colombo, L., Marconato, A., Belvedere, P.C., and Friso, C. (1980) Endocrinology of teleost reproduction: a testicular steroid pheromone in the black goby, *Gobius jozo* L. *Bollettino di Zoologia*, **47**, 355–364.

Csanyi, V. (1985) Ethological analysis of predator avoidance by the paradise fish (*Macropodus opercularis* L.) 1. Recognition and learning of predators. *Behaviour*, **92**, 227–240.

Dugatkin, L.A. and Wilson, D.S. (1993) Fish behaviour, partner choice experiments and cognitive ethology. *Reviews in Fish Biology and Fisheries*, **4**, 368–372.

Ferrari, M.C.O., Wisenden B.D., and Chivers, D.P. (2010) Chemical ecology of predator-prey interactions in aquatic ecosystems: a review and prospectus. *Canadian Journal of Zoology*, **88**, 698–724.

Fine, J.M. and Sorensen, P.W. (2005) Biologically relevant concentrations of petromyzonal sulfate, a component of sea lamprey migratory hormone, measured in stream water. *Journal of Chemical Ecology*, **31**, 2205–2210.

Fine, J.M. and Sorensen, P.W. (2010) Production and fate of the sea lamprey migratory pheromone. *Fish Physiology and Biochemistry*, **36**, 1013–1020.

Gerlach, G. and Lysiak, N. (2006) Kin recognition and inbreeding avoidance in zebrafish is based on phenotype matching. *Animal Behaviour*, **71**, 1371–1377.

Green, W.W., Mirza, R.S., and Pyle, G.G. (2011) Kin recognition and cannibalistic behaviours by adult male fathead minnows (*Pimephales promelas*). *Naturwissenshaften*, **95**, 269–272.

Heupel, M.R. and Simpfendorfer, C.A. (2008) Movement and distribution of young bull sharks *Carcharhinus leucas* in a variable estuarine environment. *Aquatic Biology*, **1**, 277–289.

Huertas, M., Hagey, L., Hofmann, A.F. et al. (2010) Olfactory sensitivity to bile fluid and bile salts in the European eel (*Anguilla anguilla*), goldfish (*Carassius auratus*) and Mozambique tilapia (*Oreochromis mossambicus*) suggests a "broad range" sensitivity not confined to those produced by conspecifics. *Journal of Experimental Biology*, **213**, 308–317.

Hunter, J.R. and Hasler, A.D. (1965) Spawning association of the redfin shiner, *Notropis umbratilis*, and the green sunfish. *Lepomis cyanellus. Copeia*, **1965**, 265–281.

Karlson, P. and Lüscher, M. (1959) "Pheromones": a new term for biologically active substances. *Nature*, **183**, 55–56.

Keefer, M.L., Caudill, C.C., Peery, C.A., and Bjornn, T.C. (2006) Route selection in a large river during the homing migration of chinook salmon (*Oncorhynchus tshawytscha*). *Canadian Journal of Fisheries and Aquatic Sciences*, **63**, 1752–1762.

Kelley, J.L. (2008) Assessment of predation risk by prey fishes, in *Fish Behaviour* (eds Magnhagen, C., Braithwaite, V.A., Forsgren, E., and Kapoor, B.G.), Science Publishers, Enfield, pp. 269–301.

Kramer, D.L. (1987) Dissolved oxygen and fish behavior. *Environmental Biology of Fishes*, **2**, 81–92.

Le Vin, A.I., Mable, B.K., and Arnold, K.E. (2010) Kin recognition via phenotype matching in a cooperatively breeding cichlid, *Neolamprologus pulcher. Animal Behaviour*, **79**, 1109–1114.

Leggett, W.C. (1977) The ecology of fish migrations. *Annual Review of Ecology and Systematics*, **8**, 285–308.

Li, W., Scott, A.P., Siefkes, M.J. *et al.* (2002) Bile acid secreted by male sea lamprey that acts as a sex pheromone. *Science*, **296**, 138–141.

Licht, T. (1989) Discrimination between hungry and sated predators: the response of guppies (*Poecilia reticulata*) from high and low predation sites. *Ethology*, **82**, 238–243.

Locatello, L., Mazzoldi, C., and Rasotto, M.B. (2002) Ejaculate of sneaker males is pheromonally inconspicuous in the black goby, *Gobius niger* (Teleostei, Gobiidae). *Journal of Experimental Biology*, **293**, 601–605.

Maes, J., Stevens, M., and Breine, J. (2007) Modelling the migration opportunities of diadromous fish species along a gradient of dissolved oxygen concentration in a European tidal watershed. *Estuarine, Coastal and Shelf Science*, **75**, 151–162.

Maruska, K.P. and Fernald, R.D. (2012) Contextual chemosensory urine signaling in an African cichlid fish. *Journal of Experimental Biology*, **215**, 68–74.

Mathis, A. and Smith, R.J.F. (1993) Chemical labelling of northern pike (*Esox lucius*) by the alarm pheromone of fathead minnows (*Pimephales promelas*). *Journal of Chemical Ecology*, **19**, 1967–1979.

Mathuru, A.S., Kibat, C., Cheong, W.F. *et al.* (2012) Chondroitin fragments are odorants that trigger fear behavior in fish. *Current Biology*, **22**, 1–7.

Mehlis, M., Bakker, T.C.M., and Frommen, J.G. (2008) Smells like sib spirit: kin recognition in three-spined stickleback (*Gasterosteus aculeatus*) is mediated by olfactory cues. *Animal Cognition*, **11**, 643–650.

Miranda, A., Almeida, O.G., Hubbard, P.C. *et al.* (2005) Olfactory discrimination of female reproductive status by male tilapia (*Oreochromis mossambicus*). *Journal of Experimental Biology*, **208**, 2037–2043.

Murphy, C.A., Stacey, N.E., and Corkum, L.D. (2001) Putative steroidal pheromones in the round goby, *Neogobius melanostomus*: olfactory and behavioral responses. *Journal of Chemical Ecology*, **27**, 443–470.

Odling-Smee, L. and Braithwaite, V.A. (2003) The role of learning in fish orientation. *Fish and Fisheries*, **4**, 235–246.

Oliveira, R.F., Almada, V.C., and Canario, A.V.M. (1996) Social modulation of sex steroid concentrations in the urine of male cichlid fish *Oreochromis mossambicus*. *Hormones and Behavior*, **30**, 2–12.

Resink, J.W., Voorthuis, P.K., Van Den Hurk, R. *et al.* (1989) Steroid glucoronides of the seminal vesicle as olfactory stimuli in African catfish, *Clarias gariepinus*. *Aquaculture*, **83**, 1–2.

Rosenthal, G.G., Fitzsimmons, J.N., Woods, K.U. *et al.* (2011) Tactical release of a sexually-selected pheromone in a swordtail fish. *PLoS ONE*, **6**, e16994.

Scholz, A.T., Horrall, R.M., Cooper, J.C., and Hasler A.D. (1976) Imprinting to chemical cues: the basis for home stream selection in salmon. *Science*, **192**, 1247–1249.

Siefkes, M.J., Scott, A.P., Zielinski, B. *et al.* (2003) Male sea lamprey, *Petromyzon marinus* L. excrete a sex pheromone from gill epithelia. *Biology of Reproduction*, **69**, 125–132.

Smith, R.J.F. (1992) Alarm signals in fishes. *Reviews in Fish Biology and Fisheries*, **2**, 33–63.

Sorensen, P.W. (1986) Origins of the freshwater attractant(s) of migrating elvers of the American eel, *Anguilla rostrata* (LeSueur). *Journal of the Environmental Biology of Fishes*, **17**, 185–200.

Sorensen, P.W. and Baker, C. (2015) Species-specific pheromones and their roles in shoaling, migration and reproduction: a critical review and Synthesis, in *Fish Pheromones and Related Cues* (eds Peter W. Sorensen and Brian D. Wisenden), John Wiley & Sons, Inc., Hoboken.

Sorensen, P.W. and Hoye, T.E. (2007) A critical review of the discovery and application of a migratory pheromone in an invasive fish, the sea lamprey, *Petromyzon marinus L. Journal of Fish Biology*, **71** (supplement D), 100–114.

Sorensen, P.W. and Scott, A.P. (1994) The evolution of hormonal sex pheromones in teleost fish: poor correlation between the pattern of steroid release by goldfish and olfactory sensitivity suggests that these cues evolved as a result of chemical spying rather than signal specialization. *Acta Physiologica Scandinavica*, **152**, 191–205.

Sorensen, P.W. and Stacey, N.E. (1999) Evolution and specialization of fish hormonal pheromones, in *Advances in Chemical Signals in Vertebrates* (eds Johnston, R.E., Müller-Schwarze, D., and Sorensen, P.W.), Kluwer Academic/Plenum Publishers, New York, pp. 15–47.

Sorensen, P.W., Fine, J.M., Dvornikovs, V. *et al.* (2005) Mixture of new sulfated steroids functions as a migratory pheromone in the sea lamprey. *Nature Chemical Biology*, **1**, 324–328.

Sorensen, P.W., Scott, A.P., Stacey, N.E., and Bowdin, L. (1995) Sulfated 17,20β-dihydroxy-4-pregen-3-one functions as a potent and specific olfactory stimulant with pheromonal actions in the goldfish. *General and Comparative Endocrinology*, **100**, 128–142.

Stacey, N. (2015) Hormonally-derived pheromones in teleost fishes, in *Fish Pheromones and Related Cues* (eds Peter W. Sorensen and Brian D. Wisenden), John Wiley & Sons, Inc., Hoboken.

Stacey, N.E and Sorensen, P.W. (2009) Hormonal pheromones in fish, in *Hormones, Brain and Behavior*, 2nd edn, vol. **2** (eds Pfaff, D.W., Arnold, A.P., Etgen, A., et al.), Elsevier Press, San Diego, pp. 639–681.

Stacey, N.E. and Sorensen, P.W. (1991) Function and evolution of fish hormonal pheromones, in *Biochemistry and Molecular Biology of Fishes*, vol. **1** (eds. Hochachka, P.L. and Mommsen, T.P.), Elsevier, Amsterdam, pp.109–135

Sweatman, H. (1988) Field evidence that settling coral reef fish larvae detect resident fishes using dissolved chemicals. *Journal of Experimental Marine Biology and Ecology*, **124**, 163–174.

Ward, A.J.W. (2015) Intraspecific social recognition in fishes via chemical cues, in *Fish Pheromones and Related Cues* (eds Peter W. Sorensen and Brian D. Wisenden), John Wiley & Sons, Inc., Hoboken.

Ward, A.J.W. and Hart, P.J.B. (2003) The effects of kin and familiarity on interactions between fish. *Fish and Fisheries*, **4**, 348–358.

Ward, A.J.W., Axford, S., and Krause, J. (2002) Mixed-species shoaling in fish: the sensory mechanisms and costs of shoal choice. *Behavioral Ecology and Sociobiology*, **52**, 182–187.

Ward, A.J.W., Axford, S., and Krause, J. (2003) Cross-species familiarity in shoaling fishes. *Proceedings of the Royal Society of London Series B*, **270**, 1157–1161.

Ward, A.J.W., Holbrook, R.I., Krause, J., and Hart, P.J.B. (2005) Social recognition in sticklebacks: the role of direct experience and habitat cues. *Behavioral Ecology and Sociobiology*, **57**, 575–583.

Webster, M.M., Goldsmith, J., Ward, A.J.W., and Hart, P.J.B. (2007) Habitat-specific chemical cues influence association preferences and shoal cohesion. *Behavioral Ecology and Sociobiology*, **62**, 273–280.

Wisenden, B.D. (2015) Chemical cues that indicate risk of predation, in *Fish Pheromones and Related Cues* (eds Peter W. Sorensen and Brian D. Wisenden), John Wiley & Sons, Inc., Hoboken.

Wisenden, B.D. and Chivers, D.P. (2006) The role of public chemical information in antipredator behavior, in *Fish Communication* (eds Ladich, F., Collins, S.P., Moller, P. et al.), Science Publisher, Enfield, pp. 259–278.

Wisenden, B.D. and Stacey, N.E. (2005) Fish semiochemicals and the evolution of *communication networks*, in Communication Networks (ed. McGregor, P.K.), Cambridge University Press, Cambridge, pp. 540–567.

Wisenden, B.D., Binstock, C.L., Knoll, K.E., *et al.* (2010) Risk-sensitive information gathering by cyprinids following release of chemical alarm cues. *Animal Behaviour*, **79**, 1101–1107.

Wyatt, T.D. (2009) Fifty years of pheromones. *Nature*, **457**, 262–263.

Wyatt, T.D. (2010) Pheromones and signature mixtures: defining species-wide signals and variable cues for identity in both invertebrates and vertebrates. *Journal of Comparative Physiology A*, **196**, 685–700.

Wyatt, T.D (2013) *Pheromones And Animal Behavior: Chemical Signals And Signature Mixtures,* 2nd edn, Cambridge University Press, Cambridge.

Chapter 8
Olfactory Discrimination
of Pheromones

Stine Lastein[1], El Hassan Hamdani[2] and Kjell B. Døving[2]

[1]University of Copenhagen, Copenhagen, Denmark
[2]University of Oslo, Oslo, Norway

8.1 INTRODUCTION

There is a common misunderstanding that one cannot distinguish between the sense of taste and the sense of smell in fishes. This presumption is typically due to a lack of knowledge about three subjects: (1) the misconception that odorants are necessarily borne by air, (2) the fact that the anatomical organization of these two chemosensory systems is actually strikingly different, and (3) because many are unaware that behavioral responses evoked by these systems are distinct.

Smell (olfaction) is mediated by the first cranial nerve, which is made up of very thin axons of olfactory sensory neurons (OSNs), and which terminate in the forebrain of the central nervous system. The sense of smell mediates a series of different behavioral patterns evoked by odorants that are related to reproduction, alarm, and habitat cues. These can all be broadly classified as pheromones (habitat cues often reflect the presence of conspecifics). In addition, the sense of smell is important for finding food. The sense of taste, on the contrary, comprises a series of taste buds situated in the mouth and over the skin of the fish body. These taste buds contain receptor cells innervated by cranial nerves that terminate in another part of the central nervous system, the most caudal region, the medulla oblongata. Fishes use the extra-oral taste buds to find food, and the oral taste buds to mediate acceptance or rejection of food items.

In this chapter, which deals with the olfactory system, we have chosen to first describe the anatomy of the olfactory system, then the physiological properties, and conclude with a description of certain behavioral reactions to odorants classified as pheromones. We shall describe its anatomical features in detail. It is reasonable to start with the olfactory cavity and the formation of the olfactory rosette where we find the OSNs. Then, we discuss how water flow is created over the sensory epithelium. The axons of the different types of OSNs project onto the relay structure, the olfactory bulb. The organization of the olfactory bulb will be described. The olfactory system of fishes is distinguished from other vertebrate sensory systems by the presence of pronounced chemotopy of the axonal projection of the different cell types onto the olfactory bulb. We shall also describe properties of the olfactory tract connecting the olfactory bulb to the telencephalon. The olfactory projections to the telencephalon will be discussed.

Fish Pheromones and Related Cues, First Edition. Edited by Peter W. Sorensen and Brian D. Wisenden.
© 2015 John Wiley & Sons, Inc. Published 2015 by John Wiley & Sons, Inc.

In the physiology section, we attempt to describe the functional properties of the different elements of the olfactory sensory system and how each of these elements work. There are different means of recording from nervous structures, and we shall describe both the mass responses and the single-neuron responses (units) of the neurons in the olfactory system. The electrophysiological properties of the OSNs will be considered in relation to the responses to pheromones. Likewise, we describe responses from the bulbar surface and the responses to single units in this structure. Of particular interest, we demonstrate a gender distinction between responses of bulbar neurons to pheromones. The olfactory tract is divided into discrete bundles, which consist of axons with different functional properties.

Pheromones were initially defined as chemical substances released by an individual and received by another individual of the same species in which they release specific behavioral or physiological reactions (Karlson and Luscher, 1959). This concept has been lately modified to be defined as "a substance or a mixture of substances released by an individual, which evokes a specific and adaptive response in conspecifics, the expression of which does not require learning" (for review, see Stacey, 2015, Sorensen (2015)). As we shall see, there are a multitude of behaviors that can be evoked by pheromones among fishes. The olfactory system mediates different behavioral responses to particular odorants. Fish detect odorants at very low concentrations of different odors, which can induce distinct types of behaviors. There are characteristic behavioral patterns including food search, but also different responses to sex odorants, habitat cues, and alarm cues. It is evident that some behavioral patterns conflict with one another, for example, alarm and reproductive behaviors. We shall also describe how habitat cues can be given (carried) by pheromones. When it comes to reproduction and reproductive behaviors, the olfactory system plays a major role in transmitting the messages between the partners to ensure successful reproductive fitness. Døving (1976) was the first to suggest that released sex hormones could function as potent sexual cues in fish. Now, it is well established that water-borne steroids and F prostaglandins (PGFs) and their conjugates are detected with great sensitivity in different fish species. Hopefully, it will be evident from this chapter that the fish olfactory system consists of three parallel pathways from the sensory epithelium to the telencephalon. Two of these pathways concern conspecifics odorants that evoke reproductive behaviors and alarm reaction, and habitat cues concerned with migration. The third pathway concerns odorants that evoke food search behaviors.

In many species, olfaction evokes behaviors that can be considered essential life processes, such as survival, escape, and reproductive success. The recordings of nervous activity in the olfactory pathways have been of great help to find the potent odorants that facilitate sexual communication. It is also of interest that most naturally occurring pheromones are not pure compounds, but mixtures. We shall briefly describe nervous responses to such mixtures.

The ray-finned fishes constitute by far the largest group of vertebrates, encompassing more than 25 000 living species. The living fishes have a long evolution as they are well adjusted to their environment, including adaptations of the olfactory system. As will be evident from this chapter, the number of species that have been investigated concerning the olfactory system is about a dozen, which is a very low number in comparison with the total number of fish species. At this time, the goldfish (*Carassius auratus*) is the best studied fish with regard to the understanding of the complexity of the fish hormonal pheromones and the olfactory system (e.g., Stacey and Sorensen, 2002, Stacey, 2015). Briefly, females release two sets of pheromones: preovulatory (steroids) and postovulatory (PGFs). These components are shown to induce physiological and behavioral responses in males. Accordingly, most of the physiological data discussed in this chapter are from studies carried on the crucian carp, *C. carassius*, a close relative of the goldfish. However, because goldfish show neither territorial nor parental behavior, other fish models seem to be required to assure a full understanding of pheromonal behavior in fish.

8.2 ANATOMY

The very large number of fish species makes it difficult to generalize about the function of a particular sensory system. However, in all species examined, one finds common elements shared by most fishes. OSNs all have an axon that terminates in the relay station, olfactory bulb. The secondary neurons mediate information to the telencephalon proper via their axons that make up the olfactory tracts. These structures are all found in the anterior part of the brain, with sensory neurons frequently gathered in a rosette that lies in front of the eyes and close to the nose tip. The gross morphology of a representative brain of the crucian carp is shown in Figure 8.1. It shows the olfactory rosettes, the short olfactory nerves running to the bulbs, and the long olfactory tracts.

In some species, such as pike and salmon, the olfactory nerves are long. This is somewhat surprising as the axons of the OSNs are very thin (0.14 μm, Døving and Gemne, 1966) and thus conduct action potentials at a slow speed (<0.1 m·s⁻¹). This means that it will take 1 s before the olfactory information reaches the olfactory bulb in a large pike with an olfactory nerve that is 10 cm long.

The long olfactory nerve means that the olfactory bulb lies close to the telencephalon proper (see Fig. 8.1). Short olfactory nerves and long tracts are characteristic of the cod, catfishes, and carps. The partition of the tract into different bundles is visible to the naked eye. The advantage such a long tract in the central nervous system to experimental biologists is evident. Details of the histological and physiological properties of the tract will be considered in later parts of this chapter. In all species examined, the OSNs lie in the olfactory epithelium and then pass (as the olfactory nerves) to the olfactory bulbs where they synapse and encode information. Information is then processed and passed to the brain via the olfactory tracts. The relation between the primary and secondary neurons is unique to the olfactory system, because there are about 1000 times more OSNs than there are secondary neurons. In this case, synaptic structures in the olfactory bulb between these neurons are of particular interest. The size of the olfactory bulbs reflects the number of OSNs and can coarsely indicate the importance of the sense of smell.

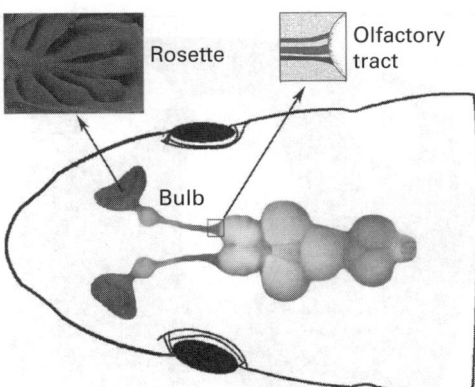

Figure 8.1. The brain of crucian carp illustrating the olfactory system. The dorsal view of the brain of a crucian carp in an outline of the head showing the olfactory rosette, the short olfactory nerves, and long olfactory tracts running from the olfactory bulbs to the telencephalon. The olfactory tracts divide as they enter the telencephalon. The lateral olfactory tract (LOT) is in green, the lateral part of the medial lateral olfactory tract (lMOT) is in red and the medial part of the medial olfactory tract (mMOT) is in blue. El Hassan Hamdani, Kjell B. Døving. The functional organization of the fish olfactory system. Progress in Neurobiology, vol. 82, issue 2. Copyright © Elsevier, 2007. (*See insert for color representation of the figure.*)

Again, it is important to mention that only a few species have been investigated. For example, the olfactory system of perciforms—the largest family of fish—has scarcely been studied. However, the brain anatomy and sensory systems of species of the perciform suborder Notothenioidei endemic to the Antarctic Ocean have been described in great detail by Eastman and Lannoo (2011).

8.2.1 Peripheral Structure

The olfactory organ in most fishes is not directly exposed to the environment. Instead, it is recessed in a cavity and protected by a flap of skin. Thus, the sensory epithelium in the olfactory cavity reflects a fundamental tradeoff; the delicate OSNs need to be protected from abrasion while simultaneously accessible to water from the external environment. Within these constraints, olfactory organs have developed two mechanisms for ventilating the sensory neurons: (1) ciliated cells that draw external water across the rosette and/or (2) accessory sacs that pump water. Based on these differences, fish are often classified as "isosmates" or "cyclosmates" (Døving *et al.*, 1977).

In isosmates, the olfactory organ consists of a series of lamellae (that contain the OSNs) on both sides of a central raphe. Because the lamellae are situated close to one another, and kinocilia on the lamellae surface beat in an organized fashion, water flows from the central region (atrium) along the corridors between the lamellae toward the outer vents. Water exits via the posterior naris (Fig. 8.2). In cyclosmates, the olfactory organ is situated in conjunction with accessory sacs. These sacs move water in and out of the olfactory cavity. Frequently, we find complex hydrodynamic features of these structures, so that incoming water flows around the olfactory organ. In some species, flaps, valves, or vents direct the water flow in the optimal direction (Zeiske *et al.*, 1976; Fig. 8.3). The relationship of these structures to pheromone reception has not been studied.

The anterior naris is frequently formed as a tube. It might be short or long and is trumpet shaped, as in ribbon eels, *Rhinomuraena* sp. Such a structure prevents or decreases the amount of water from

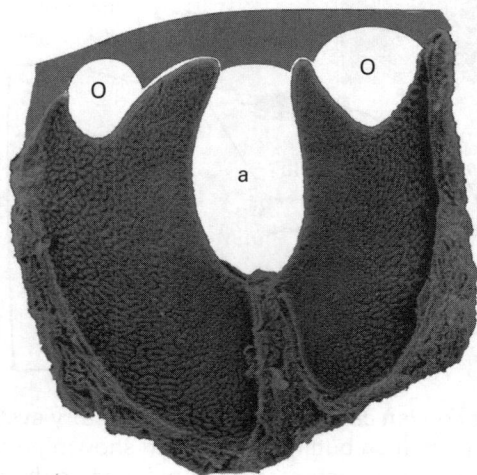

Figure 8.2.   SEM of a pair of lamellae from the olfactory organ of the cod, *Gadus morhua*. The dorsal lamella to the right is slightly smaller than the ventral one. A "top" that closes down on the organ has been drawn in red to illustrate how this organ directs water from the central atrium (a) to the peripheral outlet (o). The atrium opens to the anterior naris so that water is transported from the central region and out the posterior naris via the corridors between the lamellae. (K.B. Døving, unpublished). (*See insert for color representation of the figure.*)

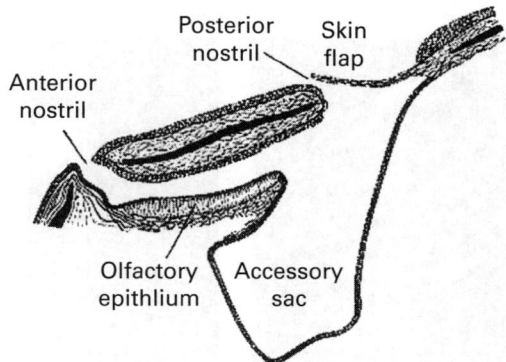

Figure 8.3. The olfactory organ of a cyprinodont. The figure illustrates the olfactory cavity with a flap at the posterior nostril that causes water to be drawn in via the anterior nostril and over the sensory epithelium when the accessory sac expands. E. Zeiske, J. Kux, R. Melinkat. Development of the olfactory organ of oviparous and viviparous cyprinodonts (Teleostei) 1. Journal of Zoological Systematics and Evolutionary Research, pp. 34–40. Copyright © 2009, John Wiley and Sons.

the boundary region of fish skin from entering the olfactory cavity—a feature that might be important as the production of mucus at the skin surface can contain substances that function as odorants. It means that the water entering the olfactory cavity will contain odorants that come from the surrounding water, not the boundary region.

The sensitivity of the olfactory organ depends on several factors. One of the most important factors is the number of OSNs that respond to a particular odorant. The total number of sensory neurons depends on the area of the sensory epithelium, and the area of olfactory epithelium increases with the size of the fish. This increase can be accomplished either by increasing the number of lamellae, or by increasing the size of the lamellae. It is interesting that the olfactory nerve layers of the olfactory bulb increases dramatically in the salmon as it starts its homeward journey (Jarrard, 1997). Is this increase in the size of the bulb an indication that the OSNs are more numerous? If so, it might also indicate that sensitivity is linked with sexual maturation and behavior.

Sensory Neurons

Three different types of sensory neurons have been described in teleosts: ciliated sensory neurons (CSNs), microvillous sensory neurons (MSNs), and crypt cells. We will now discuss the different arguments that indicate the function of these types of sensory neurons. Habitat cues and alarm substances activate CSNs that project to the medial olfactory tract (MOT). Crypt cells are implicated in the reproductive behavior of carp as they connect to the olfactory tract bundle that mediates reproductive behavior, the lateral part of the MOT. The MSNs mediate responses to food odors. The CSNs and the MSNs are easily identified in SEM (Fig. 8.5), but crypt cells are more easily identified in sections of the sensory epithelium (Hansen *et al.*, 1997). The appearance and distribution of OSNs can be seen in Figure 8.4A, B, and C.

The CSNs send longitudinal dendrites to the surface. Each dendrite protrudes slightly above the surface and forms an olfactory knob, which bears a varying number (5–10) of cilia. The MSNs have dendrites that are equipped with a large number of microvilli at the epithelial surface. Crypt cells are pear-shaped neurons located in the upper quarter of the epithelium. The conspicuous features of this type of sensory neurons are their short dendrites and that they bear microvilli and sunken cilia at the epithelial surface.

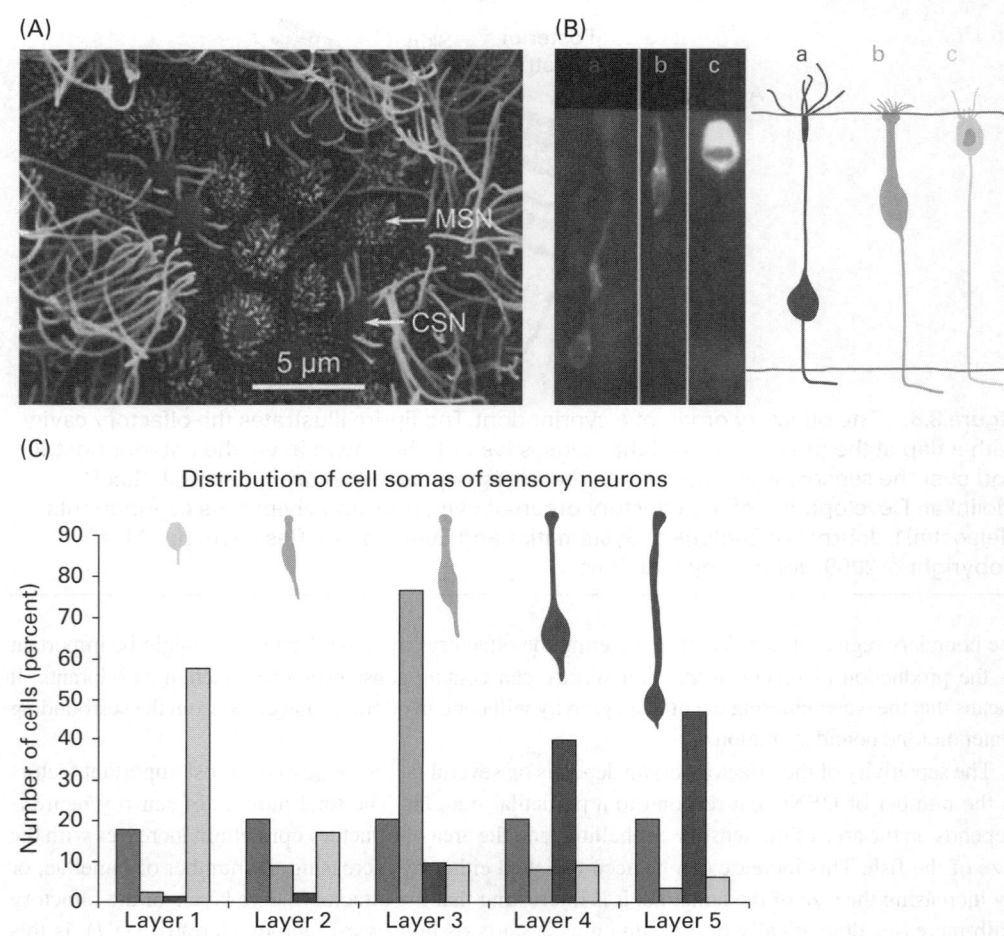

Figure 8.4. The olfactory epithelium and its sensory neurons. (A) The two types of OSNs, CSN, and MSN are seen in this SEM of the olfactory lamellae of an arctic char (*Salvelinus alpinus*). Courtesy of G. Thommesen. (B) Light microscope pictures made by DiI tracing techniques, left and cartoons right: (a) CSN; (b) MSN; (c) crypt cells. (C) Different morphologies of sensory neurons are located within different depths in the olfactory epithelium of crucian carp as revealed by retrograde staining CSNs (blue) are located near the basal lamina, MSNs (green) in the middle and crypt cells (red) at the upper most layer. The histogram demonstrates the percentage of different neurons that connect to the different parts of the olfactory tract. Red fibers connect to IMOT, green fibers to LOT, and yellow fibers to mMOT. The gray bars indicate distribution of cell somas for all cells in the epithelium. (*See insert for color representation of the figure.*)

Our understanding of OSNs increased greatly with the discovery of a large family of odorant receptors (Buck and Axel, 1991). Activated odorant receptors activate G-proteins specific to OSNs (Jones and Reed, 1989). The different G-proteins activate different downstream pathways involving adenyl cyclases or phospholipases. In catfish, it has been shown that CSNs express OR-type receptors and $G\alpha_{olf}$. Microvillous olfactory receptor neurons (ORNs) are heterogeneous, with many expressing $G\alpha_{q/11}$ whereas crypt cells express $G\alpha_{o}$. Forskolin, which interferes with $G\alpha_{olf}$/cAMP signaling, blocks responses to bile salts and markedly reduces responses to amino acids. Conversely, pharmaceuticals that interfere with $G\alpha_{q/11}$/phospholipase C signaling diminish amino acid responses but leave bile salt and nucleotide responses essentially unchanged (Hansen *et al.*, 2003).

Application of lipophilic carbocyanine dye (DiI) to the olfactory bulb revealed that the cell somas of the different categories of sensory neurons lay at different depths forming pseudostratified epithelium. For a review, see (Hamdani and Døving, 2007). CSNs have cell somas near the basal lamina; MSNs have cell somas in the middle of the epithelium; and crypt cells have cell somas near the surface. A series of neural tracing studies on the crucian carp showed a correlation between the depth of the cell soma and the projection to the olfactory bulbs and tracts. Indeed, neurons having a cell soma located near the basal lamina project to the medial part of the MOT. Moreover, CSNs project to the medial part of the olfactory bulb and connect to secondary neurons in the mMOT (Hamdani and Døving, 2002). As this part of the olfactory tract mediates alarm reactions (Hamdani *et al.*, 2000), it indicates that CSNs are activated by alarm substances. Neurons in this part of the olfactory bulb also respond to bile salts, a conclusion in line with the Thommesen's findings on salmonids (Thommesen, 1982, 1983. An anteromedial portion of the zebrafish OB is activated by bile acids (Friedrich and Korsching, 1998). Both alarm substances and bile salts are pheromonal cues involved in conspecific recognition and habitat recognition (see Sorensen and Baker, 2015; Stewart and Sorensen, 2015).

Crypt cells appear to be sensory neurons involved in detecting reproductive pheromones because they terminate on secondary neurons that make up the lateral part of the MOT (Hamdani and Døving, 2006). Ablation experiments have demonstrated that this part of the olfactory tract is involved in reproductive behavior (Weltzien *et al.*, 2003) and carries pheromonal information in carps (Kyle *et al.*, 1987; Sorensen *et al.*, 1991).

OLFACTORY NERVE

The olfactory nerve carries axons of OSNs and has been a classical preparation for studies of the conduction of nerve potentials; ahead of modern oscilloscopes (Garten, 1900) (see Fig. 8.5). There can be up to a hundred axons within one mesaxon; and although this feature would seem to favor ephapses and "cross talk" between sensory neurons, there is no evidence that this occurs. Thus, an OSN with its axon can be considered as a functional unit.

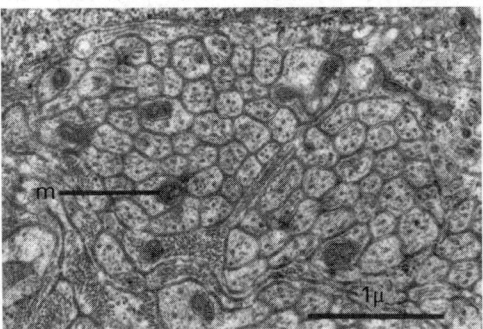

Figure 8.5. Cross section of axons in the olfactory nerve of the burbot, *Lota lota*. Electron micrograph of the olfactory nerve in burbot reveals that the axons of the OSNs have a mean diameter of 0.14 μm. Note the presence of microtubules and the enlarged diameter of the axons that contain a mitochondrion (m). Gemne, G. & Doving, K. Ultrastructural properties of primary olfactory neurons in fish (Lota lota L.), American Journal of Anatomy 126, 457–475. Copyright © John Wiley & Sons, 1969.

8.2.2 Olfactory Bulb

CELLULAR LAYERS AND ORGANIZATION

The olfactory bulb in teleost fish is a nearly spherical structure with a series of layers that receive and process olfactory information including that associated with pheromones. A sagittal section is shown in Figure 8.6. The unmyelinated fibers of the olfactory nerve are reorganized in the interface between the olfactory nerve and the olfactory bulb. The zone with glomeruli is the place for axodendritic synapses between the OSNs and the secondary neurons (mitral cells) of the olfactory bulb. The plexiform layer contains cell somas of the mitral cells. The granule cell layer contains a high density of neurons. Ependymal cells are seen lining the lateral ventricle. In the olfactory bulb, the axons of the sensory neurons terminate in specialized synaptic structures called glomeruli and make contact with secondary neurons.

The glomeruli are the main synaptic structure of the olfactory bulb; and the size of a glomerulus is thought to reflect the sensitivity to specific odorants handled by these structures. There are different cell types in the olfactory bulb with intricate synaptic connections. Very little is understood about how pheromonal information is processed, but there are strong indications that special cell types located in restricted regions of the bulb serve this function. The olfactory bulb in fishes is seemingly less organized than the mammalian counterpart, but such an interpretation may be misleading as every brain is adapted to serve the organism in the habitat in which it resides.

Bulbar neurons. The mitral cells or relay neurons described by Golgi, Ramon y Cajal, and van Gehuchten in mammals are also present in the olfactory bulb of teleosts (Golgi, 1875; van Gehuchten and Martin, 1891; Cajal, 1911). Relay neurons in teleost fishes have several forms. Examples of mitral cells are shown in Figure 8.6B. Some of these cells have only one dendrite with a dendritic arborization that occurs within the glomerulus (Fig. 8.6B(a)). Other mitral cells have several dendrites, but these enter the same glomerulus (Fig. 8.6B(b)). There are also cells that have several dendrites that terminate in different glomeruli (Fig. 8.6B(c)). A cell type seemingly unnoticed by later authors are the *ependymal mitral cells* described by Holmgren (1920). They seldom stain by the Golgi method; their somas are found at the bulbar surface and have dendrites extending into several glomeruli. Two types of mitral cells have been demonstrated in the teleost olfactory bulb (Alonso *et al.*, 1988); *Type I cells* dominate in the medial portion of the olfactory bulb. The axons arise in general from one of the dendrites close to the cell soma. *Type II cells* dominate in the lateral portion and show one to four dendrites that arborize in one dendritic field. These cells are similar to mammalian mitral cells; however, their functions are untested.

Ruffed cells were discovered by Kosaka and Hama (1979, 1981), in studies of the goldfish olfactory bulb. The ruffed cell has dendritic arborization, but seemingly lacks synaptic contacts to OSNs. The initial portion of the ruffed cell has many protrusions, which make synaptic contacts with a large number of granule cells. It is this initial portion that looks like the ruff on old-fashioned arm garments that give the cells their name. This portion, which is about 20–50 μm in diameter, has been named a nest. The function of these neurons is still unknown, but in contrast to what was assumed earlier, ruffed cells also relay neurons projecting via the olfactory tract. *Granule cells* have been defined by Kosaka and Hama (1979), as "neurons that make reciprocal synapses with the mitral cells." Such reciprocal synapses are identified in the electron microscopic slides and found most frequently in the glomerular and the nest areas. The *perinest cells* lie around the nest, and their dendrites in the nest make "symmetrical synapses" with granule cells. *Mixed-synapse cells* lie in the glomerular region and make "mixed synapses" with mitral cells, that is, mixed synapse, asymmetrical synapse, and gap junctions with mitral cells. Synaptic contacts between the different types of neurons have been schematized in Figure 8.6C in which the asymmetric synapses (excitatory) have been drawn in red and the symmetrical synapses (inhibitory) in blue. Note that the ruffed cells do not receive synaptic input from the

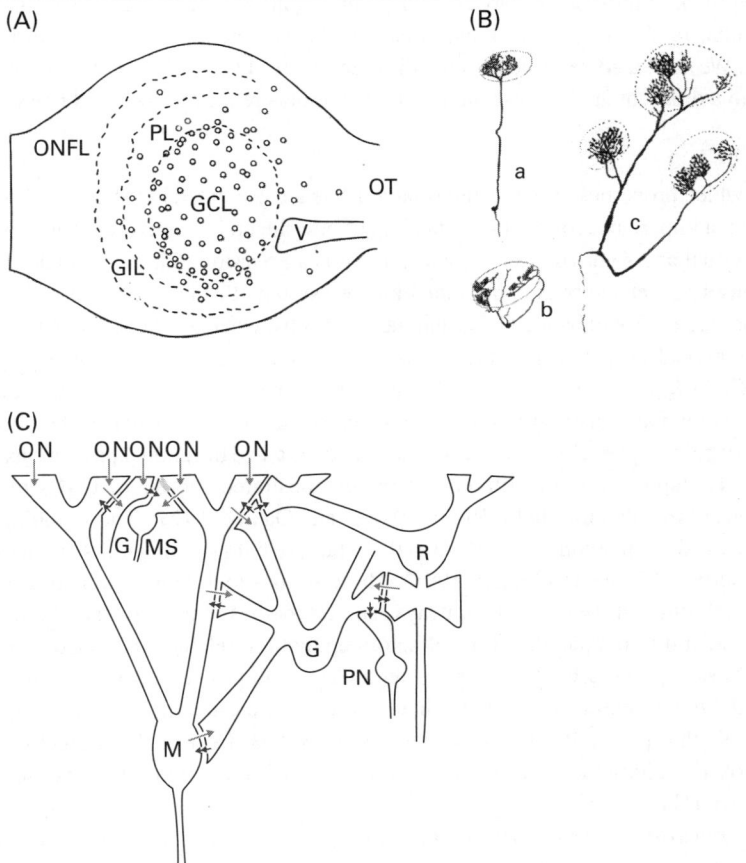

Figure 8.6.   The fish olfactory bulb and its neurons. (A) Drawing of a sagittal section of the olfactory bulb of the Mediterranean barbel *Brissus meridionalis* demonstrating the layers. The abbreviations used in this figure are GCL, granule cell layer; GIL, glomerular layer; ONFL, olfactory nerve fiber layer; OT olfactory tract; PL, plexiform layer; V, ventricle. Alonso, J.R., Lara, J., Covenas, R., Aijon, J. (1988) Two types of mitral cells in the teleostean olfactory bulb. Neuroscience research communication 3, 113–118. (B) Different types of secondary neurons in the European smelt, *Osmerus eperalanus* (a) secondary neuron with a single dendrite to glomerulus. (b) secondary neuron with several dendrites to one glomerulus. (c) secondary neuron with dendrites to several glomeruli. Holmgren, N. (1920) Zur Anatomie und Histologie des Vorder- und Zwischenhirns der Knochenfische. Acta zoologica (Stockholm). Stockholm 1, 137–315. Copyright ©The Royal Swedish Academy of Sciences, 1920. (C) The putative arrangement of the synaptic connections of the bulbar neurons in the fish. The abbreviations used in this figure are G, granule cells; M, mitral cells/mixed-synapse cell; N, nest; ON, olfactory nerve terminal; PN, perinest cell; R, ruffed cell. The asymmetric synapses (excitatory) have been colored red, and the symmetrical synapses (inhibitory) blue. Kosaka, T., Hama, K. (1982) Synaptic organization in the teleost olfactory bulb. J Physiol (Paris) 78, 707–719. (*See insert for color representation of the figure.*)

OSNs, but they make inhibitory synapses with granule cells via their gemmules (for review, see Kosaka and Hama, 1982). The interaction between the different neurons in the olfactory bulb will be discussed later. We shall also describe how the different types of OSNs project to different regions of the bulb to form a differentiated representation of the odorants related to food and pheromones.

Chemotopy

The means by which properties of the stimulus parameters are presented in the central nervous system are of fundamental importance for our understanding of how the brain works and pheromones are deciphered. In the visual and somatosensory systems, there is a point-to-point topographic representation between the sensory sheet and brain; but in the olfactory system, this is not the case. As we shall see, pheromonal cues are mediated by the CSNs and the crypt cells, whereas MSNs mediate food odors.

Labeling the axonal projections of the OSNs in a discrete region of the sensory epithelium in rainbow trout *Oncorhynchus mykiss* revealed staining of a small nerve bundle in the olfactory nerve. The abrupt resorting and redistribution of these axons at the interface between the olfactory nerve and the olfactory bulb implies that local cues control and organize the axonal projections (Riddle and Oakley, 1991). The application of different lectins demonstrated subsets of OSNs that project to restricted regions of the olfactory bulb (Riddle, Wong, and Oakley, 1993). These findings were later extended by the local application of dye (DiI) in the olfactory bulb staining OSNs with discrete morphological properties (Morita and Finger, 1998). These studies highlight the functional subdivision of the olfactory system in three parallel pathways. The cell bodies of these different types of sensory neurons are found at different depths of the olfactory epithelium (Hamdani and Døving, 2007).

In catfish, CSNs express OR-type receptors with $G\alpha_{olf/s}$ and project mostly to ventrally situated glomeruli. V2R-type receptor probes hybridize to intermediate-height receptor cells, which are apparently MSNs that project to the dorsal region of the OB. Crypt cells project to two unique glomerular territories along the ventral midline of the OB; the receptor molecules used by this cell type are unknown (Hansen *et al.*, 2003).

Prominent endocytosis in crucian carp is odorant-specific, which is to say that internalization of membranes at the peripheral part of sensory neurons is related to activation of odorant receptors. Bile salt taurolithocholate (a putative social pheromone) and the alarm agonist hypoxanthine-3-*N*-oxide stained two sets of CSNs with long dendrites (Fig. 8.7) (Døving *et al.*, 2011). A study on the olfactory organ of channel catfish, *Ictalurus punctatus*, showed that desensitization involved phosphorylation of the receptors and internalization of the ligand-bound, phosphorylated receptors by a clathrin-mediated endocytic pathway (Rankin *et al.*, 1999). It will be interesting to extend these functions to address responsiveness to pheromones and food odors in the future and how they might influence odor recognition.

Changes in Bulb Structure During Life History

As mentioned in Section 8.2.2, the volume of the outer nervous layers indicates the number of OSNs; thus, volume changes during the life history of a species probably reflect adaptive adjustments. Salmonid fishes make impressive migrations by tracking olfactory cues, which in some cases appear to be partially mediated by juvenile pheromones (Wisby and Hasler, 1954; Døving, Westerberg, and Johnsen, 1985; Stabell, 1992). In a study of chinook salmon, *O. tshawytscha*, morphological differences of the olfactory bulb were studied in different age groups and compared with telencephalon volume at each stage (Jarrard, 1997). Laminar organization and relative bulbar volumes, olfactory nerve and glomeruli layer (ONL-GL), and inner cell layer (ICL) were compared. The author labeled the primary olfactory axons "KLH-like immunoreactivity" (Riddle and Oakley, 1992). This made it possible to stain and calculate the volume of the different layers and relate it to the total volume of the bulb. The increase in volume of the various parts is shown in Figure 8.8 and demonstrates a dramatic increase in the nerve glomerular region, reflecting the increasing number of OSNs. We do

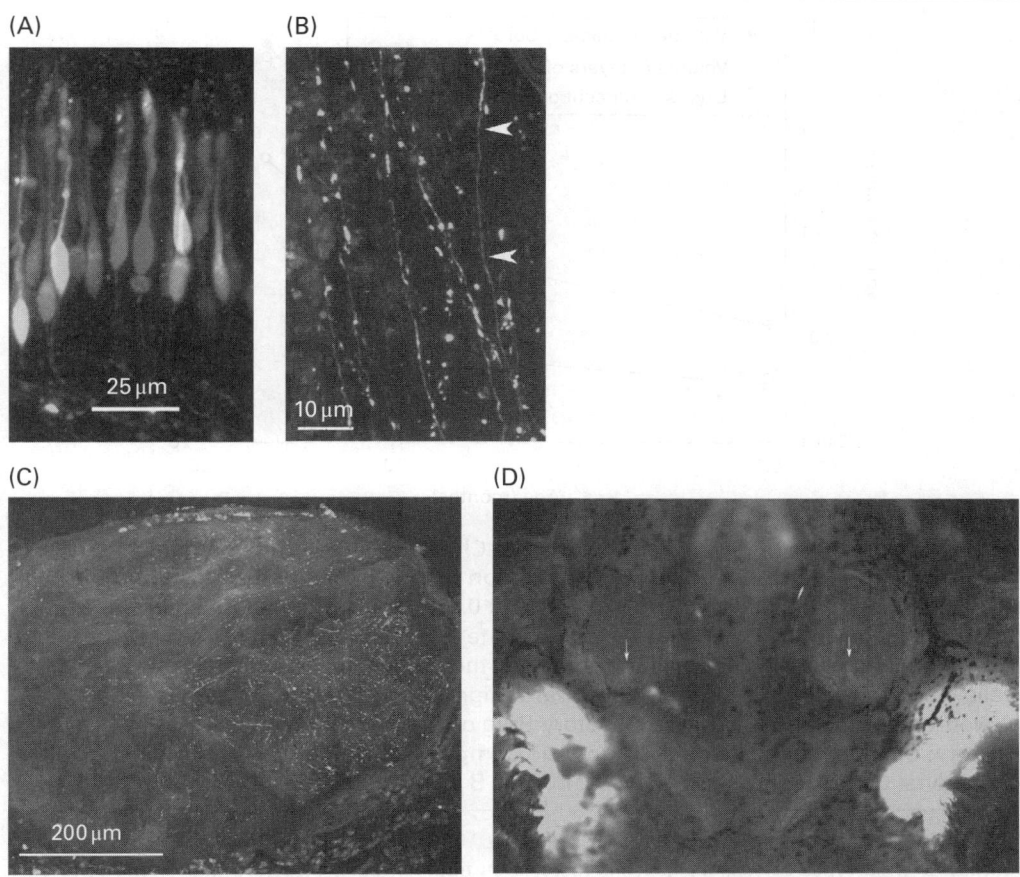

Figure 8.7. Olfactory sensory neurons visualized by endocytosis. (A) Exposing the olfactory organ of crucian carp with a bile salt taurolithocholate 10 nm together with dextran conjugated with Alexa 488 10 μm (green), thereafter a mixture of the alarm agonist hypoxanthine-3-*N*-oxide and dextran conjugated with Alexa 647 10 μm (red) caused staining of two sets of CSNs with long dendrites. The photo is a z-projection of 35 confocal images, covering a depth of 35 μm. Note the axons leaving the CSNs. Courtesy by Kenth-Arne Hansson (2011). (B) Exposing the olfactory organ of crucian carp to the bile salt taurolithocholate caused the dye to stain a subset of axons. In the olfactory nerve, the stained axons are seen to run in parallel with the long axis of the nerve. The stain is frequently seen as spots in the nerve, but regions with smooth staining are also observed (arrows). (C) Transverse section at the interface nerve and bulb. As the axons approach the interface at the anterior portion of the olfactory bulb, the axons are seen to run normal to the long axis of the nerve and aggregate in the dorsomedial portion (arrows). From Døving *et al.* (2011). (D) The CSNs responding to the bile salt taurolithocholate have been stained by ligand-selective endocytosis and surviving for 2 days after the dye has been transported to the olfactory bulb. The medial projection is seen as symmetrical point (arrows) in the olfactory bulb. From Døving *et al.* (2011). (*See insert for color representation of the figure.*)

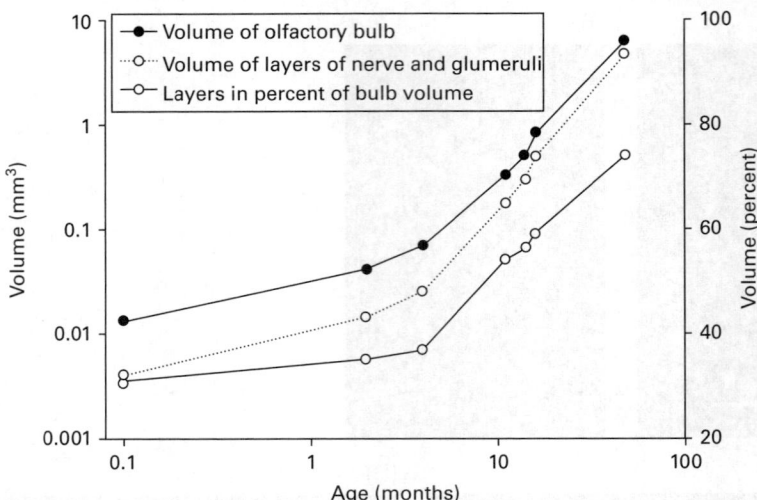

Figure 8.8. Growth of olfactory bulb layers in Chinook Salmon. The diagram shows the volume of the ONL-GL region and its relation to the volume of the olfactory bulb. Note that there are small changes in volume at 0.1, 2, and 4 months (juvenile stage, parr), but large increases at 11, 14, and 16 months (late juvenile development, smolt). The large volume at 48 months (adult) reflects the dramatic increased number of sensory cells when the salmon start their homeward migration. Data from Jarrard, H. (1997) A Neuroanatomical Study of the Olfactory Bulb of the Pacific Salmon Across Its Life History: Changes in the Relative Volumetric Composition of the Bulb and Their Possible Functional Consequences. DISS. ABST. INT. PT. B SCI. and ENG 58, 2885.

not know what types of sensory neurons increase in number, but we believe that orienting to cues from the natal river requires a high sensitivity toward certain odorants, probably of bile salt in character. It has been suggested that the imprinting needed to find the home river lies in the memory of the OSNs (Nevitt *et al.*, 1994). Such an imprinting mechanism is difficult to imagine at the time the salmon start its homeward migration in the ocean.

The pheromone hypothesis for migration of salmonids was proposed by Nordeng (1971), and additional behavior experiments support his hypothesis (Nordeng and Bratland, 2006; Nordeng, 2009). This subject will be discussed further under Section 8.4.2 as well as in Chapter 2 (Sorensen and Baker, 2015).

OLFACTORY TRACT

The connection between the olfactory bulb and telencephalon is called the olfactory tract. In many fish orders, such as cod (Gadiformes), carps (Cypriniformes), and catfishes (Siluriformes), the bulb is situated close to the sensory epithelium and the olfactory tract is long. Thus, the central nervous system becomes easily accessible to manipulation (Fig. 8.1). There are two features of the olfactory tract in fishes that can be highlighted. First, the tract is divided into bundles, and the fiber composition of the axons within different bundles is dramatically different. These findings indicate that the axons of the bundles have different functional properties. Second, the bundles enter different regions of the telencephalon, suggesting that they induce different behaviors. As we shall see in the sections that follow, this is indeed the case. Both discrete electrical stimulation of the tact bundles and ablation of the bundles demonstrate a functional division. The nomenclature used by neuroanatomists in the past century (Herrick, 1901; Sheldon, 1912), concerning the divisions of the olfactory tract, has been retained in later studies. In Figure 8.9, the different bundles have been identified as the medial part of

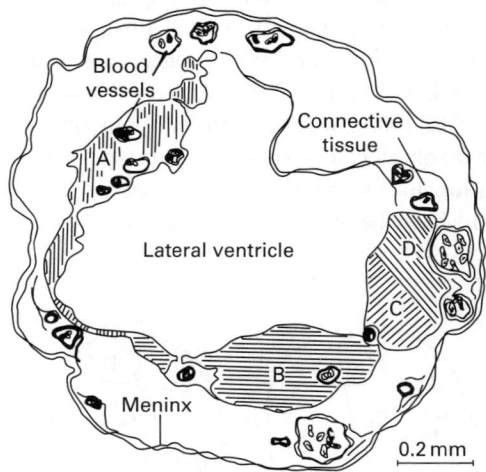

Figure 8.9. Cross section of the olfactory tract of burbot. Drawing of the olfactory tract in burbot from a cross section made close to the olfactory bulb. A, mMOT; B, IMOT; C, mLOT; and D, ILOT. Døving, K.B., Gemne, G. (1965) Electrophysiological and histological properties of the olfactory tract of the burbot (Lota lota L.). J. Neurophysiol. 28, 139–153. The American Physiological Society.

the mMOT (A); the lateral part of the medial olfactory tract lMOT (B); the medial part of the lateral olfactory tract, mLOT (C); and the lateral part of the lateral olfactory tract, lLOT (D).

Functional division of tract bundles. Studies of cod and carp fishes have shown that the LOT mediates feeding behavior. The two parts of the MOT mediate behaviors related to pheromones. The lMOT concerns behavior related to reproduction and the mMOT mediates alarm and habitat cues. Thus, a large part of the olfactory system is devoted to the processing of pheromone cues.

Nervus terminalis. In the ventral/caudal part of the olfactory bulb in fishes, there is a population of large ganglion cells with anterior processes into the olfactory nerve toward the olfactory sensory epithelium and caudal processes in the olfactory tract. These fibers form the terminal nerve, first described in 1878 (von Fritsch, 1878). The anatomy of the terminal nerve was reviewed by Holmgren (1918; for reviews, see Meek and Niewenhuys, 1998; Kawai, Oka, and Eisthen, 2009). The fibers of the *N. terminalis* project to the olfactory epithelium and bulb, but also to the retinal brain regions that control reproductive behavior. The fibers of *N. terminalis* can also form interbulbar connections. The function of this nerve is uncertain, but initial speculation that it is chemosensory has not been supported by electrophysiological recording (Fujita *et al.*, 1991). Possible mechanisms underlying the neuromodulatory functions that increase olfactory sensitivity are discussed in Kawai, Oka, and Eisthen (2009).

8.2.3 Telencephalic Projections

The extent of olfactory input to the teleost telencephalon varies greatly among species (Northcutt and Davis, 1983; Nieuwenhuys, 2009). In *Ictalurus* (von Bartheld *et al.*, 1984), *Salmo* (Northcutt and Davis, 1983), and *Gadus* (Rooney *et al.*, 1992), the olfactory terminal areas are extensive. In the weakly electric fish, *Gnathonemus petersii* (Rooney *et al.*, 1989), the olfactory terminal areas are relatively small. These differences reflect the relative importance of the olfactory sense in these species. The secondary olfactory connections of the teleosts reach the telencephalon via the medial and lateral olfactory tracts to restricted terminal fields in the ventral (areas: Vv, Vd, Vs, Vi, and AC) and dorsal (areas: Dlv, Dc, and Dp) telencephalon, whereas the dorsomedial and large parts of the dorsocentral telencephalon do not receive direct olfactory input.

It is interesting from an evolutionary viewpoint that phylogenetically disparate species such as cod (Rooney *et al.*, 1992) and goldfish (Sheldon, 1912; Oka, 1980; von Bartheld *et al.*, 1984; Levine and Dethier, 1985) show many similarities in behavioral patterns and organization of the olfactory connections in the brain. In the present context, we have decided to illustrate the projection patterns to the telencephalon in the goldfish (Oka, 1980). Because stimulation of the olfactory system, either by odorants or electrical pulses applied to the olfactory tract, causes different types of behavior, it is not surprising that different bundles of the olfactory tract terminate in different regions. The medial, lateral, and posterior terminal fields are generally found in the ventral region of the telencephalon (Fig. 8.10, red dots).

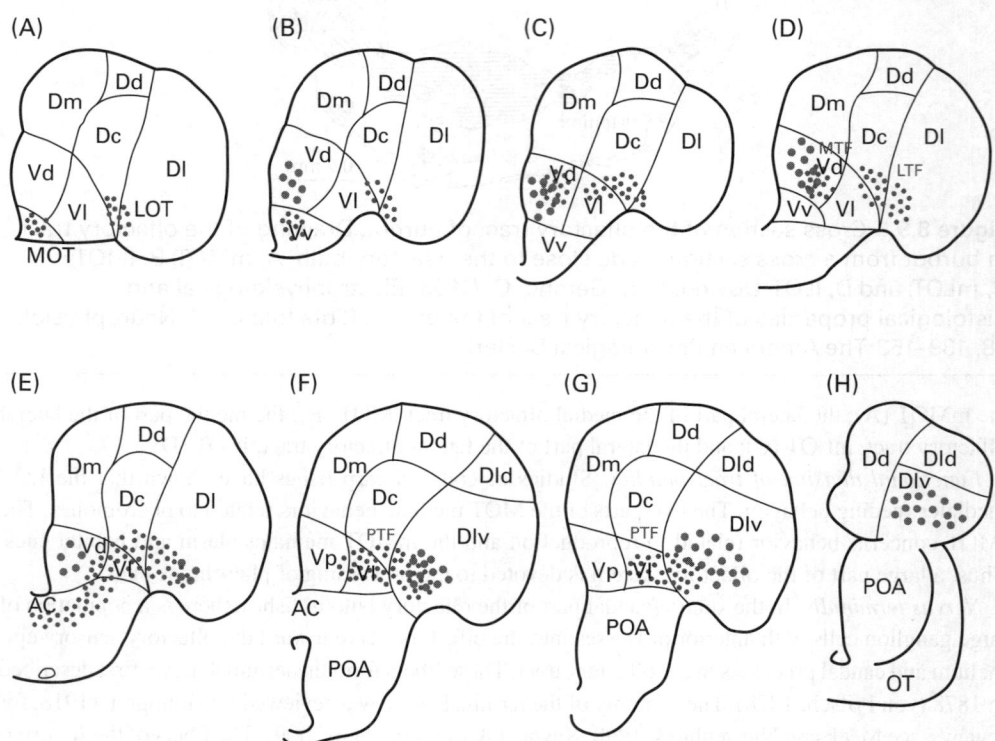

Figure 8.10. Telencephalic regions in goldfish connecting to the olfactory bulb. (A)–(H) are line drawings of frontal sections through the telencephalic hemisphere, illustrating the distribution of labeled neurons (blue dots), following primuline injections of the olfactory bulb. The sections are 150 μm apart. A, shows the area about 750 μm from the rostral end. The red dots indicate the terminal regions of degenerating axons 10 days after transection of the olfactory tract on the ipsilateral side. This figure was published in Brain Research, vol. 185, issue 2, Yoshitaka Oka, The origin of the centrifugal fibers to the olfactory bulb in the goldfish Carassius auratus. An experimental study using the fluorescent dye primuline as a retrograde tracer, pp. 215-225. Copyright © Elsevier, 1980. The abbreviations used in this figure are AC, anterior commissure; Dc, central part of area dorsalis; Dd, dorsal part of area dorsalis; Dld and Dlv, dorsal and ventral part of the lateral area dorsalis; Dm, medial part of area dorsalis; Dp, posterior part of area dorsalis; LOT, lateral olfactory tract; LTF, lateral terminal field; MTF, medial terminal field; OT, optic tract; POA, preoptic area; PTF, posterior terminal field; Vd, dorsal part of area ventralis; Vl, lateral part of area ventralis; Vp, posterior part of area ventralis; and Vv, ventral part of area ventralis. (*See insert for color representation of the figure.*)

There is a dramatic difference in the appearance of the telencephalic neurons in the terminal fields of the LOT and MOT (Ito, 1973). The neurons in the area dorsalis (D, LTF) receiving input from the LOT, have multipolar cell bodies, which give off dendrites in various directions with dense populations of dendritic spines. By contrast, neurons in the area ventralis (V, MTF) receiving input from the MOT, consist of small cells that give off a few short spine-poor or spine-free dendrites. As will be discussed in the subsequent sections, LOT mediates feeding behavior; whereas the MOT mediates social behaviors such as reproduction, alarm, and homing. As dendrite spinae have been implicated in learning and memory (Alvarez and Sabatini, 2007; Kasai *et al.*, 2010), it is interesting that the LTF receives input from the LOT that mediates feeding behavior (Døving and Selset, 1980). These findings suggest that the nervous substrate for the innate social behaviors seems fixed and remains unchanged throughout life, while feeding requires more flexible adjustments to new stimuli and to ontogenetic changes in diet. These observations are in congruence with those of the mouse olfactory system that has hard-wired neural circuits in parallel with "acquired adaptive" circuits (Kobayakawa *et al.*, 2007).

In general, the olfactory projections to the contralateral telencephalic hemisphere in teleosts are not as extensive as those to the ipsilateral side. The main projection pathway to the contralateral hemisphere is via the anterior commissure (AC) and fibers from both MOT subtracts (mMOT and lMOT) and from the LOT cross the midline. Ariëns Kappers (1906) identified fibers from both MOTs crossing the AC but reported the LOT to project only ipsilaterally within the telencephalon of the cod. He did, however, report LOT fibers crossing the AC in *Salmo salar.* In all but one of the species studied with modern methods, the LOT has been shown to cross the AC (Finger, 1975; Oka, 1980; Bass, 1981; Davis *et al.*, 1981; Ebbesson, Meyer, and Scheich, 1981; Northcutt and Davis, 1983; Prasada Rao and Finger, 1984; von Bartheld *et al.*, 1984; Levine and Dethier, 1985; Rooney *et al.*, 1989). A notable exception is the moray eel (*Gymnothorax funebris*) in which only the MOT is reported to cross the AC (Scalia and Ebbesson, 1971).

8.3 PHYSIOLOGY

8.3.1 Sensory Cells

ELECTRO-OLFACTOGRAM

In his doctoral thesis from 1956, David Ottoson described for the first time the properties of the receptor potential of the olfactory organ (in frogs) evoked by exposing the sensory epithelium to odorants (Ottoson, 1956). Because of the properties of the observed electrical activity, he named the slow response the electro-olfactogram (EOG). The wealth of information in his thesis is well worth a closer study for those interested in how to present new findings. The first recording of EOG from an aquatic vertebrate was performed on hagfish *Myxine glutinosa* (Døving and Holmberg, 1974). The olfactory lamellae of the olfactory organ extend from the dorsal roof of the cavity right in front of the brain and opens to the anterior nostril (Fig. 8.11A and B). Application of odorants evokes an EOG that is similar in form to that found in mammals (Ottoson, 1959c). The first EOG recordings from a teleost were made in 1976 (Silver *et al.*, 1976).

The EOG has found many applications; it can be used to monitor the activation of the olfactory neurons so that one can compare the peripheral input with the responses in other parts of the olfactory system (Døving, 1964). It can also be used to study the sensitivity of the olfactory organs to different odorants (Huertas *et al.*, 2010). Thommesen used the EOG to monitor the sensitivity to different odorants in various positions of the olfactory epithelium of rat (Thommesen and Døving, 1977) and fish (Thommesen, 1982, 1983). Stacey has made an extensive evaluation of the evolution of cichlids by studying the EOG responses to a large number of putative steroidal pheromones (Stacey, 2009). See also Chapter 3.

In a series of experiments, Thommesen studied the relative amplitudes of the EOGs to an amino acid and to char bile (Thommesen, 1982, 1983). Simultaneous recordings from a central and a

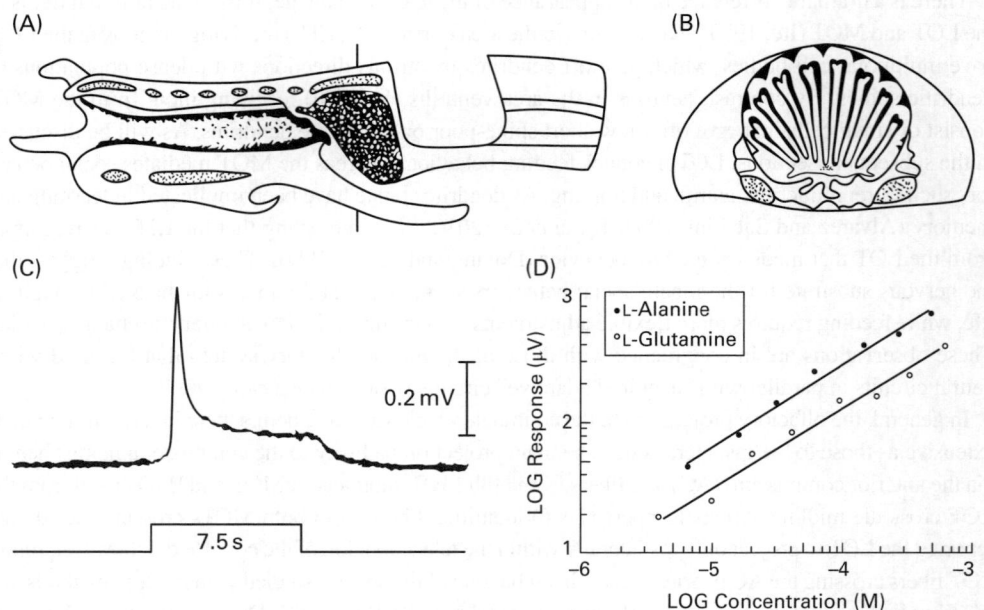

Figure 8.11. The hagfish olfactory organ and EOG. (A) The hagfish head with the brain removed. (B) The vertical line indicates the plane of cross section through the olfactory organ. Note the lamellae hanging down from the roof of the olfactory cavity. (C) An EOG from the hagfish preparation. Note the high amplitude of the initial peak. (D) Dose–response curves for two amino acids; the peak amplitude for different concentrations of L-glutamine and L-alanine. Each point is the mean of four experiments. Kjell B. Doving, Kaj Holmberg. A Note on the Function of the Olfactory Organ of the Hagfish Myxine glutinosa. Acta Physiologica, pp. 430-432. Copyright © 1974 Scandinavian Physiological Society.

peripheral electrode position demonstrated the highest amplitude for the bile at the central electrode, and the highest amplitude for the amino acid at the peripheral position (Fig. 8.12A). There was a larger response to char bile at the peripheral regions of the olfactory rosette than amino acid. The densities of CSNs were found to be highest in the peripheral parts. Thus, Thommesen proposed that the CSNs responded to bile, and the MSNs responded to amino acids. These findings were recently confirmed by the use of the ligand-specific staining of a set of CSNs by a bile salt (Døving, 2010).

Cross-adaptation studies have demonstrated that receptors sensitive to bile and L-methionine are functionally independent (Fig. 8.12B and C). Cross adaptation has also been used to demonstrate that different hormonal sex pheromones (androstenedione, 17α-hydroxy-4-pregnen-3-one and 17α,20β-dihydroxy-4-pregnen-3-one, and 17α,20β-dihydroxy-4-pregnen-3-one,) are detected by different receptors in the goldfish (Sorensen *et al.*, 1990, 1995). Eventually, these studies demonstrated that these steroids drive different biological responses. Concurrent observations of the lamellae surfaces by SEM have further shown that both types of olfactory neurons tuned to the two stimuli may be found on all lamellae, but are differentially distributed within each lamella.

Sexual dimorphism in olfactory signaling has, also, been a target of several EOG investigations. For instance, several studies reported dimorphic EOG responsiveness to different hormonal pheromones in fish (Sorensen and Goetz, 1993; Cardwell *et al.*, 1995; Stacey *et al.*, 2003). What makes males and females behave differently to pheromones is interesting from a functional viewpoint, but the neuronal mechanisms involved remain largely unknown.

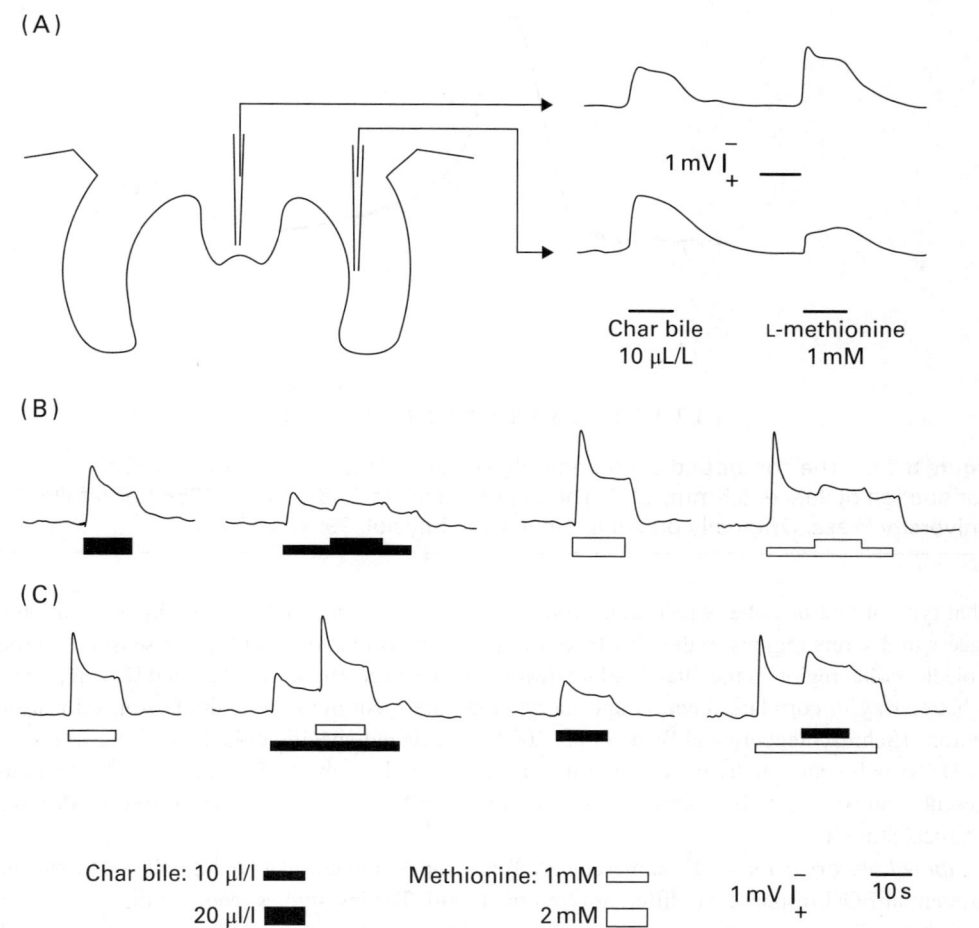

Figure 8.12. EOG recordings from the char olfactory organ. (A) Recordings from a central and a peripheral placed electrode in the olfactory organ of char. Note that the EOG amplitude was highest for bile in the central electrode position and highest for the amino acid in the peripheral electrode position (see text). (B) Cross-adaptation studies showing changes of the EOG by doubling the stimulus concentration, in each case preceded by the EOG elicited by the highest concentration. (C) Cross-adaptation studies showing the effect of adding L-methionine to a stimulation with char bile and *vice versa*, each preceded by the single EOG elicited by the added constituent. Georg Thommesen. Specificity and distribution of receptor cells in the olfactory mucosa of char (Salmo alpinus L.). Acta Physiologica, pp. 47–56. Copyright © 1982, Scandinavian Physiological Society.

FUNCTIONAL PROPERTIES OF THE OLFACTORY NERVE

The homogeneity of the axon diameters in the pike olfactory nerve is reflected in the compound action potential made by Gasser (1956) (Fig. 8.13). The peak conduction velocity is $0.06 \, \text{m} \cdot \text{s}^{-1}$.

SPECIFICITY

Several methods have been applied to describe the functional properties of the three different types of OSNs. The small size of these sensory neurons makes it difficult to record spike activity or Ca-imaging from single units; and by such *in situ* techniques, there are meager possibilities to find

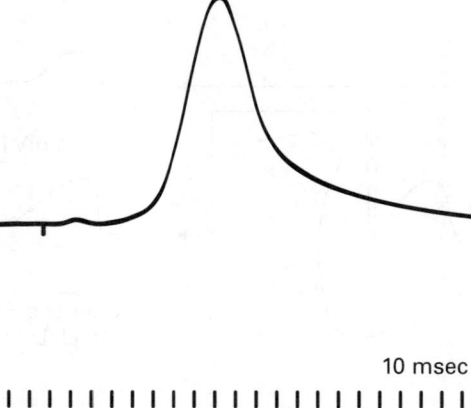

10 msec

I I I I I I I I I I I I I I I I I I I I I I I I I I I

Figure 8.13.   The compound action potential of the olfactory nerve of the pike. Conduction distance, 5.5mm, 21°C. The peak velocity is 0.06 m·s⁻¹. © 1956 Rockefeller University Press. Originally published in J. Gen. Physiol. 39: 473-496.

what type of neurons one is recording from. The indirect tracing studies made by application of tracers to discrete regions in the olfactory bulb give a gross idea of what type of sensory neurons project to what region in the olfactory bulb (Morita and Finger, 1998; Hamdani and Døving, 2007). A better way to correlate function and morphology are recordings from single isolated sensory neurons (Schmachtenberg and Bacigalupo, 2004). The ligand-specific induction of endocytosis in the OSNs will undoubtedly increase our understanding of this subject (Døving *et al.*, 2011). In the present context, we will describe properties of the different types of OSNs toward different chemical stimuli.

Ciliated sensory neurons. Thommesen (1983) was the first to correlate the density of CSNs with differential EOG responses to different types of stimuli. Tracing studies made on the crucian carp suggest that CSNs respond to bile salts and alarm substances (Hamdani and Døving, 2002). A study on ligand-specific endocytosis in OSNs showed that only a small population (<3%) of CSNs were stained by application of the bile salt taurolithocholate at 10nm. Application of an alarm agonist caused staining of another type of CSN (Døving *et al.*, 2011). A patch-clamp study on CSNs from a Cabinza grunt (*Isacia conceptionis*) showed a response to high concentration of amino acids, but bile salt was not tried (Schmachtenberg and Bacigalupo, 2004), whereas CSNs of rainbow trout responded to urine that contained bile salts (Sato and Suzuki, 2001). Channel catfish CSNs respond to bile acids (Hansen *et al.*, 2003).

Microvillous sensory neurons. Thommesen (1983) suggested that MSNs responded to amino acids. Tracing studies made on crucian carp suggest that MSNs respond to food odorants (Hamdani, Alexander, and Døving, 2001a), as food search failed to appear when the LOT was cut (Hamdani, Kasumyan, and Døving, 2001b). Direct evidence that MSNs respond to amino acids was achieved from preparations of zebrafish (Lipschitz and Michel, 2002) and rainbow trout (Sato and Suzuki, 2000, 2001). Although MSNs also respond to a mixture of amino acids, there is not yet any evidence that ciliated and MSNs respond to the putative sex pheromones (Sato and Suzuki, 2001). Data from genetic, neuroanatomical, and behavioral methods show that MSNs cause a feeding response to amino acids in zebrafish (Koide *et al.*, 2009).

Crypt cells. These cells have been implicated in reproductive behavior because they respond to sex pheromones and because the axons of these neurons project to the lMOT (Hamdani and Døving,

2006). This part of the olfactory tract mediates reproductive behavior (Weltzien *et al.*, 2003). Seasonal variation of the abundance of crypt cells in crucian carp coincides with the breeding season, implicating that these sensory neurons mediate reproductive behavior (Hamdani *et al.*, 2008). One out of 11 crypt cells responded to the combination of a postovulatory pheromone odorant Prostaglandin $F_{2\alpha}$ ($PGF_{2\alpha}$) and preovulatory phermonal odorant 17α,20β-dihydroxy-4-pregnen-3-one (17,20P), whereas 17 out of 41 responded to a mixture of eight amino acids (Vielma *et al.*, 2008).

8.3.2 Olfactory Bulb

Surface Recordings

Electrical activity evoked by stimulation of the sensory epithelium is transmitted to the olfactory bulb as action potentials in the unmyelinated axons of the sensory neurons. This activity, in turn, induces synaptic activity and nerve impulses in secondary neurons. This electrical activity can be visualized by DC recordings with electrodes placed on the surface of the olfactory bulb (Ottoson, 1959b). Electrical stimulation of the olfactory nerve of the frog evoked two potentials: one that was diminished (abolished) by antidromic pulses to the olfactory bulb and the other that was unaffected by this procedure (Ottoson, 1959b).

Differential surface recordings of the olfactory bulb of char show that amino acids stimulate neurons that project to the lateral part of the olfactory bulb and that neurons responding to bile salts project to the medial part of the olfactory bulb (Fig. 8.14) (Døving, Selset, and Thommesen, 1980).

Spontaneous Activity and Specificity

The secondary neurons in the olfactory bulb receive synaptic input from thousands of OSNs. The excitatory transmitter substance glutamate mediates synaptic activity (Murphy *et al.*, 2004). Recordings from single units in the teleost olfactory bulb have revealed two types of spike activity (Zippel *et al.*, 2000; Hamdani and Døving, 2003). Type I units are characterized by a diphasic action potential (AP) of short duration (rise time ~1 ms). Type II units displayed an AP with long duration (rise time ~1.8 ms) (Fig. 8.16). The AP of this latter unit was nearly always followed by a slow potential (SP), characteristic diphasic wave with a rise time of approximately 5 ms. The delay between AP and SP varied between 8 and 8.5 ms. Both types of units were activated by electrical stimulation of the mMOT, and the conduction velocities calculated for different APs recorded from type II units varied between 0.34 and 0.55 m·s⁻¹ (n=6) at room temperature. The appearance of the AP and the following SP indicated a particular type of unit. It has been proposed that these units represent the activity of the so-called ruffed cells (Zippel, Reschke, and Korff, 1999). The Type I units correspond presumably to mitral cells. However, because of the lack of histological identification of the units recorded, we have chosen to categorize these units as Type I and Type II.

Adaptation. As demonstrated by Ottoson (1959a), the responses of the olfactory bulb adapt very slowly. This feature was confirmed by recording the nerve action potentials from single fibers of the olfactory tract in cod. Stimulation of the olfactory epithelium revealed a steady discharge of nerve impulses to a continuous stimulation for 30 min (Fig. 8.15).

There are no overlap/cross reactions between odors with very different messages, such as food odors and sex pheromones. However, studies have shown differences in discriminatory capacity between odorants within the same class. Single-unit recording in male crucian carp showed excellent discrimination between sex pheromones released from females (Fig. 8.16B and C) (Lastein, Hamdani, and Døving, 2006). This reflects the importance for males to recognize the reproductive state of females, which is necessary to facilitate spawning. By contrast, skin odors mediating fright reactions are not that well distinguished (Lastein, Hamdani, and Døving, 2008a). When exposing fish to skin extracts from different species, the unitary discriminatory capacity varied. This might reflect that alarm

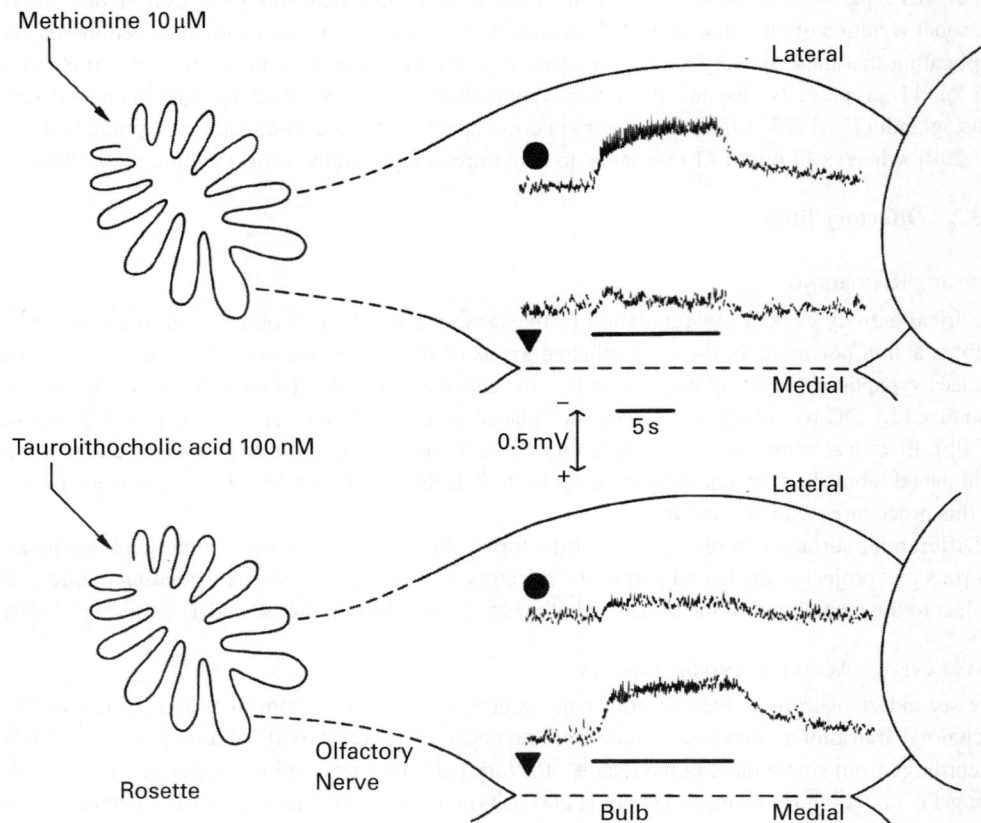

Figure 8.14. Recordings from the olfactory bulb of char. The appearance of simultaneous recordings from the medial (▼) and lateral (●) part of the right olfactory bulb of char while stimulating with an amino acid methione and a bile salt, taurolithocholic acid. Note that the response to the amino acids was mainly on the lateral part, while the bile salt caused a response mainly in the medial part of the olfactory bulb. Døving, K.B., Selset, R., Thommesen, G. Olfactory sensitivity to bile acids in salmonid fishes. Acta Physiol Scand 108, 123–131. Copyright © 1980, John Wiley & Sons, Inc.

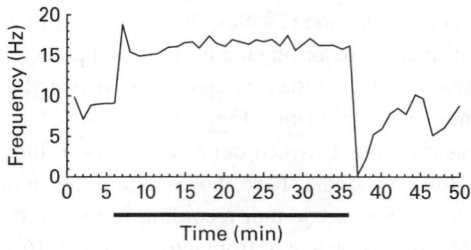

Figure 8.15. Long-term recording of spike activity in the olfactory system. Recording of the action potentials was from a single fiber of the olfactory tract in cod. Stimulus was 1 nm taurolithocholic acid. Note that the activity persisted for the entire stimulation period of 30 min. Kjøstolfsen, I. 1983. Asaptasjon I lukteorganset hos torsk (Gadus morhua L.). In Department of Biology, University of Oslo. Norway

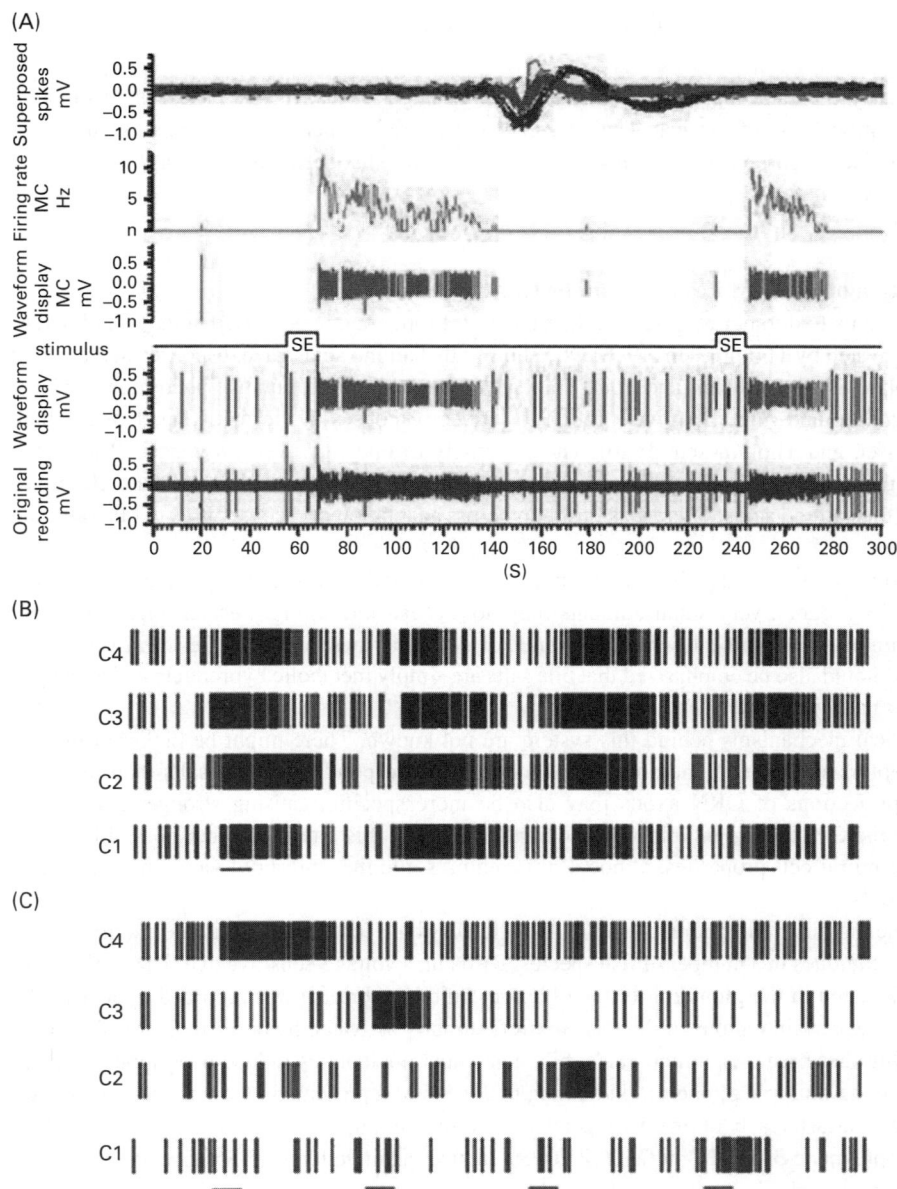

Figure 8.16. Distinction between bulbar neurons and gender differences. (A) Extracellular recordings from single units in the olfactory bulb of crucian carp. Analysis of the original recording (green) revealed two types of units. The type I unit (red) was activated by the stimulation with skin extract (SE), whereas the type II unit (blue) was inhibited during the stimulation period. The overlays shown in the upper tracing revealed the differences in appearance of the two types. El Hassan Hamdani, Kjell B. Døving. Sensitivity and Selectivity of Neurons in the Medial Region of the Olfactory Bulb to Skin Extract from Conspecifics in Crucian Carp, Carassius carassius. Chemical Senses, vol. 28, issue 3. Copyright © 2003, Oxford University Press. (B) and (C) The recordings from the single units in the olfactory bulb in crucian carp revealed dramatic differences between genders in the responses to sex pheromones. The units of the female olfactory bulb (B) did not show any discriminatory capacity. By contrast, the units in the male responded specifically to each of the four sex pheromones (A). Stine Lastein, El Hassan Hamdani, Kjell B. Døving. Gender Distinction in Neural Discrimination of Sex Pheromones in the Olfactory Bulb of Crucian Carp, Carassius carassius. Chemical Senses, vol. 31, issue 1. Copyright © 2006, Oxford University Press. (*See insert for color representation of the figure.*)

substances carry the same important message namely danger. Thus, in this case, less distinction means higher probability of survival. The difference in specificity might arise from two scenarios; it is either the sensory neurons or the wiring in the olfactory bulb that decide the outcome.

CHEMOTOPY

Representation of Odorant Responses in the Olfactory Bulb

The first notion that odorant responses are distributed in a topographic odor map in the fish olfactory bulb was suggested by Thommesen (1978) who showed that amino acids gave responses in the lateral part of the olfactory bulb of char and trout. Some years later, bile salts, putative pheromones related to kin recognition and migration, were found to induce responses in the medial part of the bulb (Døving, Selset, and Thommesen, 1980). The sensitivity to bile salts were in general 2 log units higher than that observed for the amino acid L-methionine. In other words, bile salts are 100 times more potent than amino acids. There was further no cross adaptation of the responses to bile salts and L-methionine.

The high sensitivity toward putative pheromones implies that the message is the principal function, whereas for food odors, very small amounts may not be that interesting. Further, high sensitivity toward pheromones also indicates the importance of detecting these molecules even in very small quantities. It should also be emphasized that bile salts are simply metabolic byproducts which happen to carry information about the sender.

Physiological mechanisms behind this system are not known. There might be higher affinity in odorant receptors sensitive to pheromones, or more ORNs responding to bile salts than to amino acids. The projections of ORN axons may also be more specific, causing stronger and clearer responses in the OB. Lateral inhibition could further increase this effect. Another possibility is the difference in mitral cell properties, although this requires a higher threshold value in initiation of action potentials.

The representation of odorant responses on the olfactory bulb (chemotopy) has been confirmed by a variety of techniques in a number of fish species. By using a voltage-sensitive dye (Di8-ANEPPQ), the activity induced in the glomeruli of the olfactory bulb in zebrafish was recorded optically. The results show that certain regions of the OB are preferentially activated by defined chemical odorant classes. Within these regions, amino acids, bile acids, and nucleotides induce overlapping activity patterns involving multiple glomeruli, indicating that they are represented by combinatorial activity patterns. By contrast, each of the two putative sex pheromone odorants, $PGF_{2\alpha}$ and $17\alpha,20\beta$-dihydroxy-4-pregnene-3-one-20-sulfate (17,20PS), induces a single focus of activity, at least one of which comes from a single, highly specific and sensitive glomerulus. These results confirm that the OB of cyprinid fish is organized into functional regions processing classes of odorants (Friedrich and Korsching, 1997, 1998). Amino acids induced activity in several glomeruli in the lateral OB. All nucleotides tested activated a lateral portion of the OB that overlapped with the posterior portion of the amino acid-sensitive region. Bile acids elicited activity mainly in an anterior–medial part of the OB, along with weaker activity in a posterior–lateral region. The authors concluded that in contrast to the widespread activity patterns observed with ordinary odorants, PGF and 17,20PS each induced only a single focus of activity. This is consistent with research on goldfish in that these compounds induce a spatial distribution that is more restricted than that induced by ordinary odorants. Thus, pheromone-sensitive neurons seem to be more limited in space and more specific than food neurons.

In recent studies, the chemotopy of the OB has been mapped by recording nervous activity from single units. This has been possible because many biologically relevant odors are known in this animal group. The olfactory bulb is divided into different regions based on what group of odorants, hence what type of behavior, is involved. In crucian carp, responses to skin extract were found in the medioposterior region of the bulb (Brondz, Hamdani, and Døving, 2004; Lastein, Hamdani, and Døving, 2008a). Recordings from more than 600 recording positions and 900 single units showed that a region in the ventral part of the central OB is dedicated to detection of sex pheromones, whereas the anteromedial part is involved in bile salt responses (Lastein, Hamdani, and Døving, 2008a).

There are several conclusions resulting from study on chemotopy. First, in visual and somatosensory systems, there is a point-to-point topographic representation between the sensory sheet and the brain. Such a system would be very difficult to handle as the different types of OSNs are distributed throughout the sensory epithelium. By collecting axons from all OSNs responding to a particular odorant, (or rather all OSNs expressing the same odorant receptor) and letting them converge onto a particular bulbar region, the organization is revealed. Second, the overall sensitivity of OSNs is likely enhanced because their axons are focused onto a compact region and terminate onto a small number of secondary neurons (Trotier and Døving, 1996). Third, the projection of secondary neurons onto the brain facilitates function as otherwise connections would need to be rearranged so that the messages transmitted to the animal could reach the appropriate region of the brain. For example, it is reasonable to assume that alarm reactions are initiated by activation of a particular region of the brain. All evidence indicates that the sorting of neurons connecting the olfactory system to these brain structures takes place in the olfactory bulb and is transmitted to the brain in separate channels as bundles of the olfactory tract.

The Connection between Sensory Neurons and Relay Neurons

To visualize patterns of axonal projection of OSNs to the OB, Sato, Miyasaka, and Yoshihara (2005) generated transgenic zebrafish in which spectrally distinct fluorescent proteins are expressed in the ciliated and microvillous OSNs under the control of OMP and TRPC2 gene promoters, respectively. An observation of whole-mount OB in adult double-transgenic zebrafish revealed that the ciliated OSNs project axons mostly to the dorsal and medial regions of the OB, whereas the microvillous OSNs project axons to the lateral region of the OB. A careful histological examination of OB sections clarified that the axons from the two distinct types of OSNs target different glomeruli in a mutually exclusive manner. This segregation is already established at very early developmental stages in zebrafish embryos. These findings clearly demonstrate the relationships among cell morphology, molecular signatures, and axonal terminations of the two distinct types of OSNs and suggest that the two segregated neural pathways are responsible for coding and processing of different types of odor information in the zebrafish olfactory system.

8.3.3 Olfactory Tract

Compound Action Potentials

Because the braincase of burbot, *Lota lota*, is large in relation to the brain itself, it has been possible to place bundles of the olfactory tract or the whole tract on a pair of hooked silver wires at one end to record the compound action potentials while stimulating the tract at the other end (Døving and Gemne, 1965). Thus, electrophysiological properties of the olfactory tract have been correlated to histological properties, that is, the axon diameter of nerve fibers. The bundles of the olfactory tract are situated around the lateral ventricle (Fig. 8.9). This study showed that different bundles of the tract comprise axons of different diameters and consequently have different electrophysiological properties. These findings demonstrate that axons of the secondary fibers have different properties than axons of OSNs. Remarkably, the lMOT had a large number of unmyelinated fibers between 0.5 and 1 µm in diameter (Figs. 8.17 and 8.18). The peak conduction velocity found in this part of the tract was $0.25 \, \text{m} \cdot \text{s}^{-1}$, thus, significantly higher than the conduction velocity found for the olfactory nerve . This connection is of particular interest in the present context as it relates to reproductive

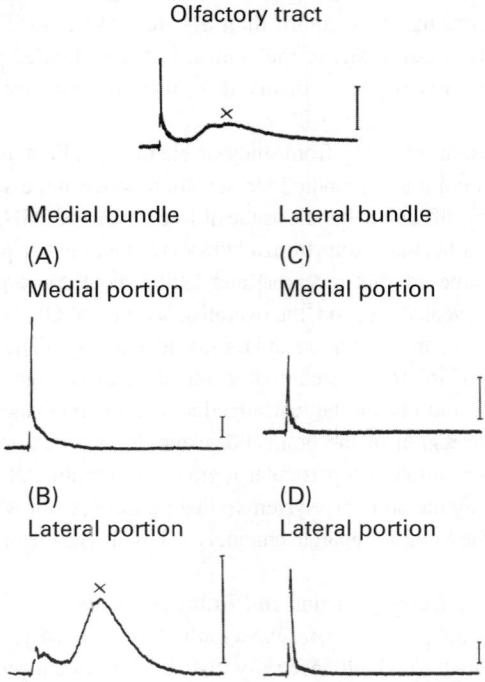

Figure 8.17.   The compound action potentials of the olfactory tract and its bundles in burbot. The top figure shows the compound action potential (CAP) of the olfactory tract, whereas figures (A)–(D) gives the CAP for the found bundles. Note that the slow component is only found in IMOT. From Døving and Gemne (1965).

behavior (see below). It is also interesting that in a range of teleosts (Rudd, *Scardinius erythrophthalmus*; bream, *Abramis brama*; tench, *Tinca tinca*; silure, *Silurus glanis*; cod, *Gadus morhua*; tree-bearded rocling, *Gaidropsarus tricirratus*; ling, *Molva molva*; blue ling, *M. yrkelange*; and lesser fork-beard, *Raniceps raninus*), the compound action potentials showed three components with peak velocities in the ranges of 0.8–4.3, 0.5–2.3, and 0.13–0.22 m·s^{-1}, which indicates that there are fibers that mediate different functional properties.

FUNCTIONAL DIVISIONS

The idea of a topographic odor map was highlighted when it was shown that stimulation of the different olfactory tract bundles induced different types of behavior in cod (Døving and Selset, 1980) (Fig. 8.19). There seems to be no rearrangement of the axons of the bundles of the olfactory tract on the way from the bulb to the telencephalon, and neurons in the medial and lateral part of the olfactory bulb have axons in the medial and lateral olfactory tract, respectively (Satou *et al.*, 1979; Dubois-Dauphin, Døving, and Holley, 1980). Findings by Døving and Selset (1980) imply that there is a functional rearrangement from the olfactory epithelium to the olfactory bulb. Thus, the cod experiments are particularly relevant because they demonstrate that primary sensory neurons of the olfactory which are randomly distributed in the olfactory epithelium project to a highly conserved and orderly arrangement in the olfactory bulb and tract.

     Stimulation of the mLOT caused the cod to swim up and down in the water and perform snapping (likely foraging behavior). Electric pulses given to the lLOT caused the cod to swim backward close

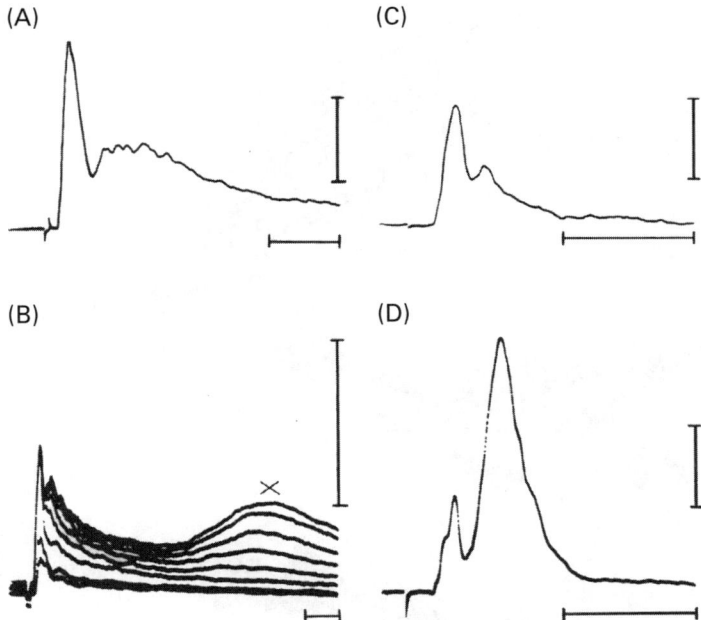

Figure 8.18. Details of the compound action potentials of the olfactory tract. Electric pulses given to the olfactory tract in burbot induce two peaks in the four bundles except lMOT, which display three peaks. Thus, there are at least two classes of fibers in the olfactory tract. (A) mMOT, conducting distance 24 mm. (B) lMOT, conducting distance 17 mm. The traces are CAP at different pulse intensities and demonstrate that the first two peaks are observed at low stimulus strength and the third (x) at the highest stimuli. (C) mLOT conduction distance, 21.5 mm. (D) lLOT conducting distance 26 mm. The time bars are all 20 ms, the vertical bars are 1 mV. From Døving and Gemne (1965).

to the bottom, thereby dragging the lower jaw barbel and the long pelvic fins along the bottom. As the stimulus strength increased the cod turned swiftly and snapped its jaws, consistent with behaviors associated with searching for food at the bottom. Stimulation of the lMOT caused cod to swim at the surface. When the stimulus strength was increased, the cod made quivering movements. Both these behaviors are seen in cod taken into aquaria during the spawning season, February through April. Finally, the stimulation of the mMOT caused the cod to lower the respiration rate, lifting the pelvic fins that made the cod lay close to the bottom, changing coloration of the sides, and displayed white spots on the back. Divers tell us that they see cod displaying such coloration among the kelp at night, and presume that the cod do this as camouflage.

Experiments connecting the different bundles of the olfactory tract to particular behaviors in other species of fishes have grossly confirmed the findings described above for burbot. Sperm release was evoked by electrical stimulation of the MOTs in male goldfish (Demski and Dulka, 1984). It was also shown that ablation of the MOT, but not the LOT, impaired reproductive behavior in goldfish (Kyle *et al.*, 1987). Within the MOT, the lMOT may be more important than the mMOT for courtship expression. In contrast to male sexual behavior, feeding responses were less affected by MOT than LOT section. With respect to male sexual behavior, these findings demonstrate differential functions for the anatomically distinct subdivisions of the goldfish olfactory tracts (Stacey and Kyle, 1983). The terminal nerve could be involved in reproductive behavior; however, experiments in male

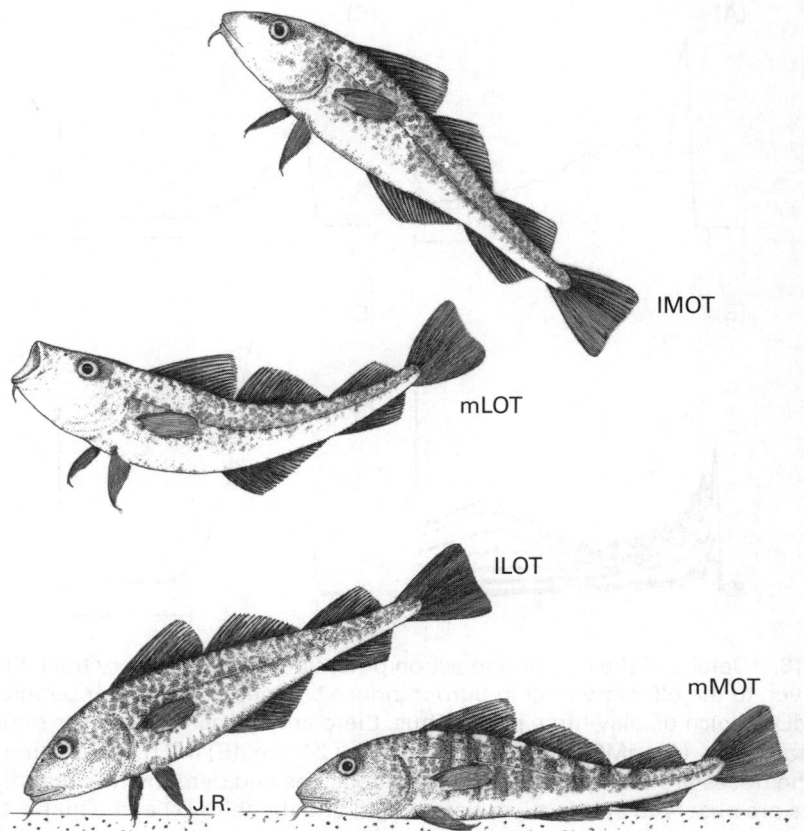

Figure 8.19. Odor-mediated behaviors in the cod. Electrical stimulation of the four bundles of the olfactory tract induces different behaviors. Stimulation of the lMOT and mMOT, that is, lateral and medial part of the MOT induce reproductive and alarm reactions respectively. Stimulation of the lLOT and mLOT, that is, lateral and medial part of the LOT induce feeding behaviors. KB Doving, R Selset. Behavior patterns in cod released by electrical stimulation of olfactory tract bundlets. Science, vol. 207, issue 4430. Copyright © 1980, The American Association for the Advancement of Science.

goldfish exposed to two sex pheromones induced activity in olfactory neurons located in the medical portion of the olfactory bulb but not in the terminal nerve cell bodies (Fujita *et al.*, 1991). Multiunit electrophysiological recordings from the olfactory tracts of goldfish have shown that the medial portion of the tract mediates responses to several sex pheromonal cues (PGF$_{2\alpha}$, 17,20βP) and amino acids while the LOT is sensitive to amino acids alone (Sorensen *et al.*, 1991).

Experiments with crucian carp where one or several bundles of the olfactory tract have been cut have also confirmed these findings. In some experiments, all bundles except one have been ablated. In dealing with ablation experiments, it is essential to understand that to have the brain bundles survive after operation, it is required that the fish remain in physiological saline. If the fish is swimming in fresh water, the function of the axons forming the tract bundles will be impaired. In a series of experiments, these precautions were taken, and it was demonstrated that each of the bundles in crucian carp carry information about particular behaviors. In sum, it has been shown that the LOT in the carp carry information about food odors, whereas lMOT mediates reproductive behavior and mMOT mediates alarm responses.

8.4 BEHAVIOR

Essential life processes are dependent on an intact olfactory system. Numerous physiological studies, combined with behavioral observations, show that nervous activation of olfactory relay neurons initiate behavior directly. It is of particular interest that all interactions with other individuals depend so heavily on chemical communication. As discussed earlier, the secondary olfactory neurons are divided into subsets, each involved in specific behaviors. Electrical stimulation of the olfactory tract bundles is a demonstration of what nervous activity means in the secondary olfactory neurons in fishes. The output of the olfactory bulb is directly related to the behavioral response to odors, and this knowledge can help us in predicting responses to odors just by observing the nervous activity.

8.4.1 Competition between Different Behaviors

Most odors that fish detect by nature are complex mixtures of multiple odorants. Studies such as those with skin extracts as stimuli are good imitators of such natural encounters. It is likely that many glomeruli are activated at the same time; often, the activation will lead to conflicting messages such as both food and danger. In this situation, the fish has to make a decision of what to do, where maximized fitness is essential. Accordingly, a response to an odor might depend on the context. Loss of fright reaction upon exposure to skin extracts is seen in some fish species when going from juvenile to adult stages (Marcus and Brown, 2003; Harvey and Brown, 2004). In crucian carp, ovulating females lose this alarm response (Lastein *et al.*, 2008b). This might be an adaptation for successful spawning, as abrasive spawning behavior might inadvertently release alarm substances. Behaviors might be incompatible, such that would require a decision-making hierarchy. Whether this takes part in higher brain centers, or is a direct cause of rewiring in the olfactory bulb, or re-organization of OSNs such as seasonal variations in appearance of crypt cells (Hamdani *et al.*, 2008), remains to be answered.

8.4.2 Habitat Cues and Migration

In 1980, it was discovered that bile salts are potent odorants for fish (Døving, Selset, and Thommesen, 1980). Bile salts are interesting mediators of olfactory information for a number of reasons: (i) The chemicals have an important function in the digestion of fat in vertebrate digestion where they function as soaps, with one lipophilic and one hydrophilic side of the molecule, (ii) This property also means that in water, bile salts adhere to substrate in water like plants, stones, gravel, and can be released from the substrate at a slow rate, and thereby function as markers in the water, (iii) Bile salts show an interesting development in the evolution of vertebrates (Haslewood, 1967a, 1967b, 1978), (iv) Bile salts show a great variability in structure (Haslewood, 1967a) and have conjugates with different amino acids, (v) Fishes show a high sensitivity to bile and bile salts in particular (Døving, Selset, and Thommesen, 1980; Thommesen, 1982; Michel and Derbidge, 1997; Lastein, Hamdani, and Døving, 2006; Rolen and Caprio, 2007; Huertas *et al.*, 2010; Zhang and Hara, 2009), and (vi) CSNs are highly and specifically sensitive to bile salts (Thommesen, 1983) and project to discrete regions of the olfactory bulb (Døving, Selset, and Thommesen, 1980; Døving *et al.*, 2011).

All these features make bile salts promising candidate pheromones. Thus, teleosts have olfactory sensitivity to a broad range of bile salts, with potential roles in both intraspecific and interspecific chemical communication. We shall raise arguments how this is used in habitat cues and fish migration based on movements of roach (*Rutilus rutilus*) and gudgeon (*Gobio gobio*) in the River Mole in England, showing that fish moved upstream or downstream from their home site, and then returned to the site of capture (Stott, 1967). Laboratory experiments indicated that the majority of gudgeons in two populations accept parts of an artificial convoluted channel as a home range after an initial period of 5 weeks and return to it when moved to another site of the channel (Table 8.1). The field and lab experiments indicate that gudgeon oriented to chemical markers adhering to the substratum.

A series of studies of the olfactory preferences of brown trout, *Salmo trutta* showed that they prefer stream water originating from their home stream over that from a neighboring stream. Moreover, they preferred water from a specific section of their home stream over home stream water taken from distant sections either upstream or downstream of their home section (Arnesen and Stabell, 1992). Of the trout displaced 200 m either upstream or downstream, 40% returned. However, very few fish lacking an olfactory organ returned to their home area within 9 weeks (Halvorsen and Stabell, 1990). The results of these experiments indicate that habitat cues might well contain conspecific pheromones, and are potentially derived from bile salt. This is most plausible, as stones, gravel, and plants have few if any substances that could function as odorants. Bile salts may function as migratory pheromones for salmonids (Døving, Selset, and Thommesen, 1980).

It has been shown that bile salts are among the many substances that function as migratory cues for the sea lamprey (*Petromyzon marinus*) (Polkinghorne *et al.*, 2001; Li *et al.*, 2002; Fine, Vrieze, and Sorensen, 2004). It has also been suggested (but not demonstrated) that bile salts might function as a migratory pheromone for the banded kokopu, (*Galaxias fasciatus*), a New Zealand diadromic species (Baker *et al.*, 2006; Sorensen and baker, 2015).

Migration in open water. It seems reasonable to the authors to consider migration of fishes as an extension of habitat preferences (Stabell, 1984). In this view, we shall extend our review to the behavior of fish during migration in open water. The behavioral mechanisms underlying homing performance of salmon in open waters were first dealt with by Westerberg (1982a, 1982b), who investigated movements of salmon in relation to its immediate environment. Westerberg also studied vertical swimming behavior of Atlantic salmon (*S. salar*) in the Baltic Sea, in a Swedish Lake system and in a Norwegian fjord. Techniques were developed to study swimming depth of fish that made it possible to continuous track individual fish within a precision of ±5 cm with a maximum recording depth of 20 m. He concluded that the salmon tended to follow the fine-structure gradients in the quasimixed surface layer or in the thermocline. Between these periods of swimming at a certain depth, the salmon made rapid excursions either down to the thermocline or up to the mixed layer at the surface. These excursions were interpreted as exploratory searches for the vertically distributed home stream odor. The downward dives were made with a vertical speed of $0.1–0.2 \, \text{m} \cdot \text{s}^{-1}$ at an angle of approximately 10°, whereas the swimming angle for the fish in the upward phase was approximately 25°. The frequency of these exploratory dives was about one per hour. Studies of the small-scale preference behavior of Pacific salmon have shown the same type of vertical movements for the species studied as previously observed for Atlantic salmon (Quinn and terHart, 1987; Ogura and Ishida, 1995; Yano *et al.*, 1997; Tanaka, Takagi, and Naito, 2000).

Table 8.1. The Distribution of Two Gudgeon Populations in an Artificial Channel.

Time	Ahead		0 h		24 h		48 h		72 h	
Area	A	B	A	B	A	B	A	B	A	B
Population A	50	0	**0**	50	**35**	15	**38**	12	**41**	9
Population B	0	50	50	**0**	15	**35**	12	**38**	11	**39**

Ahead of the experiment, two populations, each of 50 tagged gudgeon, were confined in either half of an artificial channel for 5 weeks. At 0 h, the two populations were mutually exchanged and the distribution of the two populations observed at 1, 2, and 3 days afterward. As noted, at 3 days about 80% of the gudeon had returned to their original half. The numbers back in their original areas are in bold. Data from Stott, B. (1967).

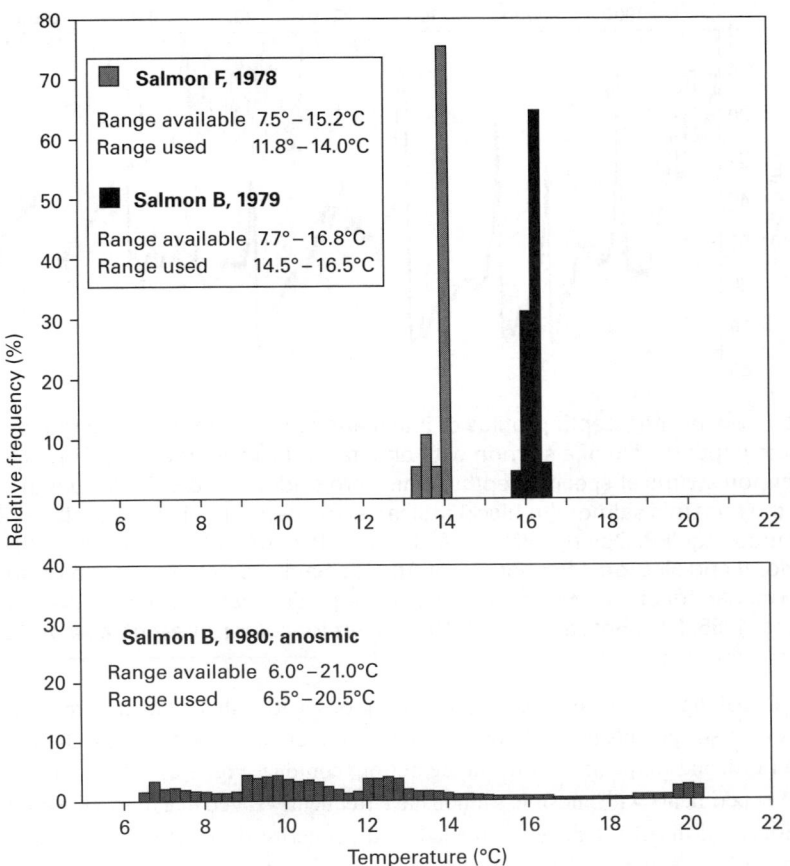

Figure 8.20. Swimming depth preferences of intact and anosmic salmon. The salmon with intact olfactory sense 1978 and 1979 preferred a restricted layer of water with a particular temperature. The anosmic salmon in 1980 did not prefer a particular layer. Westerberg, H. (Goeteborg Univ. (Sweden), Oceanografiska Inst.) (1982) Ultrasonic tracking of Atlantic salmon (Salmo salar L.), 2: Swimming depth and temperature stratification. Report - Institute of Freshwater Research, Drottningholm 60: 102–120. (*See insert for color representation of the figure.*)

Westerberg concluded that olfactory cues are probable sign stimuli for orientation in relation to water currents. His interpretation was supported by the behavior of anosmic salmon, which is dramatically different from the salmon with an intact olfactory organ. As can be seen from Figure 8.20, the intact salmon in Lake Vänern showed remarkable preferences, whereas the anosmic salmon showed no preferences and used almost the entire temperature range. Westerberg also suggested that salmon are able to detect the acceleration resulting from the relative movement of layers when crossing from one layer to the other. This suggestion seems strengthened by the observation of the keen sensitivity to infrasound in fishes (Sand and Karlsen, 1986, 2000).

Studies of salmon behavior in the Norwegian fjord system confirmed that free-swimming salmon follow a certain layer in the water for prolonged periods (Døving, Westerberg, and Johnsen, 1985). Tracking data revealed that during the surveillance period, intact salmon preferred a specific

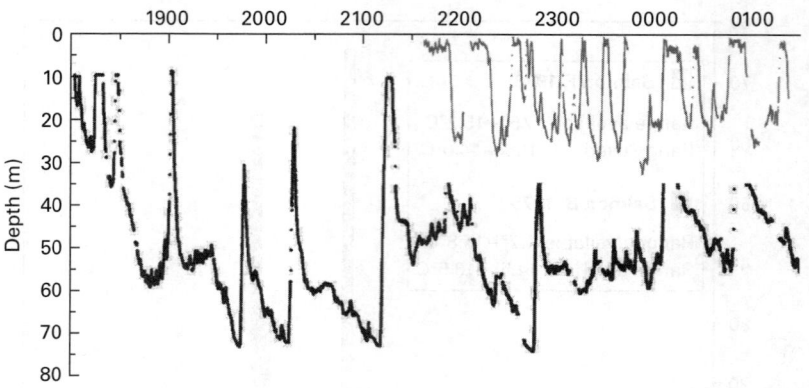

Figure 8.21. Swimming depth profiles of intact and anosmic salmon. The red trace indicates the depth profile of a salmon with an intact olfactory organ in a fjord system. Note that the salmon swims at specific depths, which are seldom below the sill depth of the fjord. However, the anosmic salmon (in black) indicates an aberrant behaviour, following the bottom contour. Kjell B. Døving, Håkan Westerberg, Peter B. Johnsen. Role of Olfaction in the Behavioral and Neuronal Responses of Atlantic Salmon, Salmo salar, to Hydrographic Stratification. Canadian Journal of Fisheries and Aquatic Sciences, vol. 42, issue 10. Copyright (c) 1985, NRC Research Press. (*See insert for color representation of the figure.*)

temperature, making few, if any excursions to regions with other temperatures (Fig. 8.21, red trace). By contrast, the anosmic salmon showed no preferences for a specific layer, made dives deep in the fjord, and frequently followed the bottom contour (Fig. 8.21, black trace).

Critics of experiments with anosmic salmon have frequently stated that due to the trauma resulting from impairment of the olfactory sense, the fish will not perform normally. One salmon tagged and followed in the Norwegian fjord system was made anosmic on one side only. This salmon behaved similar to intact fish in the tracking session, and was later captured in the River Imsa, that is, its home river. Postmortem inspection of the salmon revealed the unilateral ablation of the olfactory nerve. Thus, ablation of the olfactory nerve on one side did not cause a loss of normal behavior, nor did it influence the ability to track the home river (Døving, Westerberg, and Johnsen, 1985).

8.5 CONCLUSIONS

In this chapter, we described in detail the anatomical features and the physiological functions of the olfactory system that is essential to understand pheromone detection in fishes. The morphology of a sensory neuron is both correlated with odorant receptor identity and to its spatial termination in the olfactory bulb. Further, these labeled lines from the OSNs target the bulb, and project to the telencephalon, thus serving as important aspects and crucial for the understanding of the function of the olfactory sense. Recent and ongoing studies applying staining by endocytotic uptake of samples from the external environment will increase our understanding of the connections of these lines.

Distinct behavioral patterns evoked by different subpopulations of neurons suggest relatively hardwired connections of the neurons involved in these pathways. However, competition between different behaviors shows seasonal variation, and although some variation has been shown in peripheral structures, central plasticity cannot be excluded as one of the underlying mechanisms.

Fish are dependent on their sense of smell for all essential life processes: finding the right mating partner, avoid enemies, feeding, and migrating to spawning areas across vast oceans. Understanding the physiological mechanisms underlying olfactory-induced behaviors is essential for many aspects of fisheries management and a valuable tool for conservation of fish populations.

REFERENCES

Alonso, J.R., Lara, J., Covenas, R., and Aijon, J. (1988) Two types of mitral cells in the teleostean olfactory bulb. *Neuroscience Research Communication*, **3**, 113–118.

Alvarez, V. and Sabatini, B. (2007) Anatomical and physiological plasticity of dendritic spines. *Annual Review of Neuroscience*, **30**, 79–97.

Ariëns Kappers, C.U. (1906) The structure of the teleostean and selacianbrain. *Journal of Comparative Neurology*, **16**, 1–109.

Arnesen, A.M. and Stabell, O.B. (1992) Behaviour of stream-dwelling brown trout towards odours present in home stream water. *Chemoecology*, **3**, 94–100.

Baker, C.F., Carton, A.G., Fine, J.M., and Sorensen, P.W. (2006) Can bile acids function as a migratory pheromone in banded kokopu, *Galaxias fasciatus* (Gray)? *Ecology of Freshwater Fish*, **15**, 275–283.

Bass, A.H. (1981) Olfactory bulb efferents in the channel catfish, *Ictalurus punctatus. Journal of Morphology*, **169**, 91–111.

Brondz, I., Hamdani, E.H., and Døving, K. (2004) Neurophysiologic detector-a selective and sensitive tool in high-performance liquid chromatography. *Journal of Chromatography B*, **800**, 41–47.

Buck, L. and Axel, R. (1991) A novel multigene family may encode odorant receptors: a molecular basis for odor recognition. *Cell*, **65**, 175–187.

Cajal, R. (1911) *Histologie du système nerveux de l'homme & des vertébrés*, Maloine, Paris.

Cardwell, J.R., Stacey, N.E., Tan, E.S.P. *et al.* (1995) Androgen increases olfactory receptor response to a vertebrate ex pheromone. *Journal of Comparative Physiology A*, **176**, 55–61.

Davis, R.E., Chase, R., Morris, J., and Kaufman, B. (1981) Telencephalon of the teleost *Macropodus*: experimental localization of secondary olfactory areas and components of the lateral forebrain bundle. *Behavioral and Neural Biology*, **33**, 257–279.

Demski, L.S. and Dulka, J.G. (1984) Functional-anatomical studies on sperm release evoked by electrical stimulation of the olfactory tract in goldfish. *Brain Research*, **291**, 241–247.

Døving, K.B. (1964) Studies of the relation between the electro-olfactogram (EOG) and single unit activity in the olfactory bulb. *Acta Physiologica Scandinavica*, **60**, 150–163.

Døving, K. (1976) Evolutionary trends in olfaction, in *The Structure–Activity Relationships in Olfaction*, (ed. G. Benz), IRL Press, London, 149–159.

Døving, K.B. (2010) Smell: vertebrates, in *Encyclopedia of Animal Behavior*, vol. **3** (eds M.D. Breed and J. Moore), Academic Press, Oxford, pp. 207–215.

Døving, K.B. and Gemne, G. (1965) Electrophysiological and histological properties of the olfactory tract of the burbot (Lota lota L.). *Journal of Neurophysiology*, **28**, 139–153.

Doving, K.B. and Gemne, G. (1966) An electrophysiological study of the efferent olfactory system in the burbot. *Journal of Neurophysiology*, **29**, 665–674.

Døving, K.B. and Holmberg, K. (1974) A note on the function of the olfactory organ of the hagfish *Myxine glutinosa. Acta Physiologica Scandinavica*, **91**, 430–432.

Døving, K.B. and Selset, R. (1980) Behavior patterns in cod released by electrical stimulation of olfactory tract bundlets. *Science*, **207**, 559–560.

Døving, K.B., Dubois-Dauphin, M., Holley, A., and Jourdan, F. (1977) Functional anatomy of the olfactory organ of fish and the ciliary mechanisms of water transport. *Acta Zoologica*, **58**, 245–255.

Døving, K.B., Selset, R., and Thommesen, G. (1980) Olfactory sensitivity to bile acids in salmonid fishes. *Acta Physiologica Scandinavica*, **108**, 123–131.

Døving, K.B., Westerberg, H., and Johnsen, P.B. (1985) Role of olfaction in the behavioral and neuronal responses of Atlantic salmon, *Salmo salar*, to hydrographic stratification. *Canadian Journal of Fisheries and Aquatic Sciences*, **42**, 1658–1667.

Døving, K.B., Hansson, K.A., Backström, T., and Hamdani, E.H. (2011) Visualizing a set of olfactory sensory neurons responding to a bile salt. *Journal of Experimental Biology*, **214**, 80–87.

Dubois-Dauphin, M., Døving, K.B., and Holley, A. (1980) Topographical relation between the olfactory bulb and the olfactory tract in tench (*Tinca tinca* L.). *Chemical Senses*, **5**, 159–169.

Eastman, J.T. and Lannoo, M.J. (2011) Divergence of brain and retinal anatomy and histology in pelagic antarctic notothenioid fishes of the sister taxa Dissostichus and Pleuragramma. *Journal of Morphology*, **272**, 419–441.

Ebbesson, S.O., Meyer, D.L., and Scheich, H. (1981) Connections of the olfactory bulb in the piranha (*Serrasalmus nattereri*). *Cell and Tissue Research*, **216**, 167–180.

Fine, J.M., Vrieze, L.A., and Sorensen, P.W. (2004) Evidence that petromyzontid lampreys employ a common migratory pheromone that is partially comprised of bile acids. *Journal of Chemical Ecology*, **30**, 2091–2110.

Finger, T.E. (1975) The distribution of the olfactory tracts in the bullhead catfish, *Ictalurus nebulosus*. *Journal of Comparative Neurology*, **161**, 125–141.

Friedrich, R.W. and Korsching, S.I. (1997) Combinatorial and chemotopic odorant coding in the zebrafish olfactory bulb visualized by optical imaging. *Neuron*, **18**, 737–752.

Friedrich, R.W. and Korsching, S.I. (1998) Chemotopic, combinatorial, and noncombinatorial odorant representations in the olfactory bulb revealed using a voltage-sensitive axon tracer. *Journal of Neuroscience*, **18**, 9977–9988.

Fujita, I., Sorensen, P.W., Stacey, N.E., and Hara, T.J. (1991) The olfactory system, not the terminal nerve, functions as the primary chemosensory pathway mediating responses to sex pheromones in male goldfish. *Brain, Behavior and Evolution*, **38**, 313–321.

Garten, S. (1900) *Physiologie der marklosen Nerven*, Gustav Fischer, Jena.

Gasser, H.S. (1956) Olfactory nerve fibers. *Journal of General Physiology*, **39**, 473–496.

Golgi, C. (1875) Sulla fina anatomia dei Bulbi olfatorii. *Rivista Sperimentale di Freniatria e Medicina Legale*, **1**, 405–425.

Halvorsen, M. and Stabell, O.B. (1990) Homing behavior of displaced stream-dwelling brown trout. *Animal Behaviour*, **39**, 1089–1097.

Hamdani, E.H. and Døving, K.B. (2002) The alarm reaction in crucian carp is mediated by olfactory neurons with long dendrites. *Chemical Senses*, **27**, 395–398.

Hamdani, E.H. and Døving, K.B. (2003) Sensitivity and selectivity of neurons in the medial region of the olfactory bulb to skin extract from conspecifics in crucian carp, *Carassius carassius*. *Chemical Senses*, **28**, 181–189.

Hamdani, E.H. and Døving, K.B. (2006) Specific projection of the sensory crypt cells in the olfactory system in crucian carp, *Carassius carassius*. *Chemical Senses*, **31**, 63–67.

Hamdani, E.H. and Døving, K.B. (2007) The functional organization of the fish olfactory system. *Progress in Neurobiology*, **82**, 80–86.

Hamdani, E.H., Stabell, O.B., Alexander, G., and Døving, K.B. (2000) Alarm reaction in the crucian carp is mediated by the medial bundle of the medial olfactory tract. *Chemical Senses*, **25**, 103–109.

Hamdani, E.H., Alexander, G., and Døving, K.B. (2001a) Projection of sensory neurons with microvilli to the lateral olfactory tract indicates their participation in feeding behaviour in crucian carp. *Chemical Senses*, **26**, 1139–1144.

Hamdani, E.H., Kasumyan, A., and Døving, K.B. (2001b) Is feeding behaviour in crucian carp mediated by the lateral olfactory tract? *Chemical Senses*, **26**, 1133–1138.

Hamdani, E.H., Lastein, S., Gregersen, F., and Døving, K.B. (2008) Seasonal variations in olfactory sensory neurons—fish sensitivity to sex pheromones explained? *Chemical Senses*, **33**, 119–123.

Hansen, A., Eller, P., Finger, T.E., and Zeiske, E. (1997) The crypt cell: a microvillous ciliated olfactory receptor cell in teleost fishes. *Chemical Senses*, **22**, 694–695.

Hansen, A., Rolen, S.H., Anderson, K. *et al.* (2003) Correlation between olfactory receptor cell type and function in the channel catfish. *Journal of Neuroscience*, **23**, 9328–9339.

Hansson, K-A. (2011) *Mapping the Connectivity Within the Olfactory System in Crucian Carp, Carassiuscarassius*. MSc Thesis, University of Olso.

Harvey, M.C. and Brown, G.E. (2004) Dine or dash?: ontogenetic shift in the response of yellow perch to conspecific alarm cues. *Environmental Biology of Fishes*, **70**, 345–352.

Haslewood, G.A.D. (1967a) *Bile salt* evolution. *Journal of Lipid Research*, **8**, 535–550.

Haslewood, G.A.D. (1967b) *Bile Salts*, Methuen & Co Ltd, London.

Haslewood, G.A.D. (1978) The Biological Importance of Bile Salts, North-Holland Pub Co., Amsterdam/ New York.

Herrick, C.J. (1901) The cranial nerves and cutaneous sense organs of the North American silurid fishes. *Journal of comparative Neurology and Psychology*, **11**, 177–249.

Holmgren, N. (1918) Zur kenntnis des nervus terminalis bei Teleostieren. *Folia Neurobiologica*, **11**, 16–36.

Holmgren, N. (1920) Zur anatomie und Histologie des Vorder- und Zwischenhirns der Knochenfische. *Acta Zoologica (Stockholm)*, **1**, 137–315.

Huertas, M., Hagey, L., Hofmann, A.F. *et al.* (2010) Olfactory sensitivity to bile fluid and bile salts in the European eel (*Anguilla anguilla*), goldfish (*Carassius auratus*) and Mozambique tilapia (*Oreochromis mossambicus*) suggests a "broad range" sensitivity not confined to those produced by conspecifics alone. *Journal of Experimental Biology*, **213**, 308–317.

Ito, H. (1973) Normal and experimental studies on synaptic patterns in the carp telencephalon, with special reference to the secondary olfactory termination. *Journal für Hirnforschung*, **14**, 237–253.

Jarrard, H. (1997) A neuroanatomical study of the olfactory bulb of the pacific salmon across its life history: changes in the relative volumetric composition of the bulb and their possible functional consequences. *Dissertation Abstracts International Part B: Science and Engineering*, **58**, 2885.

Jones, D.T. and Reed, R.R. (1989) Golf: an olfactory neuron specific-G protein involved in odorant signal transduction. *Science*, **244**, 790–795.

Karlson P. and Lüscher, M. (1959) Pheromones: A new term for a class of biologically active substances. *Nature*, **183**, 55–56.

Kasai, H., Fukuda, M., Watanabe, S. *et al.* (2010) Structural dynamics of dendritic spines in memory and cognition. *Trends in Neurosciences*, **33**, 121–129.

Kawai, T., Oka, Y., and Eisthen, H. (2009) The role of the terminal nerve and GnRH in olfactory system neuromodulation. *Zoological Science*, **26**, 669–680.

Kjøstolfsen I. (1983) Adaptasjonilukteorganet hos torsk (*Gadusmorhua* L.) In: *Department of Biology*, Oslo: University of Oslo, p. 54.

Kobayakawa, K., Kobayakawa, R., Matsumoto, H. *et al.* (2007) Innate versus learned odour processing in the mouse olfactory bulb. *Nature*, **450**, 503–508.

Koide, T., Miyasaka, N., Morimoto, K. *et al.* (2009) Olfactory neural circuitry for attraction to amino acids revealed by transposon-mediated gene trap approach in zebrafish. *Proceedings of the National Academy of Sciences of the USA*, **106**, 9884–9889.

Kosaka, T. and Hama, K. (1979) Ruffed cell: a new type of neuron with a distinctive initial unmyelinated portion of the axon in the olfactory bulb of the goldfish (*Carassius auratus*) I. Golgi impregnation and serial thin sectioning studies. *Journal of Comparative Neurology*, **186**, 301–319.

Kosaka, T. and Hama, K. (1981) Ruffed cell: a new type of neuron with a distinctive initial unmyelinated portion of the axon in the olfactory bulb of the goldfish (*Carassius auratus*). III. Three-dimensional structure of the ruffed cell dendrite. *Journal of Comparative Neurology*, **201**, 571–587.

Kosaka, T. and Hama, K. (1982) Synaptic organization in the teleost olfactory bulb. *Journal of Physiology (Paris)*, **78**, 707–719.

Kyle, A.L., Sorensen, P.W., Stacey, N.E., and Dulka, J.G. (1987) Medial olfactory tract pathways controlling sexual reflexes and behavior in teleosts. *Annals of the New York Academy of Sciences*, **519**, 97–107.

Lastein, S., Hamdani, E.H., and Døving, K.B. (2006) Gender distinction in neural discrimination of sex pheromones in the olfactory bulb of crucian carp, *Carassius carassius*. *Chemical Senses*, **31**, 69–77.

Lastein, S., Hamdani, E.H., and Døving, K.B. (2008a) Single unit responses to skin odorants from conspecifics and heterospecifics in the olfactory bulb of crucian carp *Carassius carassius*. *Journal of Experimental Biology*, **211**, 3529–3535.

Lastein, S., Höglund, E., Mayer, I. *et al.* (2008b) Female crucian carp, *Carassius carassius*, lose predator avoidance behavior when getting ready to mate. *Journal of Chemical Ecology*, **34**, 1487–1491.

Levine, R.L. and Dethier, S. (1985) The connections between the olfactory bulb and the brain in the goldfish. *Journal of Comparative Neurology*, **237**, 427–444.

Li, W., Scott, A.P., Siefkes, M.J. *et al.* (2002) Bile acid secreted by male sea lamprey that acts as a sex pheromone. *Science*, **296**, 138–141.

Lipschitz, D.L. and Michel, W.C. (2002) Amino acid odorants stimulate microvillar sensory neurons. *Chemical Senses*, **27**, 277–286.

Marcus, J.P. and Brown G.E. (2003) Response of pumpkinseed sunfish to conspecific chemical alarm cues: An interaction between ontogeny and stimulus concentration. *Canadian Journal of Zoology*, **81**, 1671–1677.

Meek, J. and Niewenhuys, R. (1998) Holosteans and teleosts, in *The Central Nervous System of Vertebrates*, vol. 2 (eds R. Niewenhuys, H.J. ten Dokelaar, and C. Nicholson), Springer, Berlin, pp. 759–937.

Michel, W.C. and Derbidge, D.S. (1997) Evidence of distinct amino acid and bile salt receptors in the olfactory system of the zebrafish, *Danio rerio*. *Brain Research*, **764**, 179–187.

Morita, Y. and Finger, T.E. (1998) Differential projections of ciliated and microvillous olfactory receptor cells in the catfish, *Ictalurus punctatus*. *Journal of Comparative Neurology*, **398**, 539–550.

Murphy, G.J., Glickfeld, L.L., Balsen, Z., and Isaacson, J.S. (2004) Sensory neuron signaling to the brain: properties of transmitter release from olfactory nerve terminals. *Journal of Neuroscience*, **24**, 3023–3030.

Nevitt, G.A., Dittman, A.H., Quinn, T.P., and Moody, W.J., Jr. (1994) Evidence for a peripheral olfactory memory in imprinted salmon. *Proceedings of the National Academy of Sciences of the USA*, **91**, 4288–4292.

Nieuwenhuys, R. (2009) The forebrain of actinopterygians revisited. *Brain, Behavior and Evolution*, **73**, 229–252.

Nordeng, H. (1971) Is the local orientation of anadromous fishes determined by pheromones? *Nature (London)*, **233**, 411–413.

Nordeng, H. (2009) Natal homing in sympatric populations of anadromous Arctic char *Salvelinus alpinus* (L.): roles of pheromone recognition. *Ecology of Freshwater Fish*, **18**, 41–51.

Nordeng, H. and Bratland, P. (2006) Homing experiments with parr, smolt and residents of anadromous Arctic char *Salvelinus alpinus* and brown trout *Salmo trutta*: transplantation between neighbouring river systems. *Ecology of Freshwater Fish*, **15**, 488–499.

Northcutt, R.G. and Davis, R.E. (1983) Telencephalic organization of rayfinned fishes, in *Fish Neurobiology*, vol. **2** (eds R.E. Davis and R.G. Northcutt), The University of Michigan Press, Ann Arbor, pp. 203–236.

Ogura, M. and Ishida, Y. (1995) Homing behaviour and vertical movements of four species of Pacific salmon (*Oncorhynchus* spp.) in the central Bering Sea. *Canadian Journal of Fisheries and Aquatic Sciences*, **52**, 532–540.

Oka, Y. (1980) The origins of the centrifugal fibers to the olfactory bulb in the goldfish, *Carassius auratus*: an experimental study using the fluorescent dye primuline as a retrograde tracer. *Brain Research*, **185**, 215–225.

Ottoson, D. (1956) Analysis of the electrical activity of the olfactory epithelium. *Acta Physiologica Scandinavica*, **35**, 1–83.

Ottoson, D. (1959a) Comparison of slow potentials evoked in the frog's nasal mucosa and olfactory bulb by natural stimulation. *Acta Physiologica Scandinavica*, **47**, 149–159.

Ottoson, D. (1959b) Olfactory bulb potentials induced by electrical stimulation of the nasal mucosa in the frog. *Acta Physiologica Scandinavica*, **47**, 160–172.

Ottoson, D. (1959c) Studies on slow potentials in the rabbit's olfactory bulb and nasal mucosa. *Acta Physiologica Scandinavica*, **47**, 136–148.

Polkinghorne, C.N., Olson, J.M., Gallaher, D.G., and Sorensen, P.W. (2001) Larval sea lamprey release two unique bile acids to the water at a rate sufficient to produce detectable riverine pheromone plumes. *Fish Physiology and Biochemistry*, **24**, 15–30.

Prasada Rao, P.D. and Finger, T.E. (1984) Asymmetry of the olfactory system in the brain of the winter flounder, *Pseudopleuronectes americanus*. *Journal of Comparative Neurology*, **225**, 492–510.

Quinn, T.P. and terHart, B.A. (1987) Movements of adult sockeye salmon (*Oncorhynchus nerka*) in British Columbia coastal waters in relation to temperature and salinity stratification: ultra-sonic telemetry results, in *Sockeye Oncorhynchus nerka, Population Biology and Future Management* (eds H.D. Smith, L. Margolis, and C.C. Wood), Canadian Government Publishing Centre, Ottawa, pp. 61–77.

Rankin, M.L., Alvania, R.S., Gleason, E.L., and Bruch, R.C. (1999) Internalization of G protein-coupled receptors in single olfactory receptor neurons. *Journal of Neurochemistry*, **72**, 541–548.

Riddle, D.R. and Oakley, B. (1991) Evaluation of projection patterns in the primary olfactory system of rainbow trout. *Journal of Neuroscience*, **11**, 3752–3762.

Riddle, D.R. and Oakley, B. (1992) Immunocytochemical identification of primary olfactory afferents in rainbow trout (*Oncorhynchus mykiss*). *Journal of Comparative Neurology*, **324**, 575–589.

Riddle, D.R., Wong, L.D., and Oakley, B. (1993) Lectin identification of olfactory receptor neuron subclasses with segregated central projections. *Journal of Neuroscience*, **13**, 3018–3033.

Rolen, S.H. and Caprio, J. (2007) Processing of bile salt odor information by single olfactory bulb neurons in the channel catfish. *Journal of Neurophysiology*, **97**, 4058–4068.

Rooney, D.J., New, J.G., Szabo, T., and Ravaille-Veron, M. (1989) Central connections of the olfactory bulb in the weakly electric fish *Gnathonemus petersii*. *Cell and Tissue Research*, **257**, 492–510.

Rooney, D.J., Døving, K.B., Ravaille-Veron, M., and Szabo, T. (1992) The central connections of the olfactory bulbs in cod, *Gadus morhua* L. *Journal für Hirnforschung*, **33**, 63–75.

Sand, O. and Karlsen, H.E. (1986) Detection of infrasound in the Atlantic cod. *Journal of Experimental Biology*, **125**, 197–204.

Sand, O. and Karlsen, H.E. (2000) Detection of infrasound and linear acceleration in fishes. *Philosophical Transactions of the Royal Society of London: Series B*, **355**, 1295–1298.

Sato, K. and Suzuki, N. (2000) The contribution of a Ca(2+)-activated Cl(−) conductance to amino-acid- induced inward current responses of ciliated olfactory neurons of the rainbow trout. *Journal of Experimental Biology*, **203** (Pt 2), 253–262.

Sato, K. and Suzuki, N. (2001) Whole-cell response characteristics of ciliated and microvillous olfactory receptor neurons to amino acids, pheromone candidates and urine in rainbow trout. *Chemical Senses*, **26**, 1145–1156.

Sato, Y., Miyasaka, N., and Yoshihara, Y. (2005) Mutually exclusive glomerular innervation by two distinct types of olfactory sensory neurons revealed in transgenic zebrafish. *Journal of Neuroscience*, **25**, 4889–4897.

Satou, M., Ichikawa, M., Ueda, K., and Takagi, S.F. (1979) Topographical relation between olfactory bulb and olfactory tracts in the carp. *Brain Research*, **173**, 142–146.

Scalia, F. and Ebbesson, S.O. (1971) The central projections of the olfactory bulb in a teleost (*Gymnothorax funebris*). *Brain, Behavior and Evolution*, **4**, 376–399.

Schmachtenberg, O. and Bacigalupo, J. (2004) Olfactory transduction in ciliated receptor neurons of the Cabinza grunt, *Isacia conceptionis* (teleostei: haemulidae). *European Journal of Neuroscience* **20**, 3378–3386.

Sheldon, R.E. (1912) The olfactory tracts and centers in teleosts. *Journal of Comparative Neurology*, **22**, 177–339.

Silver, W.L., Caprio, J., Blackwell, J.F., and Tucker, D. (1976) The underwater electro-olfactogram: a tool for the study of the sense of smell of marine fishes. *Experientia*, **32**, 1216–1217.

Sorensen, P.W. (2015) Introduction to pheromones and related cues in fishes in *Fish Pheromones and Related Cues* (eds Peter W. Sorensen and Brian D. Wisenden), John Wiley & Sons, Inc., Hoboken.

Sorensen, P.W. and Baker, C. (2015) Species-specific pheromones and their roles in shoaling, migration, and reproduction: a critical review and synthesis, in *Fish Pheromones and Related Cues* (eds Peter W. Sorensen and Brian D. Wisenden), John Wiley & Sons, Inc., Hoboken.

Sorensen, P.W. and Goetz, F.W. (1993) Pheromonal function of prostaglandin metabolites in teleost fish. *Journal of Lipid Mediators*, **6**, 385–393.

Sorensen, P.W., Hara, T.J., Stacey, N.E., and Dulka, J.G. (1990) Extreme olfactory specificity of male goldfish to the preovulatory pheromone 17α,20β-dihydroxy-4-pregnen-3-one. *Journal of Comparative Physiology A*, **166**, 373–385.

Sorensen, P.W., Hara, T.J., and Stacey, N.E. (1993) Sex pheromones selectively stimulate the medial olfactory tracts of male goldfish. *Brain Research*, **558**, 343–347.

Sorensen, P.W., Scott, A.P., Stacey, N.E., and Bowdin, L. (1995) Sulfated 17,20b-dihydroxy-4-pregnen-3-one functions as a potent and specific olfactory stimulant with pheromonal actions in the goldfish. *General Comparative Endocrinology*, **100**, 128–142.

Stabell, O.B. (1984) Homing and olfaction in salmonids: a critical review with special reference to the Atlantic salmon. *Biological Reviews of the Cambridge Philosophical Society*, **59**, 333–388.

Stabell, O.B. (1992) Olfactory control of homing behaviour in salmonids, in *Fish Chemoreception* (ed. T.J. Hara), Chapman & Hall, London, pp. 249–270.

Stacey, N.E. (2009) Pheromones and reproduction, in *Reproductive Biology and Phylogeny of Fishes*, vol. **8B** (ed. B.G.M. Jamieson), Science Publishers, Enfield, pp. 94–137.

Stacey, N.E. (2015) Hormonally derived pheromones in teleost fishes, in *Fish Pheromones and Related Cues* (eds Peter W. Sorensen and Brian D. Wisenden), John Wiley & Sons, Inc., Hoboken.

Stacey, N.E. and Kyle, A.L. (1983) Effects of olfactory tract lesions on sexual and feeding behavior in the goldfish. *Physiology and Behavior*, **30**, 621–628.

Stacey, N.E. and Sorensen, P.W. (2002) Fish hormonal pheromones, in *Hormones, Brain, and Behavior*, vol. **2** (eds D.W. PfaV, A.P. Arnold, A. Etgen, *et al.*), Academic Press, New York, pp. 375–435.

Stacey, N.E., Chojnacki, A., Narayanan, A., *et al.* (2003) Hormonally derived sex pheromones in fish: exogenous cues and signals from gonad to brain. *Canadian Journal of Physiology and Pharmacology*, **81**, 329–341.

Steward, M. and Sorensen, P.W. (2015) Measuring and identifying fish pheromones, in *Fish Pheromones and Related Cues* (eds Peter W. Sorensen and Brian D. Wisenden), John Wiley & Sons, Inc., Hoboken.

Stott, B. (1967) The movements and population densities of roach (*Rutilus rutilus* L.) and gudeon (*Gobio gobio* L.) in the river mole. *Journal of Animal Ecology*, **36**, 407–423.

Tanaka, H., Takagi, Y., and Naito, Y. (2000) Behavioural thermoregulation of chum salmon during homing migration in coastal waters. *Journal of Experimental Biology*, **203** (Pt 12), 1825–1833.

Thommesen, G. (1978) The spatial distribution of odour induced potentials in the olfactory bulb of char and trout (*Salmonidae*). *Acta Physiologica Scandinavica*, **102**, 205–217.

Thommesen, G. (1982) Specificity and distribution of receptor cells in the olfactory mucosa of char (*Salmo alpinus* L.). *Acta Physiologica Scandinavica*, **115**, 47–56.

Thommesen, G. (1983) Morphology, distribution, and specificity of olfactory receptor cells in salmonid fishes. *Acta Physiologica Scandinavica*, **117**, 241–249.

Thommesen, G. and Døving, K.B. (1977) Spatial distribution of the EOG in the rat: a variation with odour quality. *Acta Physiologica Scandinavica*, **99**, 270–280.

Trotier, D. and Døving, K.B. (1996) Functional role of receptor neurons in encoding olfactory information. *Journal of Neurobiology*, **30**, 58–66.

van Gehuchten, A. and Martin, I. (1891) Le bulbe olfactif de quelques mammifères. *La Cellule*, **7**, 205–237.

Vielma, A., Ardiles, A., Delgado, L., and Schmachtenberg, O. (2008) The elusive crypt olfactory receptor neuron: evidence for its stimulation by amino acids and cAMP pathway agonists. *Journal of Experimental Biology*, **211**, 2417–2422.

von Bartheld, C.S., Meyer, D.L., Fiebig, E., and Ebbesson, S.O. (1984) Central connections of the olfactory bulb in the goldfish, *Carassius auratus*. *Cell and Tissue Research*, **238**, 475–487.

von Fritsch, G. (1878) *Untersuchungen über den feineren Bau des Fischgehirns mit besonderer Berücksichtigung der Homologien bei anderen Wirbelthierklassen*, Gutmann, Berlin.

Weltzien, F.A., Höglund, E., Hamdani, E.H., and Døving, K.B. (2003) Does the lateral bundle of the medial olfactory tract mediate reproductive behavior in male crucian carp? *Chemical Senses*, **28**, 293–300.

Westerberg, H. (1982) Ultrasonic tracking of Atlantic salmon (*Salmo salar* L.)—I. Swimming depth and temperature stratification. *Reports of the Institute of Freshwater Research, Drottningholm*, **60**, 102–115.

Westerberg, H. (1982) Ultrasonic tracking of Atlantic salmon (*Salmo salar* L.)—II. Movements in coastal regions. *Reports of the Institute of Freshwater Research, Drottningholm*, **60**, 81–101.

Wisby, W.J. and Hasler, A.D. (1954) Effect of olfactory occlusion on migrating silver salmon (*Oncorhynchus kisutch*). *Journal of the Fisheries Research Board of Canada*, **11**, 472–478.

Yano, A., Ogura, M., Sato, A. *et al.* (1997) Effect of modified magnetic field on the ocean migration of maturing chum salmon, *Oncorhynchus keta*. *Marine Biology*, **129**, 523–530.

Zeiske, E., Melinkat, R., Breucker, H., and Kux, J. (1976) Ultrastructural studies on the epithelia of the olfactory organ of cyprinodonts (Teleostei, Cyprinodontoidea). *Cell and Tissue Research*, **172**, 245–267.

Zeiske, E., Breucker, H., and Melikat, R. (1979) Gross morphology and fine structure of the olfactory organ of rainbow fish (Atheriniformes, Melanotaenidae). *Acta Zoologica*, **60**, 173–186.

Zhang, C. and Hara, T.J. (2009) Lake char (*Salvelinus namaycush*) olfactory neurons are highly sensitive and specific to bile acids. *Journal of Comparative Physiology A-Neuroethology Sensory Neural and Behavioral Physiology*, **195**, 203–215.

Zippel, H.P., Reschke, C., and Korff, V. (1999) Simultaneous recordings from two physiologically different types of relay neurons, mitral cells and ruffed cells, in the olfactory bulb of goldfish. *Cellular and Molecular Biology (Noisy-le-Grand)*, **45**, 327–337.

Zippel, H.P., Gloger, M., Nasser, S., and Wilcke, S. (2000) Odour discrimination in the olfactory bulb of goldfish: contrasting interactions between mitral cells and ruffed cells. *Philosophical Transactions of the Royal Society of London: Series B*, **355**, 1229–1232.

Chapter 9
Measuring and Identifying Fish Pheromones

Michael Stewart[1] and Peter W. Sorensen[2]

[1]National Institute of Water and Atmospheric Research Ltd, Hamilton, New Zealand
[2]University of Minnesota, St. Paul, USA

9.1   INTRODUCTION

9.1.1   Definition of "Pheromone" from a Chemical Context

The term "pheromone" when used in a biological context—as defined in earlier chapters—takes on a slightly different meaning than in a chemical context. Analytical chemistry is concerned with measuring the concentrations of individual chemicals in a specific matrix, that is, water, urine, bile, liver, etc. For this, each chemical needs to be treated as a separate entity and measured independently of others. This is usually a challenging task as a biologically active pheromone may often consist of mixtures of related chemicals, at extremely low concentrations and in the presence of interfering chemicals.

In this chapter, the terms chemical, molecule, or analyte will be used to describe a single chemical, which is either the fish pheromone of interest, or part of a pheromone mixture or "complex" (see Sorensen and Baker, 2015). In other words, when focusing on the chemistry of pheromones, we generally consider them as single compounds even if fish recognize and use them in mixtures.

9.1.2   Overview of Known Chemical Structures of Fish Pheromones

The number of chemicals excreted and used by fish for a variety of purposes that have been unambiguously determined is still very few. This is a reflection of the challenges involved in resolving a biologically relevant new chemical among the myriad of other chemicals excreted by the fish or produced by other species (e.g., microorganisms, plants) in the water. This often-complex mixture of organic chemicals is called the "matrix," and conceptually can be considered a "haystack" from which the chemical of interest is a "needle."

Steroids and prostaglandins are the predominant pheromones that have been identified from fish; however, other chemical classes have been described. The following is not an exhaustive list of chemicals that have been described from fish, but an overview of chemical classes that have been described.

Bile acids (steroid acids and sterols typically produced by the liver and often found in the bile) have been shown to be potent olfactory stimuli for many fish species (Michel and Lubomudrov,

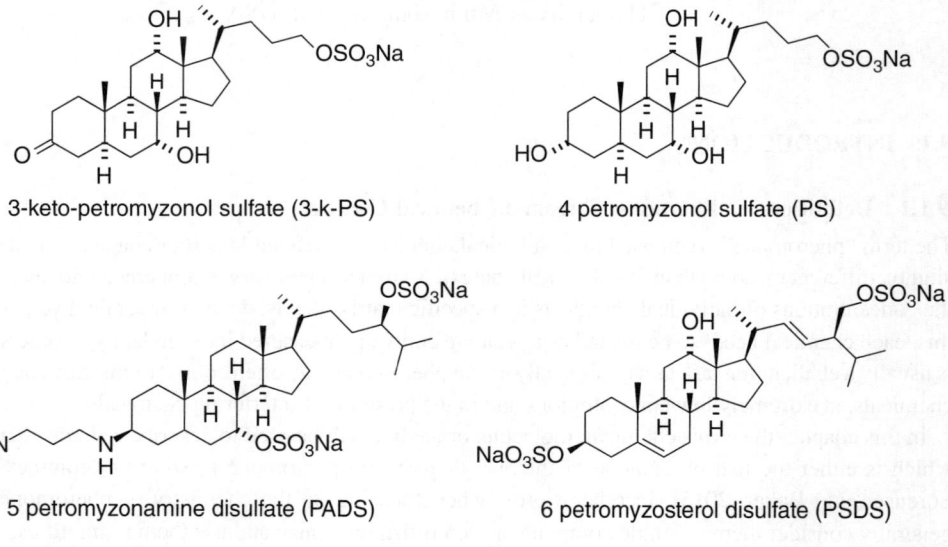

1 Taurocholic acid (TCA)

2 glycocholic acid (GCA)

1–2 Compounds

3-keto-petromyzonol sulfate (3-k-PS)

4 petromyzonol sulfate (PS)

5 petromyzonamine disulfate (PADS)

6 petromyzosterol disulfate (PSDS)

3–6 Compounds

1995; Zhang, Brown, and Hara, 2001; Baker *et al.*, 2006). They are commonly conjugated with taurine (e.g., taurocholic acid (TCA, **1**)), glycine (e.g., glycocholic acid (GCA, **2**)), and/ or sulfate or glucuronide functional groups.

Extensive studies of the sea lamprey, *Petromyzon marinus*, have identified sulfated bile acids as the important pheromones from this species. The male sex pheromone is a mixture of 3-keto-petromyzonol sulfate (3-k-PS, **3**) (Li *et al.*, 2002; Siefkes, Winterstein, and Li, 2005) and other unidentified components. Larval sea lamprey release a migratory pheromone that has three identified components: (i) petromyzonol sulfate (PS, **4**), (ii) petromyzonamine disulfate (PADS, **5**), and (iii) petromyzosterol disulfate (PSDS, **6**) (Li, Sorensen, and Gallaher, 1995; Polkinghorne *et al.*, 2001; Fine and Sorensen, 2005; Sorensen *et al.*, 2005; Hoye *et al.*, 2007).

The goldfish, *Carassius auratus*, has also been extensively studied. Steroids derived from gonadal (hormonal) biosynthetic pathways appear to be commonly used by this and many other species. Gonadal (or sex) steroids identified in the goldfish as pheromones are 17,20β-dihydroxy-4-pregnen-3-one (17,20βP, **7**), the sulfate of 17,20-βP (17,20βP-S, **8**), and androstenedione (**9**) (Scott and Sorensen, 1994).

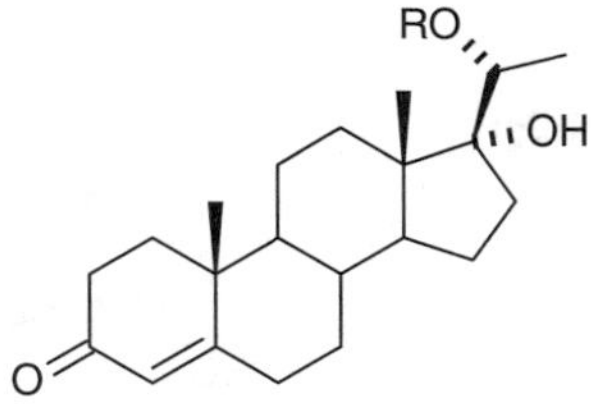

7 R=H   17,20β-dihydroxy-4-pregnen-3-one (17,20βP)

8 R=SO₃Na 17,20β-dihydroxy-4-pregnen-3-one-20-sulfate (17,20βP-S)

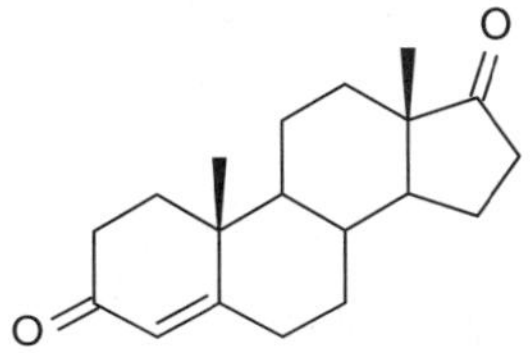

9 Androstene-3,17-dione

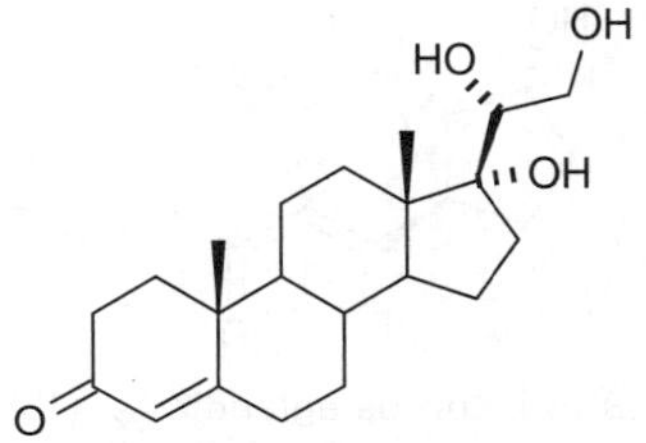

10 17,20β,21-trihydroxy-4-pregnen-3-one (20β-S)

7–10 Compounds

The closely related 17,20β,21-trihydroxy-4-pregnen-3-one (20β-S, **10**) has been identified as a maturation-inducing hormone and pheromonal precursor in the percid fish, *Gymnocephalus cernuus* (Sorensen *et al.*, 2004). Many other "hormonal pheromones" have been described in dozens of other fish species and are described by Stacey (2015).

F-series prostaglandins, PGF$_{1\alpha}$ (**11**) as well as PGF$_{2\alpha}$ (**12**) and the oxidized metabolite of PGF$_{2\alpha}$, 15-ketoprostaglandin F$_{2\alpha}$ (15 k-PGF$_{2\alpha}$; **13**) are part of the goldfish postovulatory hormonal pheromone mixture (Sorensen and Hoye, 2010). Similarly, F prostaglandins derived from ovarian tissue have been implicated as hormonal pheromones in many thousands of fishes although specific information about precise identity and production is rather scarce (Stacey, 2015; Lim and Sorensen, 2011).

Other chemical classes that have been identified with pheromonal activity in fish include L-kynurenine (**14**)—a metabolite of L-tryptophan—which was identified as a nonhormonal sex pheromone from female masu salmon (*Oncorhynchus masou*) (Yambe *et al.*, 2006). The tetrapeptide, Ile-Leu-Met-Glu (ILME, **15**) was identified from eggs of the cuttlefish, *Sepia officinalis*, as a potential pheromone involved in the transport of oocytes during egg laying (Zatylny *et al.*, 2000). Hypoxanthine-3-*N*-oxide (**16**) has been proposed as a component of the Ostariophysan alarm pheromone system with the suggestion that the nitrogen oxide functional group acts as the chief molecular trigger (Brown *et al.*, 2000). However, production of **16** by fish has yet to be confirmed using biochemical or chemical techniques.

This situation highlights the importance of not placing constraints on possible structural classes of chemicals excreted by fish, which is especially important when searching for an unknown fish pheromone, where the researcher needs to be extremely careful not to bias their search by assuming the chemical is likely to be steroid or prostaglandin based.

Subtle differences between chemicals can be extremely important in signal function. For example, 3-kPS (**3**) and PS (**4**) only differ in the oxygen on C3; being a ketone in **3** and a hydroxyl group in **4**.

11 prostaglandin $F_{1\alpha}$ (PGF$_{1\alpha}$)

12 prostaglandin $F_{2\alpha}$ (PGF$_{2\alpha}$)

13 15-ketoprostaglandin $F_{2\alpha}$ (15K-PGF$_{2\alpha}$)

14 L-kynurenine

15 Ile-Leu-Met-Glu (ILME)

16 hypoxanthine-3-*N*-oxide

11–16 Compounds

This subtle oxidative difference translates to a functional difference of **3** being the major male sex pheromone (Li *et al.*, 2002) and **4** being part of the larval migratory pheromone (Fine and Sorensen, 2005; Sorensen *et al.*, 2005) in the sea lamprey.

From a chemistry perspective, subtle changes in chemical structures of pheromone mixtures require the ability to differentiate between these sometimes extremely closely related compounds, before quantitation can be achieved. Of course, this depends on the availability of well-characterized standards. Methods to achieve this are described further on.

Synergistic, or additive effects, between chemicals have been noted in some cases (Sorensen, Vrieze, and Fine, 2003; Ferrari *et al.*, 2008), where the sum of physiological or behavioral responses of individual chemicals is less than the mixture or "whole extract." Recent studies on goldfish (Levesque *et al.*, 2011) and common carp (Lim and Sorensen, 2011) have shown that unrelated mixtures of "polar" and "nonpolar" chemicals show synergistic effects and may impart species-specificity as sex pheromones. Synergism is a biological response and is not covered in this chapter.

9.1.3 Why Quantify Fish Pheromones?

The understanding of what chemicals are excreted by fish and the resulting concentrations in natural water bodies can have various important uses. Apart from the advancement of knowledge in basic science, applied uses can be the determination of populations of threatened (Stewart and Baker, 2012) or pest species (Siefkes, Winterstein, and Li, 2005; Sorensen and Hoye, 2007), and management of these species (Sorensen, 2015). Although not strictly pheromones, the measurement of stress hormones as welfare indicators in fish (Scott and Ellis, 2007) is an important application in aquaculture.

9.1.4 Quantifying Known Fish Pheromones

Quantifying known fish pheromones can sometimes be relatively straightforward, but often it is extremely challenging. Before any attempt is made, it is necessary to have authentic chemical standards of known purity and known weight (Section 9.2.2). Techniques for analysis must be capable of differentiating between closely related relevant chemicals, and/or be capable of detecting extremely low concentrations of these chemicals, for example, in natural waters that can be 10^{-12} Molar (M), or lower at locations downstream of the fish. These extremely low concentrations usually require that some sort of concentration of the chemical(s) of interest be made from the water. Methods for achieving this are described in Section 9.2.3.

Even after a concentration step, highly sensitive detection methods are necessary to either detect or quantify the chemical of interest. Two methods that are gaining popularity are mass spectrometry (MS) (Section "Mass Spectrometry") and immunoassay (Section "Immunoassay").

9.1.5 How to Elucidate the Structure of Unknown Pheromones

Perhaps, the greatest challenge in this area is determining chemical structures of fish pheromones that are yet to be described. This can be especially complex and is usually a long-term multidisciplinary effort. Only a few examples of this seem to exist (see Section 9.1.2). As fish release pheromones primarily as a multicomponent mixture, it is necessary to determine both the number of individual chemicals that make up that mixture and the relative importance of that chemical to the pheromone mixture.

Physical separation of the chemical mixture into discrete fractions, followed by a biological assay of each to determine pheromonal activity is the approach—based on classical natural products chemistry—commonly used and is coined "bioassay-directed fractionation." This is covered in Section 9.3.3. Determination of the chemical structure is likely complicated by the low amount of material. Advanced MS techniques (Section "High Resolution Mass Spectrometry") can give valuable information on the molecule and is the tool of choice. Unambiguous determination of a new chemical structure is usually performed by nuclear magnetic resonance (NMR; see Section "Nuclear Magnetic Resonance Spectroscopy") and confirmatory synthesis (Section "Chemical Synthesis") if any ambiguity still exists. However, even with recent sensitivity advances in NMR technology, microgram-to-milligram quantities of each chemical are usually required; and for this, a large-scale, repetitive fractionation may be required to accumulate enough material.

9.2 DETECTING OR QUANTIFYING PHEROMONES WHOSE STRUCTURES ARE ALREADY KNOWN

9.2.1 Overview

A clear distinction needs to be made between detection and quantitation of pheromones. If it is only required that a chemical be detected in a matrix, then methods are only necessary to detect and confirm that chemical. If, however, information is needed about the environmental concentrations of a specific chemical, then a much more stringent and robust analytical approach is necessary to achieve this.

9.2.2   Authentic Standards

The quantitation of known chemicals requires authentic standards, for which both purity and amount provided are known. Also highly desirable are fully synthetic "surrogate" standards that have virtually the same physicochemical properties as the natural chemical, but that do not occur naturally. These are preferably deuterated (Xi *et al.*, 2011) and/or ^{13}C analogues of the chemical of interest, but non-natural isomers of the chemical of interest will suffice if isotopic standards are not feasible. If surrogate standards are not available, a "standard addition" approach may be used by spiking water samples known to be absent of pheromones (e.g., river waters lacking the species of interest) with known concentrations of standards (Fine *et al.*, 2006; Lim and Sorensen, 2011; Stewart, Baker, and Cooney, 2011). The standard analytical approach is to use these synthetic standards as spikes to ascertain the efficiency and accuracy of the concentration and quantitation step.

Often, commercially available standards are not available; and therefore, these need to be synthesized, either under contract by a commercial synthetic laboratory, or by a university laboratory. Again, unambiguous structure determination is required, with known purity a requirement.

9.2.3   Concentration Techniques

9.2.3.1   Overview

Concentration of the chemical may be required to be able to detect or quantify it. Even the most recent detection techniques are usually still less sensitive than a biological organism in detecting a pheromone.

Choosing the appropriate concentration technique and sorbent depends on both the application and the physicochemical properties of the pheromones of interest. Polarity is a term used to describe how hydrophilic (water loving) or hydrophobic (water fearing) a chemical is. This polarity is described by a partition coefficient (logP, for compounds with no ionic groups) or distribution coefficient (logD, for compounds with ionic groups) between water and a nonpolar solvent, usually octanol. LogP or logD can be estimated from chemical structure or can be measured empirically. The lower the number of log P or logD, the more hydrophilic the compound is.

9.2.3.2   Solid phase extraction (SPE)

Concentration of the chemical from water requires the right application for the right chemical. All major chromatography product suppliers have guides on the appropriate solid-phase extraction (SPE) phase to use for each chemical type, and readers are therefore urged to consult these guides if unfamiliar with these techniques.

Traditionally, silica-based reversed phase (C_{18}) SPE cartridges have been used. SPE cartridges work on the principal of chemical equilibrium between a sorbent and a solvent. In the case of reverse phase, and the solvent is water, the equilibrium for chemicals ranging from moderately polar to nonpolar is weighted toward the sorbent, and therefore the chemicals are retained by the sorbent. With the exception of highly water-soluble chemicals, large quantities of water can be passed through the SPE to concentrate the chemicals of interest. The solvent can then be modified to elute the chemical from the sorbent, by changing the equilibrium in favor of the solvent. These are useful for concentrating all but extremely polar chemicals (i.e., amino acids, nucleotides, and sugars). Less hydrophobic reverse-phase sorbents (i.e., C_8, C_3, CN, or phenyl) may be more appropriate than C_{18} for polar chemicals, where better retention may be observed. The major limitations of silica-based reversed phase SPE are that much care needs to be taken to avoid drying of the sorbent, residual silanols can cause irreversible adsorption of basic compounds; and in most cases, pH needs to be in the neutral-to-acidic range to avoid hydrolysis of the silica backbone.

More recently, sorbents that have a polystyrene and/or divinyl benzene backbone have gained importance in environmental sciences, especially for concentration of highly polar chemicals such as pharmaceuticals and pesticides. These polymeric sorbents circumvent many of the problems of silica-based sorbents. They have higher capacity, have better wetability, are tolerant to partial drying, and can be used over a wide pH range. Examples include Oasis® HLB (Waters Corporation) and Strata™-X (Phenomenex).

As most water-soluble fish pheromones have ionic, or ionizable, functional groups (sulfates, carboxylic acids, amines, etc.), they can potentially be concentrated using ion-exchange sorbents or resins. Ion-exchange sorbents can be weak or strong anion or cation exchange, or a mixed-mode hybrid with reverse phase, depending on the application required. Ion-exchange resins may be successful in removing interfering matrix material where reverse-phase separations cannot. In a recent example, a mixed-mode cation exchange sorbent was found to be superior to other phases in concentrating 3-k-PS (**3**) from stream water and removing matrix interferences (Xi *et al.*, 2011). This was counterintuitive as 3-k-PS (**3**) contains an anionic sulfate group and would not be expected to be retained by a cation exchange resin. Indeed, in distilled water this was the case, and the authors postulated that a neutral or positively charged complex of **3** must be formed with compounds in river water. These findings illustrate that, for ion exchange resins at least, a range of different phases should be tested to ascertain the most appropriate.

As stated earlier, if quantitation is the goal, recovery of the chemical(s) of interest needs to be validated, using standard surrogates or standard addition (as described in Section 9.2.2) for each concentration procedure.

9.2.3.3   Passive sampling

Passive samplers are devices that are placed in a body of water (or air for airborne chemicals) that accumulate chemicals of interest over time. The samplers are removed after a suitable time, and their contents are extracted in a similar way to SPE. Passive samplers require much validation to provide estimates of water concentrations; however, their major advantage over traditional water sampling is that the samplers are much less susceptible to short-term temporal and spatial water concentration variations, and they can be deployed at many sites concurrently, making them good candidates as tools for monitoring of fish populations via pheromone water concentrations (Stewart and Baker, 2012).

Polar Organic Chemical Integrative Samplers (POCIS) have been recently developed specifically for polar water-soluble chemicals such as pesticides and pharmaceuticals (Alvarez *et al.*, 2004). As their name suggests, POCIS accumulate chemicals over time and are especially useful where occurrence is transient or concentrations are extremely low (Alvarez *et al.*, 2005), which is the case for chemicals excreted by fish. As this technique sequesters chemicals from natural water bodies in a time-averaged manner, a useful application would be in population monitoring. POCIS have been incorporated into a method for population monitoring of the southern pouched lamprey, *Geotria australis*—which is suspected to be in population decline— in New Zealand rivers. Methodology was developed to quantify the lamprey-specific migratory pheromone, PS (**4**), in deployed POCIS and active sampling of river water. Field sampling results demonstrated that POCIS was superior to active water sampling, providing quantitative results at all sites surveyed and has promise as a monitoring tool of this (and other) species (Stewart and Baker, 2012).

9.2.3.4　Highly polar chemicals

If a chemical cannot be concentrated from water using techniques described earlier, concentration can be achieved by removal of water via evaporation.

Rotary evaporation is an established laboratory technique for removing volatile solvents. The solvent is removed by transfer from a liquid-to-gas phase and then condensed for removal. This is performed under reduced pressure to expedite the evaporation and keep temperature low to reduce chemical degradation.

If a chemical is especially labile or semivolatile, freeze-drying can be used. With this technique, water is removed by sublimation—directly from a solid to a gas—which is then condensed for removal.

Both these techniques require specialized instrumentation, which is common to most chemical laboratories. If removal of large volumes of water is required, industrial scale equipment is required.

9.2.4　Separation Techniques

9.2.4.1　Overview

The analysis of a chemical usually requires separation from the large array of co-occurring chemicals (the matrix: see section 9.1.2). This matrix can often interfere with detection techniques used to confirm the identity of the chemical. There are different separation techniques available depending on the chemical properties of the pheromones and the application required.

9.2.4.2　High Pressure Liquid Chromatography (HPLC)

Most separations of water-soluble chemicals have historically been based on reverse-phase high-pressure liquid chromatography (RP HPLC), which uses columns containing small uniform sorbent particles (3–5 μm). HPLC is characterized by highly efficient separation, reproducibility, and the ability to automate the procedure. Recently, there has been a move toward ultra HPLC (UPLC), with even smaller particle-size columns (<2 μm). This has led to improved efficiency, shorter analysis times, and higher sensitivity but is accompanied by extremely high pressures with the associated investment in specialized instrumentation required to handle this (Churchwell *et al.*, 2005; Swartz, 2005).

Buffers (e.g., phosphate or acetate) and/or pH modifiers (e.g., formic acid or triethylamine) can be added to the HPLC solvent. For chemicals with acidic or basic functional groups, buffering of the solvent phase pH is essential to give reliable retention times, good peak shape, and improved detection by MS. Of note, volatile buffers are highly desirable when coupling HPLC with MS. All major chromatography product suppliers have information on appropriate buffer systems for the application of choice.

9.2.4.3　Gas Chromatography (GC)

Gas chromatography (GC) involves the separation of chemicals based on a gas/solid phase interaction. As such, chemicals need to be volatilized, which usually limits application of this technique to relatively small, moderately nonpolar chemicals. GC is the common method of separation for insect pheromones, which as they are transported by air, are by definition much more likely to be volatile. On the contrary, water-soluble chemicals—such as those excreted by fish—frequently have polar functional groups to increase solubility in water, but which greatly decreases their volatility. In the case of steroids, these may be hydroxyl groups or amines (OH or NH_2, classed as free steroids) and/or ionic groups (e.g., sulfate, glucuronide, taurine which are classed as conjugates).

However, polar conjugates may be either selectively or nonselectively removed (see Section "Chemical and Biochemical Modification, Section 9.3.5.2"), and separation achieved by GC. Derivatization of OH or NH$_2$ groups on free steroids is still necessary (See Becker, Galili, and Degani (1992); Van Weerd *et al.* (1991) for examples). However, GC has become less popular due to the versatility of RP HPLC as a separation tool for water-soluble chemicals.

9.2.4.4 Thin Layer Chromatography (TLC)

Thin-layer chromatography (TLC) works on the same principal as HPLC, however with much lower resolution. TLC can give rapid qualitative data but with the inability to automate and requires intensive labor. When chemicals from fish are in high concentration, as in bile or liver or possibly holding tank water extracts, TLC can be combined with UV detection (if pheromones have a UV chromophore) or specific stain development (i.e., ninhydrin for amines) and used as a useful qualitative tool. However, for extremely low environmental concentrations of fish pheromones—as experienced in natural waterbodies—a much more sensitive detection technique is required. Metabolites from the incorporation of radioactive precursors into pheromone biosynthesis can be traced via this technique (Scott and Canario, 1992; Vermeirssen and Scott, 1996). Alternatively, strips of sorbent may be scraped off, extracted with solvent and analyzed by MS.

9.2.5 Detection Techniques for Measuring Known Pheromones

9.2.5.1 Overview

MS and immunoassay (IA) are two extremely sensitive techniques for detecting and quantifying environmental concentrations of organic chemicals. Other techniques that have some use are fluorescence and ultraviolet spectroscopy.

9.2.5.2 Mass Spectrometry (MS)

MS has the ability to detect and quantify extremely low concentrations of chemicals in complex matrices. A mass spectrometer creates charged molecules in the gas phase and effectively "filters out" those chemicals of interest from the matrix. As such, MS is ideally suited to water-soluble chemicals due to functional groups that are either ionic (i.e., sulfate and carboxylic acid) or for which ionization can be induced (i.e., amine and phenol).

Two relatively recent developments in MS have made it much more suitable for detection and quantitation of fish pheromones. "Soft ionization" techniques and probes (ESI, APCI, MALDI, etc.) deliver the chemical of interest to the mass spectrometer largely intact, providing more chemical information and sensitivity. More recently, various combinations of tandem MS (MS/MS) provide greater filtering capacity of the analyte of interest, which allows for lower detection limits and greater selectivity in detection of compounds at extremely low concentrations. Two recent examples use tandem MS to achieve extremely low quantitation limits of lamprey pheromones in stream waters (Stewart, Baker, and Cooney, 2011; Xi *et al.*, 2011).

9.2.5.3 Fluorescence Spectrometry

Fluorescence spectroscopy is extremely sensitive and in some cases perhaps more so even than MS; however, it is restricted to those chemicals that are naturally fluorescent or for which fluorescence can be induced. The best application of fluorescence spectroscopy is when extremely low limits of detection are required.

However, fluorescence spectroscopy does not have the selectivity or versatility of MS; therefore, care should be taken to reduce and avoid inclusion of interfering fluorescent chemicals in the matrix. Two examples of fluorescence spectroscopy are induced fluorescence and fluorescent tags. Polkinghorne *et al.* (2001) used induced fluorescence to analyze selectively for lamprey bile acids. By using a custom column containing 3α-hydroxysteroid dehydrogenase, bile acids with a

17 BDETS　　　　　　　　　　　　　　　　　　　　　18

17–18 Compounds

3α-hydroxy functional group were oxidized selectively, resulting in the reduction of NAD to NADH. NADH production was recorded on a fluorescence detector.

Attachment of a fluorescent tag as a second method for inducing fluorescence has been achieved. The fluorescent tag 1,2-benzo-3,4-dihydrocarbazole-9-ethyl-*p*-toluenesulfonate (BDETS; **17**) was attached to bile acids for sensitive detection (down to 1.3×10^{-14} M) in serum (You *et al.*, 2004). A direct application to fish pheromones has not been found; however, a procedure using 1-anthroyl nitrile (Goto *et al.*, 1983)—a fluorescent tag selective to 3α hydroxy steroids—was successfully used to form a fluorescent derivative (**18**) of petromyzonol sulfate (**4**) with a 1×10^{-13} M detection limit (Stewart and Baker, unpublished data).

9.2.5.4　Ultraviolet (UV) Spectrometry

Ultraviolet (UV) spectroscopy is very versatile but restricted as a detection technique. The chemical of interest needs to have a chromophore that absorbs UV radiation. UV can be combined with TLC (using a lamp with wavelength 254 or 365 nm) or post-HPLC elution as single wavelength or multiple wavelength (diode array) detectors. Diode array detectors also incorporate the visible spectrum (400–800 nm) and are useful if a characteristic UV spectrum can be obtained (usually from 250 to 600 nm). However, even if a chemical has a unique UV spectrum, the extinction coefficient is usually too low for UV to be applicable when detecting environmental concentrations of water-soluble chemicals. As such, UV detection is restricted to higher concentrations as would be experienced from fish organs or fluids, that is, liver, bile, and urine.

9.2.5.5　Immunoassay

Immunoassay uses indirect procedures (competitive binding to antibodies) to measure miniscule quantities of compounds and represents a reasonable alternative to direct chemical measurement because it can be both quick and inexpensive. However, if not conducted with appropriate controls, misidentification of compounds is possible. Nevertheless, various types of immunoassays have been successfully developed to measure fish pheromones on numerous occasions. Two techniques exist, radioimmunoassay (RIA) and enzyme-linked immunosorbent assay (ELISA, ELIZA, or EIA). This next section introduces immunoassay methods and describes their application in the quantitation of fish pheromones.

9.2.5.5.1　Radioimmunoassay (RIA)

RIA was developed in the 1970s for measuring insulin in human blood but has since been widely deployed to measure innumerable types of chemical structures in various types of fluids. The first step in the use of a RIA is to develop antibodies for the compound of interest (e.g., the antigen). This is a complex, somewhat trouble prone, process and is described further below. The second step requires synthesis of a radiolabeled antigen to serve as a tracer. In RIA, known amounts of tracer are added to samples containing unknown quantities of the compound of interest (in our case, the pheromone) in the presence of antibodies. The radiolabeled and unlabeled antigens then bind to the antibody (or antibodies) in ratios that reflect their relative abundance. The excess product is then washed away, and the amount of radiation bound to it is measured and compared with a standard curve. This value should reflect the quantity of antigen (or antigen-like compounds; see below) present in the sample. The main strength of RIA is that once established, and all conditions for its implementation have been met (see below), it can be very simple and inexpensive to run as dozens of samples can be measured concurrently. With a high-quality antibody, sensitivity can also be in the picogram range (i.e., similar to MS). Further, sample purity need not always be high because radioactive labels do not suffer from background matrix effects. A few pheromonal structures are not ideally suited to analysis by MS (e.g., androstenedione, a pheromone in the goldfish (Sorensen *et al*., 2005, unpublished) and therefore are more suited to measurement using immunoassay. Major disadvantages of RIA are the need to produce antibodies and labeled antigens, and the need for specialized equipment to measure and manage radiation (both operationally and administratively).

Establishing and validating a RIA

Establishing a credible RIA is a complex process that is briefly described here. Aside from the common need for concentration of samples (following the procedures described earlier), RIA requires radiolabeled and unlabeled authentic standards as well as antibodies for them. Tritium (^{3}H) and ^{14}C are most often used as labels. Synthesis of radiolabeled compounds adds significantly to the cost of RIAs, but only trace amounts (microgram quantities) are typically needed and the half-life is long (thousands of years); therefore, a single synthesis can be sufficient for decades. Probably, the single greatest challenge with developing an RIA is producing an antibody (or antibodies) that are both specific and sensitive enough to the antigen. Two types of antibodies can be developed. Polyclonal antibodies are derived by injecting animals (e.g., rabbits) with conjugated forms of the antigen in the hope that they will naturally produce antibodies that can be purified from their blood. Usually, multiple antibodies are produced. Monoclonal antibodies are produced in a similar fashion using hybridoma cell culture. Although more difficult and expensive, this technique can ultimately produce very large quantities of a single (and very pure) antibody. In any case, both techniques are part "art" and part "skill" as scientists must trigger production of an antibody to small molecules (i.e., pheromones) that are not always antigenic. Often technology companies take lead roles in this complex process. Various conjugates are often attached to the antigen to trigger an immune response, but this can influence antibody specificity. Certain natural conjugates (e.g., sulfates) can also cleaved by the animal upon injection; therefore, it has not proven practical to measure them directly by immunoassay. In general, antibodies only recognize (bind) to specific portions of antigens, and this can present a problem if binding is not restricted to portions of the antigen (pheromone) that are highly distinctive.

Because of the risk of nonspecificity, RIAs must be validated. Thus, before deploying an RIA, it is essential to determine its sensitivity and specificity in natural samples. Several steps are involved. In general, the sensitivity of the antibody is determined first. This involves determining the relative binding of unlabeled and labeled authentic antigen to polyclonal or monoclonal antibodies. Different quantities of all three are generally tested in clean solvent to optimize the assay and evaluate its potential sensitivity. The second step typically entails determining antibody specificity. This is determined using synthetic standards of analogues of the antigen that might be expected to be naturally present in samples. Here, experience is important as some analogues may need to be synthesized and decisions made about relevant concentrations to test as antibodies are rarely 100% specific. A final step involves validating the assay in natural samples (i.e., fluids) to ensure that no unanticipated interference or nonspecific binding occurs. Typically, fluids (water) from fish believed not to produce pheromone are extracted and then known amounts of unlabeled standard added to it for this set of assays. This yields a standard curve. In general, binding is relatively linear (on a log scale), but nonspecific reactions and saturation can occur; therefore, it is critical to describe the range of concentrations over which the antigen is accurately measured in a linear and predictable manner. Each antibody will have its own curve, and this typically requires samples to be carefully diluted so that values measured can be extrapolated to the standard curve for quantification.

A key question when interpreting data generated by immunoassay (and especially when using a commercial kit) is how extensive validation has been. A credible and well-validated RIA has the potential to measure hundreds or even thousands of samples at low cost with high sensitivity and specificity. All assays must include blanks and controls.

Using RIAs

Dozens of studies have successfully deployed RIAs and yielded what appear to be reasonable values. The first study to use RIA appears to have been that of Dulka *et al.* (1987) who used RIA to measure the hormonal pheromone, 17,20βP (**7**), which is both a sex hormone and pheromone in many fish species. Female goldfish were found to release many hundreds of nanograms of this hormonal steroid per hour before ovulation, seemingly proving its function as pheromone. It is fortunate that the goldfish was used as a model for sex pheromone function because this species uses many relatively common hormonal products as sex pheromones, and RIAs already existed for many of them. Alexander Scott and his colleagues. (see Scott and Canario, 1992; Scott and Ellis, 2007) have since been responsible for many of the first studies of hormonal pheromone release in goldfish as well as various cyprinids, salmonids, and even some flatfish. Of special note was their work on measuring various conjugates that they measured by cleaving the moieties of interest before analysis and then measuring using specific antisera. Various F prostaglandins have also been measured in goldfish water and urine (Sorensen *et al.*, 1988) using RIA. Although antibody specificity is typically somewhat less to PGFs than steroids, the measured values have recently been validated by MS and ELISA (Lim and Sorensen, 2011; see below).

9.2.5.5.2 Enzyme-linked immunosorbent Assays (ELISAs)

Enzyme linked immunosorbent assays or ELISAs, like also rely on antibodies to measure the compounds; but instead of using radiolabeled antigen, they use various types of biochemical tags (typically fluorescent), which they then visualize using various types of enzyme system antibodies. Sensitivity and specificity is fundamentally the same as for RIA as are most of ELISA's strengths (ease of use and low cost) and weaknesses (antibody nonspecificity). Additional factors are that no radiation is used, therefore special facilities are not needed; however, the use of complex biochemical systems to visualize bound antigen opens up the possibility of interference from the chemical matrix. Several types of ELISAs exist (indirect, sandwich, competitive, etc.), and we will not review them here other than to mention that most ELISAs use two antibodies—the first of which binds to the antigen, and the second (which typically has a chemical tag) binds to the former, which in turn binds to the well so the product can be visualized.

Developing ELISAs

Like an RIA, an ELISA requires authentic standards and antibodies for that standard. In addition, ELISAs typically use at least one other antibody to react with the first antibody, and this must be developed along with the enzyme system to achieve visualization. These procedures are lengthy and complex and almost always carried out by commercial laboratories. Because the labeled antigen that is used as a tracer is usually visualized (sometimes with fluorescence), special care must be taken to address the possibility of nonspecific interference by the compounds found in the matrix. Interference (which has not yet been reported) could conceivably lead to artificially low or high measurements.

Using ELISAs

Efforts have been made to implement ELISAs to measure sea lamprey bile acids (Yun *et al.*, 2002). Recently, F prostaglandins were measured by ELISA in carp-release waters and validated by MS measurements (Lim and Sorensen, 2011). ELISA proved to be especially valuable in the latter study as it could distinguish between some prostaglandin metabolites that could not be differentiated by the MS techniques used. Commercially available ELISAs have proven to have similar sensitivities as MS and relatively easy to implement once validated (which follows the same protocols used for RIA). It is best suited for using standard products as development is expensive and tedious, especially if large numbers of measurements are needed. Hormonal pheromones are often very good candidates.

9.3 IDENTIFYING PHEROMONES WHOSE STRUCTURES ARE UNKNOWN

9.3.1 Overview

The discovery of chemicals excreted from fish is a complicated process. Consequently, there are still not many examples of new chemicals described from fish (see Section 9.1.2). However, with improvements in analytical technologies, especially toward sensitivity, these challenges are not insurmountable.

9.3.2 The Question of What Is a Pheromone

As stated earlier, a pheromone can be a single chemical component, but is most usually a mixture of related compounds. Fish excrete many other chemicals that do not function as a pheromone; therefore, it is essential to be able to assess the pheromonal activity and importance of individual compounds excreted. For this, physiological and behavioral responses to the pheromone (either as a chemical or as a chemical mixture) need to be established and used to guide isolation and identification.

9.3.3 Bioassay-Directed Fractionation

The procedure of separating out biologically relevant chemicals from a mixture has been around for decades and was initially developed in the field of bioactive natural product discovery. Bioactivity-directed fractionation is an iterative process whereby crude mixtures are physically separated into "fractions" of semipure mixtures, which are subsequently analyzed for a biological response. Bioactive "fractions" are then re-fractionated—usually following a different but complementary separation procedure—and each "subfraction" reassayed. This process is repeated until an endpoint is reached, which can be the identification of the impure bioactive chemical or the purification of the bioactivity into a homogenous (pure) chemical. In either case, confirmation of bioactivity should be validated against a synthetically derived identical chemical, to confirm that the bioactivity observed is the candidate compound and not an unidentified minor impurity.

A comprehensive example of bioassay-directed fractionation of fish pheromones from concept to confirmation (Sorensen *et al.*, 2005; Hoye *et al.*, 2007; Fine and Sorensen, 2008) has been provided in the sea lamprey (*P. marinus*). Concentration and separation of the pheromones was achieved by solid-phase extraction and HPLC, using EOG assays and behavioral trials to identify the active pheromones. Structural identification of the active pheromones was achieved by MS and ^{1}H nuclear magnetic resonance, with associated pattern recognition. Finally, unambiguous identification of both chemical structure and biological relevance was achieved by synthesis of the two novel pheromones— PADS (**5**) and PSDS (**6**). Structure identification of **5** and **6** is covered in more detail in Section 9.3.5.4.

9.3.4   Pheromone Class Determination

As stated in Section 9.1.2, preconceptions about the structure class (or classes) of pheromone that may be excreted from a species that has not been extensively studied cannot be made. However, *potential* structural classes could be suggested from related species that have been examined. This information can be combined with other questions to build up a picture about the chemical such as the following:

1. Is the chemical a large biomolecule or a small molecule?
2. Is the chemical retained on reverse-phase SPE?
3. Is elution on SPE (or retention time on HPLC) pH dependent?
4. How does the chemical behave on different ion-exchange resins?
5. What authentic standards are detected by EOG?
6. Is there any cross reactivity in RIA experiments with known compounds?
7. Can a molecular weight be established with MS, and can this be related to known standards?
8. What molecular fragmentations can be achieved by hyphenated MS?
9. What information can be obtained from NMR experiments?

Establishing if the pheromone is a biomolecular polymer (i.e., a protein or large peptide) or a small molecule can be achieved rapidly by molecular weight cutoff membranes. The approximate size of the pheromone (closely related to molecular weight) can often be established by filtration through a membrane of defined nominal molecular weight cutoff, that is, 1000 Da. If the molecule is retained by the membrane, then it is likely a biopolymer; if it is not, then it is a small molecule. Of note, there are no described fish pheromones that are biopolymers; however, this does not preclude them as candidates in the pursuit of a new pheromone. More refined molecular weight determinations can be made via gel permeation techniques, where separation is based on size and can be compared with suitable standards.

If the analyte is not retained on reverse-phase SPE, then it is unlikely to have a steroidal backbone or to be a prostaglandin. Even steroids with multiple highly polar functional groups, that is, PADS (**5**) and PSDS (**6**), are retained on these sorbents.

Highly polar functional groups have either acidic or basic properties. Glucuronides are acidic, whereas amines are basic. The elution on SPE, or the retention time on reverse-phase HPLC, will be affected by manipulation of the solvent pH. Parallel fractionations of a crude extract (SPE or HPLC) with either acidic (e.g., formic acid, pKa 3.8) or basic (e.g., triethylamine, pKa 11.0) pH conditions, followed by appropriate bioassay can provide useful information. However, of note, the pheromone may contain both acidic and basic functional groups, such as in PADS (**5**), which can complicate the issue.

Further evidence for basic or acidic functional groups can be obtained by ion-exchange chromatography. Acidic functional groups can be retained on anion exchange resins, whereas basic functional groups may be retained on cation resins. However, caution needs to be exercised. As stated earlier, Xi *et al.* (2011) recently reported that 3-k-PS (**3**) was extracted from stream water more efficiently on a mixed-mode cation exchange resin than a mixed-mode anion exchange resin. The inference from these results is that 3-k-PS contains a basic functional group—such as an amine—that may act as a cation. However, 3-k-PS contains a sulfate, which would be expected to be retained by anion exchange resins.

Biological investigations can give information as to potential pheromone candidates. Extremely helpful, and occasionally, enabling information has also come from electro-olfactogram (EOG) recording. In fact, the first sex pheromones identified in the goldfish were isolated and eventually identified using this technique alone, as the olfactory system can be extremely specific and typically

19 5α-cyprinol sulfate

19 Compounds

expresses a few olfactory receptors that have extreme sensitivity and specificity (Stacey and Sorensen, 2009). For example, careful structure-activity studies of olfactory ligands derived from hormonal products tentatively point to several dozen novel steroid structures with pheromonal function (see Stacey, 2014). To date, all hormonal products that have been found to be detected by EOG with notable sensitivity and specificity by fishes have eventually been found to have pheromonal function (Sorensen and Caprio, 1998). EOG experiments can be performed on multiple standards of various structural classes in a short time period. For example, Baker *et al.* (2006) recorded EOG measurements for 30 bile acid standards against the banded kokopu, *Galaxias fasciatus*. Four exhibited strong responses, of which one of those, 5α-cyprinol sulphate (**19**) was subsequently detected in the gall bladder. Eom, Jung, and Park (2009) measured EOG responses for 19 prostaglandins as sex pheromones in the salamander, *Hynobius leechii*, with follow-up LC/MS/MS analysis and behavioral tests to identify the dominant pheromone.

Of note, the chemical may not be novel and may have been identified from other sources. If it is suspected that the chemical is not novel and that this chemical can be sourced—for example, from a commercial supplier—then confirming the structure is greatly simplified. Sometimes, immunoassays can also assist greatly in this process if the structures are commonplace. Indeed, some level of antibody nonspecificity may even be helpful.

9.3.5   Structure Identification

9.3.5.1   Overview

Obtaining the purified fish pheromone is a challenging task in itself, however is only part of the way to establishing the chemical identity of an unknown pheromone. This section discusses some of the traditional methods that are used to solve the chemical structure.

9.3.5.2   Chemical and Biochemical modification

The pheromone may be chemically or biochemically manipulated to obtain information about the chemical structure. This can be useful when there is a very low amount of material, as success of each modification can be monitored by MS. It is important that the pheromone is of the highest purity possible; therefore, impurities do not pose complications in the monitoring procedure.

Chemical modifications can include oxidation and reduction, hydrolysis (cleavage of functional groups) and alkylation (e.g., methylation and acetylation). However, chemical modifications can be potentially misleading, unless information is available about functional groups present in the molecule.

More selective biochemical modifications may also be undertaken such as the use of glucuronidases or sulfatases, which are used to cleave glucuronides and sulfates, respectively. The loss of sulfate or glucuronic acid (either single or multiple) can be readily monitored by MS. Vermeirssen and Scott (1996) used both chemical and biochemical modifications to analyze conjugated steroids in rainbow trout, *Oncorhynchus mykiss*. Sulfate groups were cleaved by acid hydrolysis and glucuronide groups were cleaved by β-glucuronidase enzyme.

9.3.5.3　High resolution mass spectrometry (HRMS)

High resolution MS (HRMS) can give indicative molecular formulae—the proportion of each element in the molecule (usually CHONS)—for an unknown molecule. Generally, the smaller the molecule, the fewer the available number of permutations of molecular formulae. HRMS is used in combination with other spectroscopic information to solve the structure. Yambe *et al.* (2006) used time of flight (TOF) MS to establish a molecular formula of $C_{10}H_{12}N_2O_3$ for the unknown pheromone from the urine of masu salmon, *O. masou*. Characteristic spectroscopic data included the observations that a spot on TLC absorbed UV light at 365 nm (suggesting a conjugated aromatic moiety) and was positive to ninhydrin spray (suggesting an amine functional group). UV maxima at 260 and 370 nm supported a conjugated aromatic system. Two-dimensional (2D) NMR experiments suggested kynurenine as the gross structure, Marfey's analysis (Marfey, 1984) was used to solve the stereochemistry, and final confirmation of L-kynurenine (**14**) provided by comparison with an authentic standard.

High-resolution MS has been incorporated into the back end of tandem mass spectrometers, designated quadrupole time of flight (qTOF). Using qTOF, fragmentation of the molecule can be performed at the front end of the instrument and each fragment subsequently measured in high resolution, providing valuable information on which functional groups may be present (as they cleave), or more importantly what may be the core structure.

9.3.5.4　Nuclear magnetic resonance spectrometry

Nuclear magnetic resonance (NMR) spectroscopy is a technique that is essential in the unambiguous confirmation of a new chemical (with the notable exception of structure determination by single-crystal X-ray diffraction). Like MS, NMR spectroscopy has developed rapidly and has found various applications. In the study of organic small molecules, proton (^{1}H) and carbon (^{13}C) NMR experiments—and 2D combinations of the two—have been invaluable in a "jigsaw" approach to solving chemical structures.

Information on the chemical structure class, functional groups present, position of attachment of functional groups, (relative) stereochemistry at chiral centers, and preferred structural conformations can be obtained through NMR spectroscopy. Concepts such as chemical shift (a nuclei-specific shift away from the applied magnetic field) and ^{1}H–^{1}H spin coupling (through bond effects) are especially useful. Comparisons of regions of the NMR spectrum can be made with authentic standards and information built up about the molecule via this method. This was used to great effect in the discovery of the chemical structures of PADS (**5**) and PSDS (**6**) (Hoye *et al.*, 2007), which is described later in this section.

NMR was initially designed for fields in organic chemistry such as natural products and synthetic chemistry, where milligram quantities (10^{-3} g) of material are routinely available. Recent technological advancements, especially with low-volume capillary probes (see Schroeder and Gronquist (2006) for a review) and cryogenic probes (low temperature), in combination with inverse detection (^{1}H) 2D methods and hyphenation with HPLC (see, e.g., Exarchou *et al.* (2005)) have meant that the chemical structure can be achieved on low microgram quantities. With their low volume (around 2 µl), capillary probes are ideally suited for hyphenation techniques such as LC/NMR and LC/SPE/NMR (Exarchou *et al.*, 2005).

These highly sensitive techniques are especially useful when pheromones can only be sourced at physiological concentrations, and where there is still an extensive concentration effort

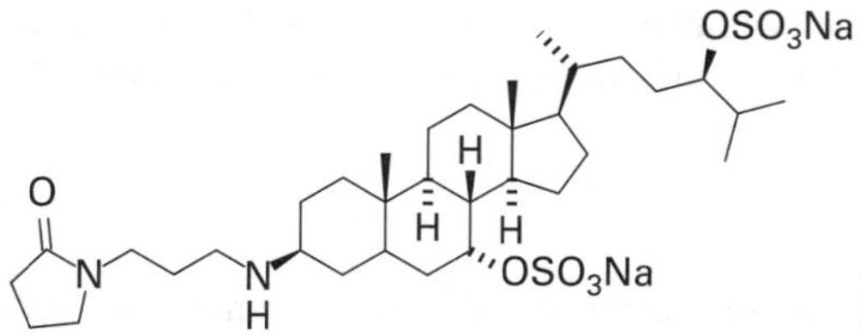

5 petromyzonamine disulfate (PADS)

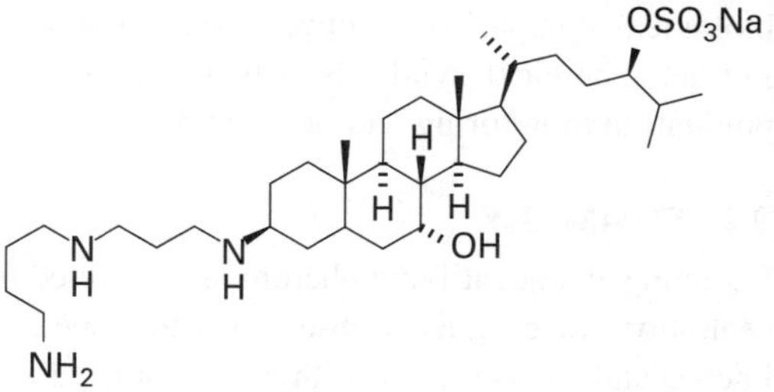

20 squalamine

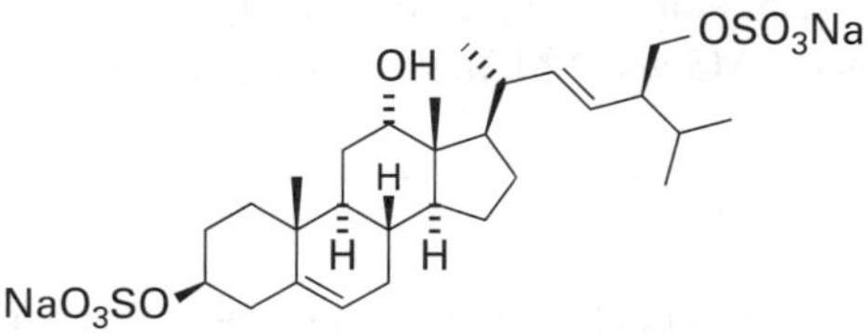

6 petromyzosterol disulfate (PSDS)

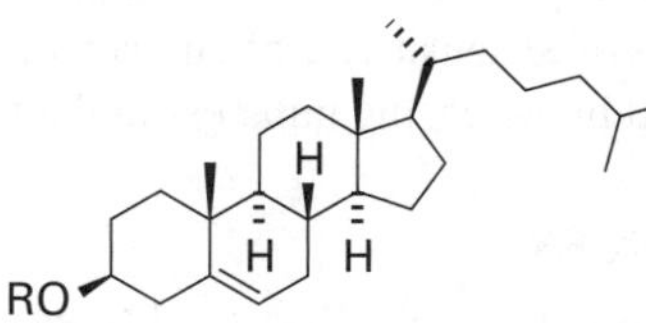

21   R = H   Cholesterol
22   R = SO$_3$Na   Cholesterol sulfate

5–6, 20–22   Compounds

required to obtain these chemicals. For example, Sorensen *et al.* (2005) processed 8,000 liters of lamprey larval holding water to obtain submilligram quantities of PADS (**5**) and PSDS (**6**). Even with traditional NMR technology, it was possible to solve the structures of PADS (**5**) and PSDS (**6**) on submilligram quantities using the process of "pattern matching" of the ^{1}H NMR spectra of the natural compounds with authentic synthetic standards to gain information on functional groups. The full structure elucidation of these complex structures is an excellent model of using this approach and underlines the multidisciplinary approach that is usually necessary. PADS (**5**) has a similar structure to squalamine (**20**); and once this had been established, it was possible to narrow the structural possibilities of **5**, with final confirmation made via synthesis of model *partial* structures and finally the full molecule of PADS (**5**). PSDS (**6**) showed similarities in its ^{1}H NMR spectra to cholesterol (**21**), with further investigations revealing cholesterol sulfate (**22**)—an excellent match for the A and B ring systems. Confirmation of the side chain of **6** was made via 2D NMR experiments and comparison with substructures in the literature, with final confirmation by total synthesis of **6**.

Purity must also be high (about 90%), to avoid spurious resonances in the NMR spectra, which can complicate, or even prevent, structural assignment.

9.3.5.5 Chemical synthesis

There are often significant challenges in unambiguous structure determination, especially where very low amounts of pheromone are available or the chemical is sufficiently complex that ambiguities—such as stereochemical uncertainties—are present. If any ambiguity remains, chemical synthesis of the putative pheromone(s) is necessary. For example, doubt may exist over stereochemistry of a chiral center, or conformation of substituents on a ring system. If necessary, all viable options may need to be synthesized to find the correct structure. Physicochemical and biological comparisons between the natural and synthetic pheromone(s) can then be made. Further, chemical synthesis may be the only viable route to production of pheromones for population monitoring and/or control.

9.4 SUMMARY

Detecting and quantifying pheromones excreted from fish is complicated by the extremely low concentrations of chemicals observed, and sometimes subtle differences between the pheromone of interest and others present with no pheromonal activity. However, recent advancements in analytical technology, especially toward improved *sensitivity* and *selectivity*, have made this task somewhat easier, if not yet routine. There are still few examples of pheromones unambiguously identified from fish, owing to the challenges involved in acquiring sufficient material and performing appropriate biological assays. However, with improvements in separation techniques (especially UPLC) and greatly improved sensitivity of characterization methods (MS and NMR), we can expect structures to be solved in increasing numbers in the future.

REFERENCES

Alvarez, D.A., Petty, J.D., Huckins, J.N. *et al.* (2004) Development of a passive, in situ, integrative sampler for hydrophilic organic contaminants in aquatic environments. *Environmental Toxicology and Chemistry*, **23**, 1640–1648.

Alvarez, D.A., Stackelberg, P.E., Petty, J.D. *et al.* (2005) Comparison of a novel passive sampler to standard water-column sampling for organic contaminants associated with wastewater effluents entering a New Jersey stream. *Chemosphere*, **61**, 610–622.

Baker, C.F., Carton, A.G., Fine, J.M., and Sorensen, P.W. (2006) Can bile acids function as a migratory pheromone in banded kokopu, *Galaxias fasciatus* (Gray)? *Ecology of Freshwater Fish*, **15**, 275–283.

Becker, D., Galili, N., and Degani, G. (1992) GCMS-indentified steroids and steroid glucoronides in gonads and holding water of *Trichogaster trichopterus* (Anabantidae, Pallas 1770). *Comparative Biochemistry and Physiology Part B: Comparative Biochemistry*, **103**, 15–19.

Brown, G.E., Adrian, J.C., Smyth, E. *et al.* (2000) Ostariophysan alarm pheromones: laboratory and field tests of the functional significance of nitrogen oxides. *Journal of Chemical Ecology*, **26**, 139–154.

Churchwell, M.I., Twaddle, N.C., Meeker, L.R., and Doerge, D.R. (2005) Improving LC–MS sensitivity through increases in chromatographic performance: comparisons of UPLC–ES/MS/MS to HPLC–ES/MS/MS. *Journal of Chromatography B*, **825**, 134–143.

Dulka, J.G., Stacey, N.E., Sorensen, P.W., and Van Der Kraak, G.J. (1987) A sex steroid pheromone synchronizes male-female spawning readiness in goldfish. *Nature*, **325**, 251–253.

Eom, J., Jung, Y.R., and Park, D. (2009) F-series prostaglandin function as sex pheromones in the Korean salamander, *Hynobius leechii*. *Comparative Biochemistry and Physiology—Part A: Molecular & Integrative Physiology*, **154**, 61–69.

Exarchou, V., Krucker, M., Van Beek, T.A. *et al.* (2005) LC–NMR coupling technology: recent advancements and applications in natural products analysis. *Magnetic Resonance in Chemistry*, **43**, 681–687.

Ferrari, M.C.O., Vavrek, M.A., Elvidge, C.K. *et al.* (2008) Sensory complementation and the acquisition of predator recognition by salmonid fishes. *Behavioral Ecology and Sociobiology*, **63**, 113–121.

Fine, J.M. and Sorensen, P.W. (2005) Biologically relevant concentrations of petromyzonol sulfate, a component of the sea lamprey migratory pheromone, measured in stream water. *Journal of Chemical Ecology*, **31**, 2205–2210.

Fine, J.M. and Sorensen, P.W. (2008) Isolation and biological activity of the multi-component sea lamprey migratory pheromone. *Journal of Chemical Ecology*, **34**, 1259–1267.

Fine, J.M., Sisler, S.P., Vrieze, L.A. *et al.* (2006) A practical method for obtaining useful quantities of pheromones from sea lamprey and other fishes for identification and control. *Journal of Great Lakes Research*, **32**, 832–838.

Goto, J., Saito, M., Chikai, T. *et al.* (1983) Studies on steroids: CLXXXVII. Determination of serum bile acids by high-performance liquid chromatography with fluorescence labeling. *Journal of Chromatography B: Biomedical Sciences and Applications*, **276**, 289–300.

Hoye, T.R., Dvornikovs, V., Fine, J.M. *et al.* (2007) Details of the structure determination of the sulfated steroids PSDS and PADS: new components of the sea lamprey (*Petromyzon marinus*) migratory pheromone. *Journal of Organic Chemistry*, **72**, 7544–7550.

Levesque, H., Scaffidi, D., Polkinghorne, C., and Sorensen, P. (2011) A multi-component species identifying pheromone in the goldfish. *Journal of Chemical Ecology*, **37**, 219–227.

Li, W., Sorensen, P., and Gallaher, D. (1995) The olfactory system of migratory adult sea lamprey (*Petromyzon marinus*) is specifically and acutely sensitive to unique bile acids released by conspecific larvae. *The Journal of General Physiology*, **105**, 569–587.

Li, W., Scott, A.P., Siefkes, M.J. *et al.* (2002) Bile acid secreted by male sea lamprey that acts as a sex pheromone. *Science*, **296**, 138–141.

Lim, H. and Sorensen, P. (2011) Polar metabolites synergize the activity of prostaglandin $F_{2\alpha}$ in a species-specific hormonal sex pheromone released by ovulated common carp. *Journal of Chemical Ecology*, **37**, 695–704.

Marfey, P. (1984) Determination of D-amino acids. II. Use of a bifunctional reagent, 1,5-difluoro-2,4-dinitrobenzene. *Carlsberg Research Communications*, **49**, 591–596.

Michel, W.C. and Lubomudrov, L.M. (1995) Specificity and sensitivity of the olfactory organ of the zebrafish, *Danio rerio. Journal of Comparative Physiology A: Neuroethology, Sensory, Neural, and Behavioral Physiology*, **177**, 191–199.

Polkinghorne, C.N., Olson, J.M., Gallaher, D.G., and Sorensen, P.W. (2001) Larval sea lamprey release two unique bile acids to the water at a rate sufficient to produce detectable riverine pheromone plumes. *Fish Physiology and Biochemistry*, **24**, 15–30.

Schroeder, F.C. and Gronquist, M. (2006) Extending the scope of NMR spectroscopy with microcoil probes. *Angewandte Chemie International Edition*, **45**, 7122–7131.

Scott, A.P. and Canario, A.V.M. (1992) 17α,20β-Dihydroxy-4-pregnen-3-one 20-sulphate: a major new metabolite of the teleost oocyte maturation-inducing steroid. *General and Comparative Endocrinology*, **85**, 91–100.

Scott, A.P. and Ellis, T. (2007) Measurement of fish steroids in water—a review. *General and Comparative Endocrinology*, **153**, 392–400.

Scott, A.P. and Sorensen, P.W. (1994) Time course of release of pheromonally active gonadal steroids and their conjugates by ovulatory goldfish. *General and Comparative Endocrinology*, **96**, 309–323.

Siefkes, M.J., Winterstein, S.R., and Li, W. (2005) Evidence that 3-keto petromyzonol sulphate specifically attracts ovulating female sea lamprey, *Petromyzon marinus. Animal Behaviour*, **70**, 1037–1045.

Siefkes, M.J., Winterstein, S.R., and Li, W. (2005) Evidence that 3-keto petromyzonol sulphate specifically attracts ovulating female sea lamprey, *Petromyzon marinus. Animal Behaviour*, **70**, 1037–1045.

Sorensen, P.W. (2015) Applications of pheromones in invasive fish control and fisheries conservation. in Fish Pheromones and Related Cues (P.W. Sorensen and Brian D. Wisenden), John Wiley and Sons, Inc. Hoboken.

Sorensen, P.W. and Baker, C. (2015) Specis-specific pheromones and their roles in shoaling, migration and reproduction: a critical review and synthesis, in *Fish Pheromones and Related Cues*, (eds Peter W. Sorensen and Brian D. Wisenden), John Wiley & Sons, Inc., Hoboken.

Sorensen, P.W. and Hoye, T.R. (2007) A critical review of the discovery and application of a migratory pheromone in an invasive fish, the sea lamprey *Petromyzon marinus* L. *Journal of Fish Biology*, **71**, 100–114.

Sorensen, P.W. and Hoye, T.R. (2010) Pheromones in vertebrates, in *Comprehensive Natural Products II* (eds M. Lew and L. Hung-Wen), Elsevier, Oxford, pp. 225–262.

Sorensen, P.W., Hara, T.J., Stacey, N.E., and Goetz, F.W. (1988) F prostaglandins function as potent olfactory stimulants that comprise the postovulatory female sex pheromone in goldfish. *Biology of Reproduction*, **39**, 1039–1050.

Sorensen, P.W., Vrieze, L.A., and Fine, J.M. (2003) A multi-component migratory pheromone in the sea lamprey. *Fish Physiology and Biochemistry*, **28**, 253–257.

Sorensen, P.W., Murphy, C.A., Loomis, K. *et al.* (2004) Evidence that 4-pregnen-17,20β,21-triol-3-one functions as a maturation-inducing hormone and pheromonal precursor in the percid fish, *Gymnocephalus cernuus*. *General and Comparative Endocrinology*, **139**, 1–11.

Sorensen, P.W., Fine, J.M., Dvornikovs, V. *et al.* (2005) Mixture of new sulfated steroids functions as a migratory pheromone in the sea lamprey. *Nature Chemical Biology*, **1**, 324–328.

Stacey, N. (2015) Hormonally-derived pheromones in teleost fishes, in *Fish Pheromones and Related Cues* (eds Peter W. Sorensen and Brian D. Wisenden), John Wiley & Sons, Inc., Hoboken.

Stacey, N.E. and Sorensen, P.W. (2009) Hormonal pheromones in fish, in *Hormones, Brain and Behavior*, vol. 1 (eds D.W. Pfaff, A.P. Arnold, A.M. Etgen *et al.*), Elsevier Press, San Diego, pp. 639–681.

Stewart, M. and Baker, C. (2012) A sensitive analytical method for quantifying petromyzonol sulfate in water as a potential tool for population monitoring of the Southern Pouched Lamprey, *Geotria Australis*, in New Zealand streams. *Journal of Chemical Ecology*, **38** (2), 135–144.

Stewart, M., Baker, C.F., and Cooney, T. (2011) A rapid, sensitive, and selective method for quantitation of lamprey migratory pheromones in river water. *Journal of Chemical Ecology*, **37**, 1203–1207.

Swartz, M.E. (2005) UPLC™: an introduction and review. *Journal of Liquid Chromatography & Related Technologies*, **28**, 1253–1263.

Van Weerd, J.H., Sukkel, M., Lambert, J.G.D., and Richter, C.J.J. (1991) GCMS-identified steroids and steroid glucuronides in ovarian growth-stimulating holding water from adult African catfish, *Clarias gariepinus*. *Comparative Biochemistry and Physiology Part B: Comparative Biochemistry*, **98**, 303–311.

Vermeirssen, E.L.M. and Scott, A.P. (1996) Excretion of free and conjugated steroids in rainbow trout (*Oncorhynchus mykiss*): evidence for branchial excretion of the maturation-inducing steroid, 17,20β-dihydroxy-4-pregnen-3-one. *General and Comparative Endocrinology*, **101**, 180–194.

Xi, X., Johnson, N.S., Brant, C.O. *et al.* (2011) Quantification of a male sea lamprey pheromone in tributaries of Laurentian Great Lakes by liquid chromatography—tandem mass spectrometry. *Environmental Science & Technology*, **45**, 6437–6443.

Yambe, H., Kitamura, S., Kamio, M. *et al.* (2006) l-Kynurenine, an amino acid identified as a sex pheromone in the urine of ovulated female masu salmon. *Proceedings of the National Academy of Sciences*, **103**, 15370–15374.

You, J., Shi, Y., Ming, Y. *et al.* (2004) Development of a sensitive reagent, 1,2-benzo-3,4-dihydrocarbazole-9-ethyl-*p*-toluenesulfonate, for determination of bile acids in serum by HPLC with fluorescence detection, and identification by mass spectrometry with an APCI source. *Chromatographia*, **60**, 527–535.

Yun, S.-S., Scott, A.P., Siefkes, M.J., and Li, W. (2002) Development and application of an ELISA for a sex pheromone released by the male sea lamprey (*Petromyzon marinus* L.). *General and Comparative Endocrinology*, **129**, 163–170.

Zatylny, C., Gagnon, J., Boucaud-Camou, E., and Henry, J. (2000) ILME: a waterborne pheromonal peptide released by the eggs of *Sepia officinalis*. *Biochemical and Biophysical Research Communications*, **275**, 217–222.

Zhang, C., Brown, S., and Hara, T. (2001) Biochemical and physiological evidence that bile acids produced and released by lake char (*Salvelinus namaycush*) function as chemical signals. *Journal of Comparative Physiology B: Biochemical, Systemic, and Environmental Physiology*, **171**, 161–171.

Chapter 10
Effects of Pollutants on Olfactory Detection and Responses to Chemical Cues Including Pheromones in Fish

K. Håkan Olsén

Södertörn University, Huddinge, Sweden

10.1 INTRODUCTION TO OLFACTORY-MEDIATED RESPONSES INDUCED BY CHEMICAL CUES

Chemical information by molecules released from heterospecific and conspecific individuals can give information about the presence of food items, predators, competitors, mates, and the recognition of kin (e.g., Olsén, 1992, 1999; Müller-Schwarze, 2006; Stacey and Sorensen, 2006). Numerous species of fish are dependent on the olfactory sense to detect sex pheromones during the reproduction (for review, see Liley, 1982; Stacey and Sorensen, 2005). Other chemical senses (e.g., gustation) appear to have little, if anything, to do with the detection of conspecific cues (Sorensen and Caprio, 1998). This chapter deals with the possible effects of pollutants on detection and responses to both intraspecific, that is pheromones, and interspecific olfactory cues in fish.

Most of the information about sex pheromones in fish is based on studies of goldfish (*Carassius auratus*) and salmonid fishes, but these models have proven broadly applicable to many other species (Stacey and Sorensen, 2009). Sex pheromone production in goldfish is closely connected to the production of sex hormones (for review, see Stacey and Sorensen, 2002, 2006; Stacey *et al.*, 2003). In the species studied thus far that use sex pheromones, several have been found to detect sex steroid hormones (either in the free form or connected to a glucuronide or a sulfate group) and/or F prostaglandins (reviewed by Stacey and Sorensen, 2009). These hormonally derived sex pheromones are often referred to as "hormonal pheromones," and their production is described by Stacey (2015). Goldfish and the closely related crucian carp (*C. carassius*) use both steroids and prostaglandins (Sorensen, Hara, and Stacey, 1991; Bjerselius and Olsén, 1993; Kobayashi, Sorensen, and Stacey, 2002) and are sensitive to the free hormone 17,20β-dihydroxy-4-pregnen-3-one (17,20β-P) and prostaglandin $F_{2\alpha}$ (PGF$_{2\alpha}$). Goldfish, crucian carp, and common carp (*Cyprinus carpio*) seem to use very similar pheromone systems (see Kobayashi, Sorensen, and Stacey, 2002; Lim and Sorensen, 2011).

Fish Pheromones and Related Cues, First Edition. Edited by Peter W. Sorensen and Brian D. Wisenden.
© 2015 John Wiley & Sons, Inc. Published 2015 by John Wiley & Sons, Inc.

217

Exposure to 17,20β-P has a priming effect that increases the milt volume in both goldfish and crucian carp (Bjerselius, Olsén and Zheng, 1995a, 1995b). In crucian carp, this has also been shown during studies in net pens in their natural environment (Olsén *et al.*, 2004; Olsén, Sawisky, and Stacey, 2006), and there are indications that males are not primed by heterospecific ovulated females although they release the same hormonal sex pheromones (Stacey *et al.*, 2011).

In salmonid fishes, both endocrinological and behavioral effects by female odors have also been demonstrated in males. Male rainbow trout (*Oncorhynchus mykiss*) are attracted to water carrying the scent of an ovulated female and show endocrine stimulation by female odors (Emanuel and Dodson, 1979; Olsén and Liley, 1993). The identity of the active molecule(s) is, however, not yet known despite great efforts have been made to isolate and identify the pheromone (Scott, Liley, and Vermeirssen, 1994). Results of a recent study with masu salmon (*O. masou masou*) stress that pheromones may not always be hormones or their metabolites, and they implicate roles for an amino acid (Yambe *et al.*, 2006). Atlantic salmon and brown trout females release odors with priming or releasing effects. Both ovarian fluid and urine contain odors that stimulate hormonal levels and increase milt volumes. $PGF_{2\alpha}$ is the priming pheromone or at least one important part of it (Waring, Moore, and Scott, 1996; Olsén *et al.*, 2000, 2001, 2002; Moore *et al.*, 2002). Urine, but not ovarian fluid, also contains molecules of unknown identity, which attracts males (Olsén *et al.*, 2000, 2001, 2002).

In salmonid fishes, the olfactory sense is also important for the homing of adult fish to their natal river to spawn (e.g., Wisby and Hasler, 1954; Cooper and Hirsch, 1982; Hasler and Scholz, 1983), and many believe pheromones are involved in this process (e.g., Nordeng, 1977; Døving, Selset, and Thommesen, 1980; Stabell, 1984). If the olfactory sense is affected, then the homing behavior should be less accurate as the fish will have a problem recognizing their home river (see Bertmar and Toft, 1969; Bertmar, 1979). The identity and source of the home river odors are still not known, but the odors seem to be learned before the fish leave the river (e.g., Scholz *et al.*, 1976; Morin, Dodson, and Dore, 1989). There is also compelling evidence that sockeye salmon (*O. nerka)*, and probably other salmonids, have a very precise homing ability as they return to their natal site at 4 years of age (Quinn, Stewart, and Boatright, 2006). There are suggestions that returning adults orient to odors released by juvenile fish of their natal population (e.g., Nordeng, 1977; Stabell, 1984). Salmonid fishes are able to discriminate between water scented by individuals from the same population and a different population (Olsén, 1986a, 1986b, 1986c; Courtenay *et al.*, 2001) but also between unfamiliar relatives and nonrelatives from the same population (Quinn and Busack, 1985; Olsén, 1989; Winberg and Olsén, 1992; Brown, Brown, and Crosbie, 1993; Olsén and Winberg, 1996; Olsén *et al.*, 1998). Bile acids have been suggested as migratory pheromones that guide spawning salmonids back to their home river (Døving, Selset, and Thommesen, 1980; Lasteine *et al.*, 2015). Bile acids from sea lamprey (*Petromyzon marinus*) larvae have also now been shown to function as pheromones that attract mature adults upstream to spawn (Bjerselius *et al.*, 2000; Sorensen *et al.*, 2005). In the sea lamprey, specific bile acids released from males may also act as female attractants (Yun, Scott, and Li, 2003; Siefkes, Winterstein, and Li, 2005).

10.2   POSSIBLE EFFECTS OF POLLUTANTS ON CHEMICAL COMMUNICATION IN FISH

Streams, lakes, and the oceans act as sinks for various pollutants where the olfactory sense of fish can be exposed directly to these chemicals. Some of these molecules are known to interfere with the function of the primary olfactory receptor cells (Lürling and Scheffer, 2007). Pollutants can affect one or more components of chemical communication [herein defined as "an action on the part of one organism (or cell) that alters the probability pattern of behavior in another organism (or cell) in an adaptive fashion" (Wilson, 1975)] by acting on either the producer or

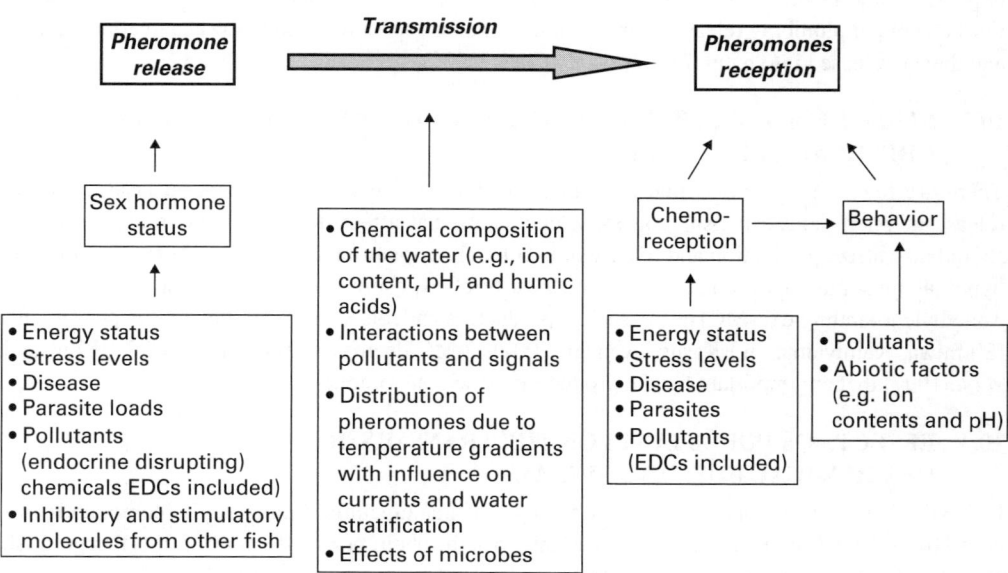

Figure 10.1. Various biotic and abiotic factors can influence the three links of the sex-pheromone communication. In addition to effects on the sender and receiver, the signal molecule(s) can be affected during the transmission step. T. Breithaupt & M. Thiel. Chemical Communication in Crustaceans. Copyright 2010, Springer Business and Science Media.

releaser of a pheromone or the receiver individual that detects the signal. Pheromonal function may also be affected by biotic and abiotic factors, such as bacteria, organic materials, and pH (see Fig. 10.1).

One aspect that has not received much attention thus far is the effects of endocrine disruptors on levels of endogenous hormones/sex pheromones regulating reproductive interactions. The sensitivity and ability of the olfactory system to sex pheromones, at the receptor cell level or at higher levels, is itself under hormonal control (e.g., Cardwell *et al.*, 1995; Yambe and Yamazaki, 2000, 2001; Yambe *et al.*, 2003). Behavioral and endocrine responses of males to female pheromones are dependent on male sex steroid hormones and chemicals affecting the hormone levels. Endocrine disrupting chemicals (EDCs) may interfere with sex pheromone perception and communication between the sexes. EDCs such as estrogenic chemicals and various pharmaceutical drugs commonly occur in sewage treatment effluents (Fent, Weston, and Caminada, 2006).

Pesticides present in surface runoff and wind transport enter streams and lakes (e.g., Environment Agency, UK, 1998, 2003; Barata *et al.*, 2007). For example, more than 50% of the samples taken from various streams in agricultural areas in Sweden contain any of 38 different pesticides (Hessel *et al.*, 1997; Kreuger, 1998, 2000; Kreuger, Peterson, and Lundgren, 1999; Törnquist *et al.*, 2005). The concentration of these compounds is correlated with the level of use in the local area (Kreuger, 1998). Pesticides are designed and synthesized to be biologically active at low concentrations. Water-borne concentrations often affect nervous and endocrine function of nontarget organisms, for example, on fish and other vertebrates (e.g., Bonde, Comhaire, and Ombelet, 1998).

Metals, especially copper, are also very toxic to the olfactory sense of fish. Copper compounds are used as pesticides, against fungi in wood preservative, in antifouling paints, and as an algicide (e.g., Le Jeune *et al.*, 2006; Garcia-Valcárcel and Tadeo, 2007; Gatidou and Thomaidis, 2007). Further, copper plumbing releases significant amounts of copper that enters sewage treatment plants, and thence released into natural water ways (Clark, 2001).

10.3 EFFECTS OF POLLUTANTS ON THE PRODUCTION OF CONSPECIFIC CHEMICAL CUES AND SIGNALS

There are few, if any, studies that have directly studied effects of pollutants on production and release of pheromones in fish. Logically, chemicals that affect the sex hormone system in fish should also affect production and release of hormonal sex pheromones if those chemicals affect the hypothalamus–pituitary–gonad axis. In mice, a brief exposure to the polyaromatic hydrocarbon 3-methylchoranthrene (3-MC) decreased the production and release of male specific sex pheromones (Shiraiwa, Kamiyama, and Kashiwayanagi, 2007). 3-MC decreases testosterone levels (Konstandi *et al.*, 1997) that are important for the production of sex pheromones.

10.4 EFFECTS OF POLLUTANTS ON THE TRANSMISSION OF CHEMICAL CUES AND SIGNALS

Hubbard, Barata, and Canario (2002) observed that manufactured humic acids (compounds produced by chemical companies from coal) might block the pheromonal effects of 17,20β-P in goldfish, probably by adsorbing the molecule. Humic acids are known to adsorb hydrophobic substances such steroid pheromones (Mesquita, Canario, and Melo, 2003). In experiments with swordtail fish (*Xiphophorus birchmanni*), Fisher, Wong, and Rosenthal (2006) found that females exposed to manufactured humic acids also lost their strong preference for water scented by conspecific males over heterospecific males (*X. malinche*) both during and after the exposure. The same kind of effect was observed with zebrafish (*Brachydanio rerio*) that lost their ability to discriminate between conspecific and heterospecific urinary chemical cues (Fabian *et al.*, 2007). However, more recent work by Fine and Sorensen (2010) has called into question the precise interpretation of these results by re-examining the effects of natural humic acids (previous studies had used high levels of manufactured humics that also had very low pHs) on the sea lamprey olfactory sense and found that it does not appear to be affected by natural stream humics at natural concentrations. Nevertheless, these studies do reinforce the concept that pollutants found in waters may be interfering with olfactory detection and communication simply because these systems are so sensitive and open to exposure.

The modulating effect of ammonium–ammonia on the attraction of Arctic char (*Salvelinus alpinus*) to conspecific juvenile odors could have been caused by interactions between the chemical cues and reactive ammonia (Olsén, 1986a, 1986b, 1986c). A recent review on insect chemical signals suggests that air pollutants have a similar effect in terrestrial environments (McFredrick *et al.*, 2009), particularly during attraction to mates and social aggregations that depend on complex mixtures of compounds.

10.5 EFFECTS OF POLLUTANTS ON OLFACTORY DETECTION OF CHEMICAL CUES AND SIGNALS

Toxic effects of pollutants on the olfactory sense in fish have been reviewed by Tierney *et al.* (2010). The authors review the effects of various pollutants on the electrophysiological responses with electroencephalogram (EEG) and electro-olfactogram (EOG) recording to amino acids, but also to bile acids and prostaglandins, and effects on the behavior to food and skin extracts.

10.5.1 General Introduction to the Olfactory System

There is tremendous variation in the structure of olfactory organs among fishes. In general, there are two openings: (i) the nares that are present on each side of the head. Water enters the olfactory chamber through the anterior naris opening and leaves the chamber through the posterior naris.

(ii) An olfactory rosette with lamellae develops from a central axis or middle raphe in each of the two pits (e.g., Hara, 1986; Olsén, 1993; Kudo *et al.*, 2009). Olfactory receptor cells are present on the lamellae. With age, the lamellae get bigger in size; and in some fish, secondary lamellae develop which increase surface area. Nonsensory cells with kinocilia are import for the transport of water through the olfactory chamber (Døving *et al.*, 1977). In the olfactory epithelium, it is possible to distinguish three different types of receptor cells: (i) microvillar, (ii) ciliated, and (iii) crypt receptor cells (Hansen and Zielinski, 2005; Hamdani *et al.*, 2008). It has also been suggested that these types of receptor cells mediate responses to different chemical classes and hence different behaviors (Døving and Lastein, 2009; Lastein *et al.*, 2015). Olfactory receptor cells are bipolar neurons developed from basal cells in the olfactory epithelium. The dendrites of these cells are exposed directly to the water and hence damage from pollutants. Axons of olfactory neurons terminate in glomeruli in the olfactory bulbs and connect to mitral cells that join to olfactory tracts for transport of signals further to nuclei in the brain. Olfactory neurons with the same receptor type end up in the same glomeruli (Mori, Nagao, and Yoshihara, 1999). Olfactory receptor neurons have the capacity to regenerate damaged cells (e.g., Zippel, 1993). For further detailed information about the structure of the olfactory system, see Lastein *et al.* (2015).

10.5.2 Effects of Metals on Olfactory Function

Several studies have indicated that metals interfere with olfactory function in fish. Sutterlin (1974) reviewed and discussed early studies concerning pollutant effects on the chemical sense in aquatic organisms. Most electrophysiological studies on pollutant effects on olfactory receptor cell function were done with amino acids as odorants as they relate to foraging. However, there is no reason to believe that pollutant disturbances on responses to pheromones would be fundamentally different.

Early studies by Hara and collaborators have shown with electrophysiological techniques (EOG, EEG) that low concentrations of copper and some other heavy metals such as mercury and silver reduce or completely block the olfactory response to amino acids that fish use to detect food (e.g., Hara *et al.*, 1976, 1983). Sandhal *et al.* (2004) demonstrated with EOG and EEG that 7 days' exposure to 10 or 20 µg Cu/L reduced or inhibited responses to amino acid L-serine or to the bile acid taurocholic acid in juvenile coho salmon (*O. kisutch*) by 50% (compared with controls). In a recent study, a 10-min exposure of the olfactory epithelium in goldfish to a high concentration of copper sulfate (100 µM) followed by EOG measurements to amino acids, catecholamines, bile acids, the hormonal pheromones 17,20β-dihydroxy-4-pregnen-3-one-20-sulfate, and $PGF_{2\alpha}$ revealed that the olfactory responses recovered after 3 days with the amino acids, but it took 28 days to fully recover the response to the prostaglandin (Kolmakov *et al.*, 2009). The study also gave some support to that microvillous receptor cell types are connected to amino acid detection, and ciliated receptor cells are involved in sex pheromone and bile acid detection. Some of the effects stated earlier might be due to Cu depression of transcription of the genes important for olfactory signal transduction (Tilton *et al.*, 2011).

Olfactory receptor neurons can be a transport route of metal ions and organic molecules to the olfactory bulbs and the brain in vertebrates, fish included, with high risks of severely disturbing effects on the function of the CNS (e.g., Tomlinson and Esiri, 1983; Hastings and Evans, 1991; Evans and Hasting, 1992; Tallkvist *et al.*, 1998, 2002; Henriksson, Tallkvist, and Tjälve, 1999; Rouleau *et al.*, 1999; Persson, Larsson, and Tjälve, 2002; Persson, Henriksson, and Tjälve, 2003). Behavioral studies need to be conducted to investigate the disturbing effects on the olfactory system.

In a study using EOG, Bjerselius *et al.* (1993) found that increasing the Ca^{2+} in solutions with copper ($CuCl^-$), reduced the immediate negative effects of copper exposure in juvenile Atlantic salmon, *Salmo salar*. These authors also found a positive effect of Ca^{2+} on the recovery of the response to L-alanine after copper exposure. The Ca^{2+} seems to protect the olfactory receptors from the effects of Cu^{2+}. It is also known that Ca^{2+} is important for the olfactory receptor response (e.g., Suzuki, 1982;

Restrepo *et al.*, 1990; Gautam *et al.*, 2007). Lately, it has been shown that Ca^{2+} could be an olfactory stimulus of its own (Hubbard and Canario, 2007).

Silver nanoparticles affect EOG responses to amino acid L-alanine (Bilberg *et al.*, 2011). Exposure to a low concentration of nanoparticles (0.45 µg/l) enhanced the response, whereas exposure to high concentration (45 µg/l) decreased the response. The authors suggested that the effects were due to a combination of silver particles and silver ions.

10.5.3   Effects of Organic Pesticides on Olfactory Function

Several studies have shown effects of pesticides on the olfactory function in fish. Tierney and collaborators studied the changes of coho salmon parr in response to amino acids, both with EOG and with behavior tests. The fungicide iodocarb (IPBC) and the herbicide glyphosate (the active ingredient of the commonly used Roundup®) decreased the olfactory response to L-serine at relatively low concentrations (Tierney *et al.*, 2006b), but the other three pesticides tested (chlorothalonil, endosulfan, 2,4-dichlorophenoxyacetic acid) only had effects at high concentrations. Sandhal *et al.* (2004) did not find any effects of the pyrethroid esfenvalerate on EOG or EEG responses to either an amino acid or a bile acid, but the pesticide evoked irregular postsynaptic activity in the olfactory bulbs. It is not known if the postsynaptic effects by esfenvalerate had any effects on the ability of the fish to detect the odors and behave in a proper way.

In a study with coho salmon, a 30-min exposure of the olfactory rosette to the insecticide carbofuran at 200 µg/l, but not to the lower concentrations tested, significantly reduced the AChE activity in the olfactory rosette (but with no change in the olfactory bulb or in the rest of the brain) (Jarrard, Delaney, and Kennedy, 2004). Exposure to two other carbamates—IPBC and mancozeb—resulted in a significant increase in the enzyme activity in the brain at 0.005 and 2.2 µg/l, respectively. All three carbamates reduced EOG response to the amino acid L-serine (10^{-5} M). The authors speculated that the effects of carbamates might have been through increased mucus production in the olfactory epithelium due to AChE inhibition resulting in an increased diffusion distance of the odorants.

10.5.4   Effects of pH on Olfactory Function

Olfactory responsiveness to amino acids is dependent on the pH of the amino acid solution. Hara (1982) found that the olfactory bulb response was highest near the isoelectric point of the amino acid tested. The bulbar response to fish skin mucus was also highly pH-dependent similar to amino acids. Thus, the pH of the water can either affect the olfactory molecules themselves (Leduc, Kelly, and Brown, 2004) or have direct effects on the olfactory receptors (Thommesen, 1983). A combination of aluminum and decreased pH depressed the electrophysiological response of the olfactory sense in rainbow trout to the amino acid L-serine (Klaprat, Brown, and Hara, 1988). Low pH has negative effects on the detection of female odors in male Atlantic salmon (Moore, 1994).

10.5.5   Effects of Endocrine Disrupting Chemicals

Olfactory sensitivity to sex pheromones is in at least some fish dependent on the production of sex hormones. By treating juvenile male *Puntius* species with methyltestosterone olfactory sensitivity to the female sex pheromone 15keto-PGF$_{2\alpha}$ was induced, and the treated males demonstrated sex-specific pheromone-dependent behaviors (Cardwell *et al.*, 1995). These effects with methyltestosterone have been also observed in juvenile fish of other cyprinids (Belanger, Pachkowski, and Stacey, 2010). Chemicals affecting sex hormone production should affect the olfactory perception of pheromones. Studies on effects of EDCs on olfactory function in fish are lacking. Future studies could reveal that chemicals with effects on the sex hormone status will affect pheromone detection and reproduction in fish.

10.6 EFFECTS OF POLLUTANTS ON BEHAVIORAL RESPONSES OF FISH TO CHEMICAL CUES

Blaxter and Hallers-Tjabbes (1982) have compiled and discussed some of the early studies about effects of pollutants on senses and behaviors in aquatic organisms. A recent review dealt with the effects of anthropogenic contamination on all kinds of behavior (Sloman and Wilson, 2006). A minor part of the studies referred to were dealing with disturbances of olfaction and connected behaviors, that is, responses to alarm pheromone and food odors.

10.6.1 Effects of Copper

Several fishes use chemical alarm cues to warn them of the risk of predators. The odors are either released by damaged conspecifics or released in the faces by the predator preying on the fish species (e.g., Brown, Chivers, and Smith, 1995; Brown, 2003; Ferrari *et al.*, 2007). Detection of predators is under strong selection as it is important to the prey fish to be able to detect, avoid the predator, and assess the risk of being in a certain environment. Wisenden (2015) discusses these cues but not whether and how their detection is influenced by pollutants, therefore it is covered here. Exposing Colorado pikeminnow (*Ptychocheilus lucius*) to copper for 24 h reduced the fishes' ability to show a behavior, fright response to conspecific skin extract (Beyers and Farmer, 2001), and this negative effect increased with concentration. Interestingly, the authors found that after copper exposure (66 μg/l), the fishes had a reduced response to skin extract after exposure for 24 h than after a 96 h. The authors suggested that the fishes after 96 h had developed a physiological adaptation and increased tolerance to copper. In a field study, McPherson, Mirza, and Pyle (2004) observed that Iowa darters (*Etheostoma exile*) from a contaminated lake (mainly copper, but also some zinc and nickel) did not, in contrast to fish from a clean lake, avoid conspecific skin extract containing fright odors.

Young rainbow trout exposed to copper since early stage (22 μg/l during about 40 weeks) did not discriminate between two currents with pure water or water conditioned by heterospecifics and water scented by water from their own tank containing conspecific odors (Saucier, Astic, and Rioux, 1991). The fish showed some recovery of their olfactory discrimination ability after 2 or 10 weeks in clean water.

10.6.2 Effects of Organic Pesticides

In experiments with chinook salmon (*O. tshawytscha*), the organophosphate insecticide, diazinon, significantly inhibited exposed fish fright responses to skin extract; instead of decreasing their swimming and foraging behavior, these fish continued their high activity after exposure to the pesticide at 1 and 10 μg/l for 2 h and recovery for 1 h (Scholz *et al.*, 2001). The authors also gave some preliminary results which indicated that a less number of adult fish exposed to 10 μg/l homed to the place they were caught, a hatchery, 2 weeks earlier. The fright responses to skin extract (dashing behavior and freezing) were different in juvenile coho salmon after a 0.5-h exposure to the carbamate fungicide IPBC (Tierney *et al.*, 2006a). The immediate dashing behavior to skin extract was decreased after exposure to all IPBC concentration tested (1, 10, and 100 μg/l). The time-freezing behavior was not changed after 1 μg/l, decreased after exposure to 10 μg/l, but was increased after 100 μg/l when compared with unexposed control fish.

Tierney *et al.* (2007) observed that a 30-min exposure to three different pesticides, 1 μg/l IPBC, 1 μg/l atrazine, or 100 μg/l AI Roundup, eliminated attraction to 10^{-7} M L-histidine. Interesting Roundup gave an EOG response, and the juvenile rainbow trout avoided concentrations over 10 mg/l. Saglio *et al.* (2003) also have found that previously demonstrated attraction of goldfish to mixtures of four L-amino acids (glycine, alanine, valine, taurine) found in conspecific urine decreased after an 8-h exposure to prochloraz (fungicide) or carbofuran (insecticide), thereby indicating that the function of the olfactory sense was affected. Exposure to the herbicide nicosulfuron did not significantly decrease the attraction to the amino acid mixture.

10.6.3 Effects of pH

Decreases in the pH of water have been shown to affect the responses of fishes to alarm cues (Brown *et al.*, 1982; Leduc, Kelly, and Brown, 2004; Leduc *et al.*, 2007, 2008). It is yet to be shown whether these effects are direct (i.e., they influence olfactory receptor function) or merely alter the chemistry of the cues (Thommesen, 1978). Recently, Heuschele and Candolin (2007) showed that increased pH made stickleback females more interested in conspecific male odors. They spent a significant amount of time with the male-scented water when the pH was 9.5 compared to 8. Acidification of the oceans have raised concerns about changes in olfactory discrimination abilities of relevant chemical cues such as about suitable settlement sites in fish larvae (Munday *et al.*, 2009). A slight decrease in pH, from 6.8 to 6.4 within a period of 20 min, decreased nest digging significantly in females of landlocked sockeye salmon, hime salmon (*O. nerka*) (Kitamura and Ikuta, 2000). The reason behind the impaired behavior is probably that the fish detected an increase in CO_2 in water containing carbonate salts after addition of H^+ and that chemical senses are involved (Höglund, 1961). It is not known if the reproductive behaviors and endocrine responses to female odors were affected in males.

10.6.4 Other Pollutants

Lower and Moore (2007) exposed juvenile of Atlantic salmon to low concentrations of a brominated flame retardant (hexabromocyclododecane, HBCD; water concentration of about 10 ng/l) for 30 days over the peak period of smoltification. The olfactory responses of these fish to smolt urine were tested on a weekly basis using EOG. Exposure to HBCD reduced the responses to the smolt urine to about 40% of the original baseline responses. HBCD exposure did not change the fishes' osmoregulatory function as estimated by the gill $Na^+ K^+$ ATPase activity. It has been suggested that declining return rates of Atlantic salmon return to home river may be attributable to spraying of a carbamate insecticide—4-nonylphernol (4-NP) mixture during the smoltification period a few years earlier (Fairchild *et al.*, 1999). The authors suggested that the estrogenic 4-NP was the reason behind the decline, but disturbance of the olfactory sense during learning of home river cues might have also caused the problem. In a previous study, chlorinated paraffins (C_{12} and C_{16}) used as flame retardants were found to be selectively accumulated in the olfactory organ of rainbow trout (Darnerud *et al.*, 1989). In another study, juvenile Arctic char were found to no longer be attracted to water scented by relatives after exposure to a surfactant, linear alkylbenzene sulphonate (LAS) used in detergents (Olsén and Höglund, 1985). Water scented by conspecifics changed from attractive to repellent in juvenile Arctic char when ammonia and ammonium ions (NH_4Cl) were added to the current. They preferred to be in the unscented half (Olsén, 1986b, 1986c). In a recent study, Ward *et al.* (2008) showed that low concentrations of the surfactant 4-nonylphenol impaired the tendency of juvenile banded killifish (*Fundulus diaphanus*) to be attracted to conspecific scented water while increasing distances between shoaling fishes.

10.7 EFFECTS OF POLLUTANTS ON THE BEHAVIOR AND ENDOCRINOLOGICAL RESPONSIVENESS OF REPRODUCTIVELY ACTIVE CONSPECIFICS

The possible impacts of pollutants on hormonal (primer) and behavior (releaser) effects of sex pheromone detection have seemingly not been studied. Instead, we describe some possible effects in Figure 10.2.

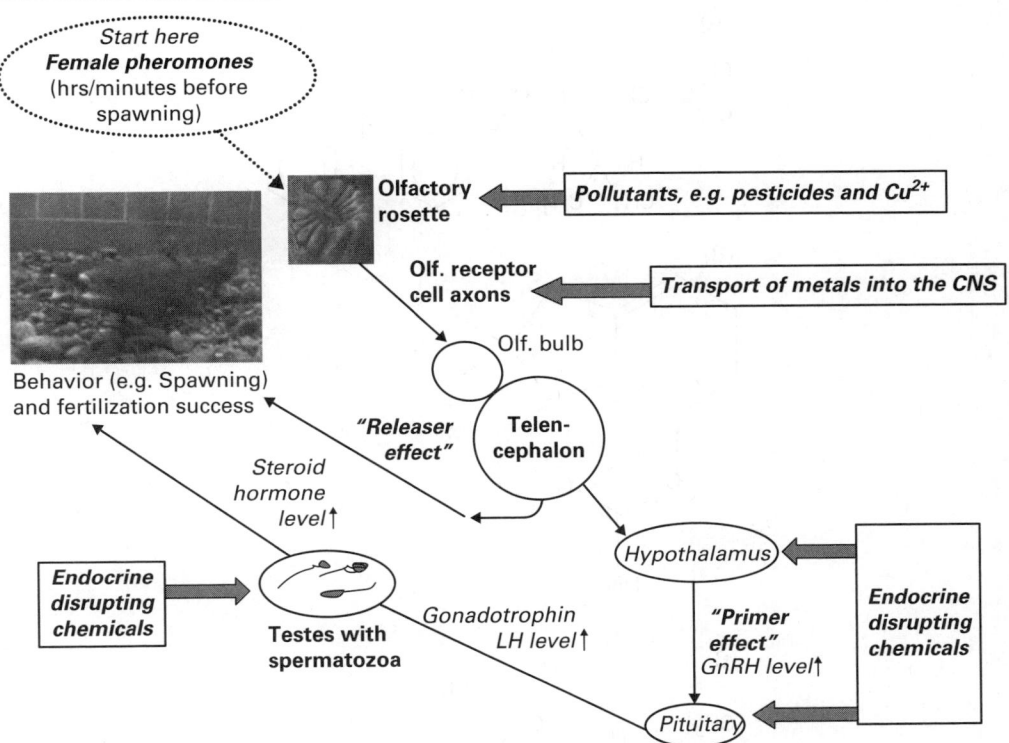

Figure 10.2.   Possible negative effects of pollutants on various parts of chemical communication, from detection of pheromones to the resulting behavior and physiological responses. T. Breithaupt & M. Thiel. Chemical Communication in Crustaceans. Copyright 2010, Springer Business and Science Media. (*See insert for color representation of the figure.*)

10.7.1   Copper

In a recent study, brown trout mature male parr were exposed to copper sulfate before being exposed to the female priming pheromone $PGF_{2\alpha}$ or being placed into a large stream tank with a small group of mature adult fish and untreated parr (Jaensson and Olsén, 2010). Male parr exposed during 4 days to Cu concentrations of 10 or 100 µg/l had significantly lower volumes of expressible milt after exposure to $PGF_{2\alpha}$ than the controls. The results from the stream tank showed that copper exposed males spent less time with nest digging females and courted females less frequently. The exposed males also swam less time upstream compare to the intact fish. The groups were followed for only a day; therefore, long-term ramifications are not known.

10.7.2   Organic Pesticides

Studies of the effects of pesticides on the olfactory sense in spawning fish have focused on salmonids. A number of common pesticides have been shown to affect pheromone mediated endocrine function and reproduction in salmon even at low concentrations (see below; Fig. 10.3). There is a good possibility that decreases in the sensitivity of pesticide exposed fish to primer and releaser pheromones causes strong decreases in their spawning willingness and success. For instance, exposure of mature male salmon parr to environmental levels of certain common pesticides inhibited

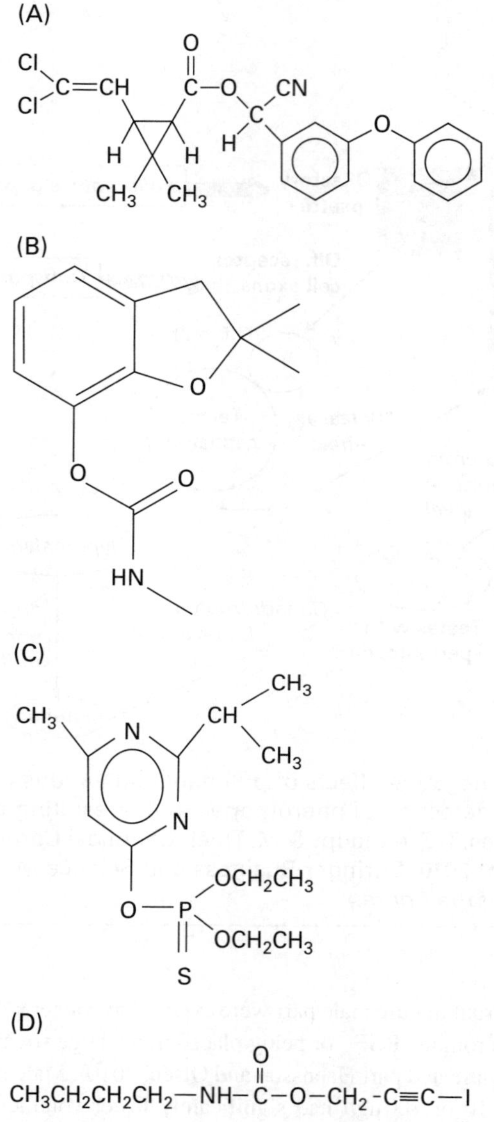

(A)

(B)

(C)

(D)

$$CH_3CH_2CH_2CH_2-NH-\overset{\overset{\displaystyle O}{\|}}{C}-O-CH_2-C\equiv C-I$$

Figure 10.3. Molecular structures of four pesticides with effects on the function of the olfactory sense and olfactory-mediated behaviors in fish. The insecticides cypermethrin (A), carbofuran (B), the organophosphate insecticide diazinon (C), and the carbamate fungicide iodocarb (IPBC) (D). Adapted from KEMI, Swedish Chemicals Agency, Sundbyberg, Stockhom, Sweden.

the olfactory detection of the female reproductive priming pheromone, which is thought to be involved in the synchronization of spawning between the two sexes (Moore and Waring, 1996, 1998; Waring and Moore, 1997). Similarly, exposing the olfactory epithelium to the organophosphate diazinon for 30 min decreased the EOG responses to L-serine and $PGF_{2\alpha}$ at a concentration as low as 1 µg/l (Moore and Waring, 1996). These authors also observed that milt volumes and blood plasma levels of sex hormones (17,20ß-P, LH) were no longer elevated to the same extent when fish were exposed to ovulated females' urine after 120-h exposure to 0.3 µg/l or higher concentration of this pesticide. Later, Moore and Waring (1998) demonstrated reduced priming effects of female urine in mature Atlantic salmon parr (milt volumes, blood plasma levels of sex steroids) after exposure to the herbicide atrazine at the same low levels that gave reduced EOG responses to $PGF_{2\alpha}$ (0.04 µg/l, or above). The study also showed that the *in vitro* release of sex steroids from the testes of atrazine exposed parr was affected. Atrazine and the related simazine were later tested individually or in mixtures of both (Moore and Lower, 2001), and it was shown that the herbicides had additive effects.

It has recently been demonstrated that environmental levels of the synthetic pyrethroid insecticide *cypermethrin* (0.1 µg/l) may interfere with salmon reproduction directly by reducing fertilization rates (Moore and Waring, 2001). This study was done *in vitro* where the water had been contaminated by the pesticide. The authors did not test the ability of sperms from pesticide-exposed males to fertilize eggs. However, they demonstrated, however, that *cypermethrin* concentrations lower than 0.1 µg/l abolished the endocrine effects of the female priming pheromone (Moore and Waring, 2001). No studies of the reproductive behavior were performed.

In a recent study with spawning groups of brown trout in a big stream tank, males exposed to 1 µg/l *cypermethrin* for 4 days spent less time with the nest digging female and courted her less frequently than control males exposed the control (Jaensson *et al.*, 2007). The exposed males also had lower plasma levels of 11-KT and volumes of strippable milt. In a priming experiment, males pre-exposed to *cypermethrin* had lower blood plasma levels of 11-KT and 17,20β-P after 5 h of exposure to ovarian fluids, indicating that the olfactory sense was affected. Surprisingly, exposed males were present when the female spawned indicating involvement of other senses other than olfaction in attracting males. *Cypermethrin* is used to treat salmonid fish in aquaculture for sea lice (*Lepeophtheirus salmonis*), a commonly occurring copepod parasite (Ernst *et al.*, 2001), and also in dips to prevent and treat ticks, lice, and scrab on sheep (Moore and Waring, 2001). The insecticide is also used in forestry for the treatment of spruce seedlings against fine weevil (Torstensson, Börjesson, and Arvidsson, 1999); therefore, it is possible that streams are contaminated.

10.7.3 Effects of Endocrine Disrupting Compounds

Very few studies have directly examined the possible effects of EDCs on pheromone-mediated reproductive behaviors in fish. Bjerselius *et al.* (2001) exposed mature goldfish males during 24–28 days to the female 17ß-estradiol via the food or via the ambient water, and studied their behavior toward females injected with the hormone $PGF_{2\alpha}$ that acts as a releaser pheromone after it and its metabolite are released into the water. All aspects of the male reproductive behavior were significantly reduced or absent after estrogen exposure, and the males had lost their spawning tubercles and their testes decreased in size. Food search behavior was not negatively affected. Estrogenic chemicals have also affected male reproductive behaviors in other fish species (e.g., Gray, Teather, and Metcalfe, 1999; Bayley *et al.*, 1999; Kristensen *et al.*, 2005; Lavelle and Sorensen, 2011). It is, however, not possible to determine if the exposure decreased the olfactory sensitivity for sex pheromones or if changes in androgen levels had direct effects on certain brain areas important for male specific behaviors (see Hallgren, Linderoth, and Olsén, 2006).

10.8 SUMMARY AND FUTURE DIRECTIONS

Fish are highly dependent on their olfactory sense to mediate foraging, detection of predators, kin recognition, and identification of sex partners. Yet, these systems are highly susceptible to interference from pollutants. Olfactory receptor cells are exposed directly to the environment. Thus, there are many opportunities for pollutants to disturb the olfactory detection of chemical signals and pheromones, blocking or disrupting the detection of this information. Olfactory receptor neurons can transport metals, molecules, and viruses into the olfactory bulbs and perhaps to higher brain centers.

Pollutants can also affect the production of chemical signals. EDCs can affect the production of pheromones or detection of the same. There is not much information published about pollutant effects on chemical communication in fish. Most of what we know is based on studies that use electrophysiological techniques measuring responses of olfactory receptor cells (EOG) to amino acids, L-serine, in particular. Most of the few studies done concerning detection disturbances of pheromones have been done with salmonid fish, mostly male Atlantic salmon, and endocrine priming sex pheromones. Further, the majority of the chemical tested are metals (mostly copper salts) and various organic pesticides. Some of the pesticides, especially insecticides, have general effects on the nerve system. Studies on effects of pollutants on chemical communication and behavior in fish are scare. Information about kinship is one aspect of olfactory communication that has hardly been studied in connection to pollutant effects. Sibling specific odors have been suggested to be used during social interactions and shoaling in salmonids, but also during mate choice (Olsén, 1999). Recent studies have indicated that odors originating from the MHC molecules, probably specific peptides, may be included in the chemical information transmitted (Olsén *et al.*, 1998; Milinski *et al.*, 2005). Such peptides should easily be broken down and metabolized by microorganisms and the information is lost. It can be worth looking into the possible effects of various pharmaceutical drugs that have been shown to affect olfaction and taste in humans (Doty and Bromly, 2004) and are present in effluents from sewage treatment plants (e.g., Waiser *et al.*, 2011). Studies of anthropogenic effects on chemical communication in fish, and other aquatic organisms as well, are still in their infancy and demand more attention (Barry, 1999; Lürling and Scheffer, 2007; Olsén, 2011).

ACKNOWLEDGMENT

This work was supported by The Foundation for Baltic and Eastern European studies (Östersjöstiftelsen)

REFERENCES

Barata, C., Damasio, J., Lopez, M.A. *et al.* (2007) Combined use of biomarkers and in situ bioassays in *Daphnia magna* to monitor environmental hazards of pesticides in the field. *Environmental Toxicology and Chemistry*, **26**, 370–379.

Barry, M. (1999) Chemical communication in planktonic organisms: environmental contaminants can mimic the effects of natural chemical signals. *SETAC – Europe News*, **10**, 6–8.

Bayley, M., Larsen, P.F., Baekgaard, H., and Baatrup, E. (1999)The effects of vinclozolin, an anti-androgenic fungicide, on male guppy secondary sex characters and reproductive success. *Biology of Reproduction*, **69**, 1951–1956.

Belanger, R.M., Pachkowski, M.D., and Stacey, N.E. (2010) Methyltestosterone-induced changes in electro-olfactogram responses and courtship behaviors of cyprinids. *Chemical Senses*, **35**, 65–74.

Bertmar, G. (1979) Home range, migrations and orientation mechanisms of the River Indalsälven trout, *Salmo trutta* L. *Report Institute of Freshwater Research, Drottningholm*, **58**, 5–26.

Bertmar, G. and Toft, R. (1969) Sensory mechanisms of homing in salmonid fish. I. Introductory experiments on the olfactory sense in grilse of Baltic salmon (*Salmo salar*). *Behaviour*, **35**, 235–241.

Beyers, D.W. and Farmer, M.S. (2001) Effects of copper on olfaction of Colorado pikeminnow. *Environmental Toxicology and Chemistry*, **20**, 907–912.

Bilberg, K., Døving, K.B., Beedholm, K., and Baatrup, E. (2011) Silver nanoparticles disrupt olfaction in crucian carp (*Carassius carassius*) and Eurasian perch (*Perca fluviatilis*). *Aquatic Toxicology*, **104**, 145–152.

Bjerselius, R. and Olsén, K.H. (1993) A study of the olfactory sensitivity of crucian carp (*Carassius carassius*) and goldfish (*Carassius auratus*) to 17 20ß-dihydroxy-4-pregnen-3-one and prostaglandin F$_{2\alpha}$. *Chemical Senses*, **18**, 427–436.

Bjerselius, R., Winberg, S., Winberg, Y., and Zeipel, K. (1993) Ca^{2+} protects olfactory receptor function against acute Cu(II) toxicity in Atlantic salmon. *Aquatic Toxicology*, **25**, 25–138.

Bjerselius, R., Olsén, K.H., and Zheng, W.B. (1995a) Behavioral and endocrinological response of mature male goldfish to the sex hormone 17, 20ß-dihydroxy-4-pregnen-3-one in the water. *Journal of Experimental Biology*, **198**, 747–754.

Bjerselius, R., Olsén, K.H., and Zheng, W.B. (1995b) Endocrine, gonadal and behavioral responses of male crucian carp to the hormonal pheromone 17, 20ß-dihydroxy-4-pregnen-3-one. *Chemical Senses*, **20**, 221–230.

Bjerselius, R., Li, W., Teeter, J.H. *et al.* (2000) Direct behavioral evidence that unique bile acids released by larval sea lamprey (*Petromyzon marinus*) function as a migratory pheromone. *Canadian Journal of Fisheries and Aquatic Sciences*, **57**, 557–569.

Bjerselius, R., Lundstedt-Enkel, K., Mayer, I., and Olsén, K.H. (2001) Male goldfish reproductive behaviour and physiology are severely affected by exogenous exposure to 17β-estradiol. *Aquatic Toxicology*, **53**, 139–152.

Blaxter, J.H.S. and Hallers-Tjabbes, C.C.N. (1982) The effect of pollutants on sensory systems and behaviour of aquatic animals. *Netherlands Journal of Aquatic Ecology*, **26**, 43–58.

Bonde, J.P., Comhaire, F., and Ombelet, W. (eds) (1998) Environmental influences on spermatogenesis and sperm quality. *Middle East Fertility Society Journal*, **3**, 1–53.

Brown, G.E. (2003) Learning about danger: chemical *alarm* cues and local risk assessment in prey fishes. *Fish and Fisheries*, **4**, 227–234.

Brown, G.E., Brown, J.A., and Crosbie, A.M. (1993) Phenotype matching in juvenile rainbow trout. *Animal Behaviour*, **46**, 1223–1225.

Brown, G.E., Chivers, D.P., and Smith R.J.F. (1995) Localized defecation by pike: a response to labelling by cyprinid alarm pheromone? *Behavioral Ecology and Sociobiology*, **36**, 105–110.

Brown, S.B., Evans, R.E., Thompson, B.E., and Hara, T.J. (1982) Chemoreception and aquatic pollutants, in *Chemoreception in Fishes. Developments in Aquaculture and Fisheries Science*, vol. **8** (ed. T.J. Hara), Elsevier, Amsterdam, Oxford, New York, pp. 363–394.

Cardwell, J.R., Stacey, N.E., Tan, E.S.P. *et al.* (1995) Androgen increases olfactory receptor response to a vertebrate sex pheromone. *Journal of Comparative Physiology, A*, **176**, 55–61.

Clark, R. (2001) *Marine Pollution*. Oxford University press, Oxford, UK, 248 pp.

Cooper, J.C. and Hirsch, P.J. (1982) The role of chemoreception in salmonid homing, in *Chemoreception in Fishes. Developments in Aquaculture and Fisheries Science*, vol. **8** (ed. T.J. Hara), Elsevier, Amsterdam, Oxford, New York, pp. 343–362.

Courtenay, S.C., Quinn, T.P., Dupuis, H.M. *et al.* (2001) Discrimination of family-specific odours by juvenile coho salmon: roles of learning and odour concentration. *Journal of Fish Biology*, **58**, 107–125.

Darnerud, P.O., Lund, B.-O., Brittebo, E.B., and Brandt, I. (1989) 1,2-Dibromoethane and chloroform in the rainbow trout (*Salmo gairdneri*): studies on the distribution of nonvolatile and irreversibly bound metabolites. *Journal of Toxicology and Environmental Health*, **26**, 209–221.

Doty, R.L. and Bromley, S.M. (2004) Effects of drugs on olfaction and taste. *Otolaryngologic Clinics of North America*, **37**, 1229–1254.

Døving, K.B. and Lastein, S. (2009) The alarm reaction in fishes-odorants, modulations of responses, neural pathways. *Annals of the New York Academy of Sciences*, **1170**, 413–423.

Døving, K.B., Dubois-Dauphin, M., Holley, A., and Jourdan, F. (1977) Functional anatomy of the olfactory organ of fish and the ciliary mechanism of water transport. *Acta Zoologica (Stockholm)*, **58**, 245–255.

Døving, K.B., Selset, R., and Thommesen, G. (1980) Olfactory sensitivity to bile acids in salmonid fishes. *Acta Physiologica Scandinavica*, **108**, 121–131.

Emanuel, M.E. and Dodson, J.J. (1979) Modification of the rheotropic behavior of male rainbow trout (*Salmo gairdneri*) by ovarian fluid. *Journal of Fisheries Research Board Canada*, **36**, 63–68.

Environment Agency (1998) *Report on the Welsh Sheep Dip Monitoring Programme*. Environment Agency, Cardiff, Wales, UK.

Environment Agency (2003) *Pesticides 2002. The Annual Report of the Environment Agency Pesticide Monitoring Programme*. Environment Agency, Wallingford, Oxon, UK.

Ernst, W., Jackman, P., Doe, K. *et al.* (2001) Dispersion and toxicity to non-target aquatic organisms to pesticides used to treat sea lice on salmon in net pen enclosures. *Marine Pollution Bulletin*, **42**, 433–444.

Evans, J. and Hastings, L. (1992) Accumulation of Cd(II) in the CNS depending on the route of administration: intraperitoneal, intratracheal, or intranasal. *Fundamental and Applied Toxicology*, **19**, 275–278.

Fabian, N.J., Albright, L.B., Gerlach, G. *et al.* (2007) Humic acid interferes with species recognition in zebrafish (*Danio rerio*). *Journal of Chemical Ecology*, **33**, 2090–2096.

Fairchild, W.L., Swansburg, E.O., Arsenault, J.T., and Brown, S.B. (1999) Does an association between pesticide use and subsequent declines in catch of Atlantic salmon (*Salmo salar*) represent a case of endocrine disruption? *Environmental Health Perspectives*, **107**, 349–357.

Fent, K., Weston, A.A., and Caminada, D. (2006) Ecotoxicology of human pharmaceuticals. *Aquatic Toxicology*, **76**, 122–159.

Ferrari, M.C.O., Brown, M.R., Pollock, M.S., and Chivers, D.P. (2007) The paradox of risk assessment: comparing responses of fathead minnows to capture-released and diet-released alarm cues from two different predators. *Chemoecology*, **17**, 157–161.

Fine, J.M. and Sorensen, P.W. (2010) Production and fate of the sea lamprey migratory pheromone. *Fish Physiology and Biochemistry*, **36**, 1013–1020.

Fisher, H.S., Wong, B.B.M., and Rosenthal, G.G. (2006) Alteration of the chemical environment disrupts communication in a freshwater fish. *Proceedings of the Royal Society of London, Series B: Biological Sciences*, **273**, 1187–1193.

Garcia-Valcárcel, A.I. and Tadeo, J.L. (2007) Evaluation of laboratory assays for the assessment of leaching of copper and chromium from ground-contact wood. *Environmental Toxicology and Chemistry*, **26**, 2115–2121.

Gatidou, G. and Thomaidis, N.S. (2007) Evaluation of single and joint toxic effects of two antifouling biocides, their main metabolites and copper using phytoplankton bioassays. *Aquatic Toxicology*, **85**, 184–191.

Gautam, S.H., Otsuguro, K.-I., Ito, S. *et al.* (2007) T-type Ca channels mediate propagation of odor-induced Ca transients in rat olfactory receptor neurons. *Neuroscience*, **144**, 702–713.

Gray, M.A., Teather, K.L., and Metcalfe, C.D. (1999) Reproductive success and behavior of Japanese medaka (*Oryzias latipes*) exposed to 4-tert-octylphenol. *Environmental Toxicology and Chemistry*, **18**, 2587–2594.

Hallgren, S.L.E., Linderoth, M., and Olsén, K.H. (2006) Inhibition of cytochrome p450 brain aromatase reduces two male specific sexual behaviours in the male Endler guppy (*Poecilia reticulata*). *General and Comparative Endocrinology*, **147**, 323–328.

Hamdani, E.H., Lastein, S., Gregersen, F., and Døving, K.B. (2008) Seasonal variations in olfactory sensory neurons - Fish sensitivity to sex pheromones explained? *Chemical Senses*, **33**, 119–123.

Hansen, A. and Zielinski, B.S. (2005) Diversity in the olfactory epithelium of bony fishes: development, lamellar arrangement, sensory neuron cell types and transduction components. *Journal of Neurocytology*, **34**, 183–208.

Hara, T.J. (1982) Structure-activity relationships of amino acids as olfactory stimuli, in *Chemoreception in Fishes*. Developments in Aquaculture and Fisheries Science, vol. **8** (ed. T.J. Hara), Elsevier, Amsterdam, Oxford, New York, pp. 343–362.

Hara, T.J. (1986) Role of olfaction in fish behavior, in *The Behaviour of Teleost Fishes* (ed. T.J. Pitcher), Croom Helm, London, Sydney, pp. 152–176.

Hara, T.J., Law, Y.M.C., and Macdonald, S. (1976) Effects of mercury and copper on the olfactory response in rainbow trout. *Journal of the Fisheries Research Board of Canada*, **33**, 1568–1573.

Hara, T.J., Brown, S.B., and Evans, R.E. (1983) Pollutants and chemoreception in aquatic organisms, in *Aquatic Toxicology* (ed. J.O. Nriagu), Wiley, New York, pp. 247–306.

Hasler, A.D. and Scholz, A.T. (1983) *Olfactory Imprinting and Homing in Salmon. Investigation Into the Mechanisms of the Imprinting Process.* Springer-Verlag, Berlin.

Hastings, L. and Evans, J.E. (1991) Olfactory primary neurons as a route of entry for toxic agents into the CNS. *NeuroToxicology*, **12**, 707–714.

Henriksson, J., Tallkvist, J., and Tjälve, H. (1999) Press the escape key to close transport of manganese via the olfactory pathway in rats: dosage dependency of the uptake and subcellular distribution of the metal in the olfactory epithelium and the brain. *Toxicology and Applied Pharmacology*, **156**, 119–128.

Hessel, K., Kreuger, J., and Ulén, B. (1997) Kartläggning av bekämpningsmedelrester i yt-, grund- och regnvatten i Sverige 1985–1995. Resultatfrånmonotoringochriktadprovtagning. Division of Water Quality Management, Swedish University of Agricultural Sciences. *Research Report Ekohydrology*, **42**, 72 (In Swedish).

Heuschele, J. and Candolin, U. (2007) An increase in pH boosts olfactory communication in sticklebacks. *Biology Letters*, **3**, 411–413.

Höglund, L.B. (1961) The reaction of fish in concentration gradients. *Report of the Institute of Freshwater Research, Drottningholm*, **43**, 1–147.

Hubbard, P.C. and Canario, A.V.M. (2007) Evidence that olfactory sensitivities to calcium and sodium are mediated by different mechanisms in the goldfish *Carassius auratus*. *Neuroscience Letters*, **44**, 90–93.

Hubbard, P.C., Barata, E.N., and Canario, A.V.M. (2002) Possible disruption of pheromonal communication by humic acid in the goldfish, *Carassius auratus*. *Aquatic Toxicology*, **60**, 169–183.

Jaensson, A. and Olsén, K.H. (2010) Effects of copper on olfactory mediated endocrine responses and reproductive behaviour in mature male brown trout (*Salmo trutta* Linneaus) parr to conspecific females. *Journal of Fish Biology*, **76**, 800–817.

Jaensson, A., Scott, A., Moore, A. *et al.* (2007) Effects of a pyretroid pesticide on endocrine responses to female odours and reproductive behaviour in male parr of brown trout (*Salmo trutta* L.). *Aquatic Toxicology*, **81**, 1–9.

Jarrard, H.E., Delaney, K.R., and Kennedy, C.J. (2004) Impacts of carbamate pesticides on olfactory neurophysiology and cholinesterase activity in coho salmon (*Oncorhynchus kisutch*). *Aquatic Toxicology*, **69**, 133–148.

Kitamura, S. and Ikuta, K. (2000) Acidification severely suppresses spawning of hime salmon (land-locked sockeye salmon, *Oncorhynchus nerka*). *Aquatic Toxicology*, **51**, 107–113.

Klaprat, D.A., Brown, S.B., and Hara, T.J. (1988) The effect of low pH and aluminum on the olfactory organ of rainbow trout, *Salmo gairdneri*. *Environmental Biology of Fishes*, **22**, 69–77.

Kobayashi, M., Sorensen, P.W., and Stacey, N.E. (2002) Hormonal and pheromonal control of spawning behavior in the goldfish. *Fish Physiology and Biochemistry*, **26**, 71–84.

Kolmakov, N.N., Hubbard, P.C., Lopes, O., and Canario, A.V.M. (2009) Effect of acute copper sulfate exposure on olfactory responses to amino acids and pheromones in goldfish (*Carassius auratus*). *Environmental Science & Technology*, **43**, 8393–8399.

Konstandi, M., Pappas, P., Johnson, E. *et al.* (1997) Modification of reproductive function in the rat by 3-methylcholanthrene. *Pharmacological Research*, **35**, 107–111.

Kreuger, J. (1998) Pesticides in stream water within an agricultural catchment in southern Sweden, 1990-1996. *Science of the Total Environment*, **216**, 227–251.

Kreuger, J. (2000) Övervakning av bekämpningsmedel i vatten från ett avrinningsområde i Skåne. Division of Water Quality Management, Swedish University of Agricultural Sciences. *Ekohydrologi*, **54**, 52 (In Swedish).

Kreuger, J., Peterson, M., and Lundgren, E. (1999) Agricultural inputs of pesticide residues to stream and pond sediments in a small catchment in southern Sweden. *Bulletin of Environmental Contamination and Toxicology*, **62**, 55.

Kristensen, T., Baatrup, E., and Bayley, M. (2005) 17 alpha -ethinylestradiol reduces the competitive reproductive fitness of the male guppy (*Poeciliareticulata*). *Biology of Reproduction*, **72**, 150–156.

Kudo, H., Shinto, M., Sakurai, Y., and Kaeriyama, M. (2009) Morphometry of olfactory lamellae and olfactory receptor neurons during the life history of chum salmon (*Oncorhynchus keta*). *Chemical Senses*, **34**, 617–624.

Lastein, S., Hmadani, E.H., and Døving, K.B. (2015) Olfactory discrimination of pheromones, in *Fish Pheromones and Related Cues* (eds Peter W. Sorensen and Brian D. Wisenden), John Wiley & Sons, Inc., Hoboken.

Lavelle, C., and Sorensen, P.W. (2011) Behavioral responses of adult male and female fathead minnows to a model estrogenic effluent and its effects on exposure regime and reproductive success. *Aquatic Toxicology*, **101**, 521–528.

Leduc, A.O.H.C., Kelly, J.M., and Brown, G.E. (2004) Detection of conspecific alarm cues by juvenile salmonids under neutral and weakly acidic conditions: laboratory and field tests. *Oecologia*, **139**, 318–324.

Leduc, A.O.H.C., Roh, E., Breau, C., and Brown, G.E. (2007) Effects of ambient acidity on chemosensory learning: an example of an environmental constraint on acquired predator recognition in wild juvenile Atlantic salmon (*Salmo salar*). *Ecology of Freshwater Fish*, **16**, 385–394.

Leduc, A.O.H.C., Lamaze, F.C., McGraw, L., and Brown, G.E. (2008) Response to chemical alarm cues under weakly acidic conditions: a graded loss of antipredator behaviour in juvenile rainbow trout. *Water, Air, & Soil Pollution*, **189**, 179–187.

Le Jeune, A.H., Charpin, M., Deluchat, V. *et al.* (2006) Effect of copper sulphate treatment on natural phytoplanktonic communities. *Aquatic Toxicology*, **80**, 267–280.

Liley, N.R. (1982) Chemical communication in fish. *Canadian Journal of Fisheries and Aquatic Sciences*, **39**, 22–35.

Lower, N. and Moore, A. (2007) The impact of a brominated flame retardant on smoltification and olfactory function in Atlantic salmon (*Salmo salar* L) smolts. *Marine and Freshwater Behaviour and Physiology*, **40**, 267–284.

Lürling, M. and Scheffer, M. (2007) Info-disruption: pollutants and the transfer of chemical information between organisms. *Trends in Ecology and Evolution*, **22**, 374–379.

McFredrick, Q.S., Fuentes, J.D., Roulston, T. *et al.* (2009) Effects of air pollution on biogenic volatiles and ecological interactions. *Oecologia*, **160**, 411–420.

McPherson, T.D., Mirza, R.S., and Pyle, G.G. (2004) Responses to wild fishes to alarm chemicals in pristine and metal-contaminated lakes. *Canadian Journal of Zoology*, **82**, 694–700.

Mesquita, R.M.R.S., Canario, A.V.M., and Melo, E. (2003) Partition of fish pheromones between water and aggregates of humic acids. Consequences for Sexual signaling. *Environmental Science & Technology*, **37**, 742–746.

Milinski, M., Griffiths, S., Wegner, K.M. *et al.* (2005) Mate choice decisions of stickleback females predictably modified by MHC peptide ligands. *Proceedings of the National Academy of Sciences, USA*, **102**, 4414–4418.

Moore, A. (1994) An electrophysiological study on the effects of pH on olfaction in mature male Atlantic salmon (*Salmo salar*) parr. *Journal of Fish Biology*, **45**, 493–502.

Moore, A. and Lower, N. (2001) The impact of two pesticides on olfactory-mediated endocrine function in mature male Atlantic salmon (*Salmo salar* L.) parr. *Comparative Biochemistry and Physiology B*, **129**, 269–276.

Moore, A. and Waring, C.P. (1996) Sublethal effects of the pesticide Diazinon on olfactory function in mature male Atlantic salmon (*Salmo salar* L.) parr. *Journal of Fish Biology*, **48**, 758–775.

Moore, A. and Waring, C.P. (1998) Mechanistic effects of a triazine pesticide on reproductive endocrine function in mature male Atlantic salmon (*Salmo salar* L.) parr. *Pesticide Biochemistry and Physiology*, **62**, 41–50.

Moore, A. and Waring, C.P. (2001) The effects of a synthetic pyrethroid pesticide on some aspects of reproduction in Atlantic salmon (*Salmo salar*). *Aquatic Toxicology*, **52**, 1–12.

Moore, A., Olsén, K.H., Lower, N., and Kindahl, H. (2002) The role of F-series prostaglandins as reproductive priming pheromones in the brown trout. *Journal of Fish Biology*, **60**, 613–624.

Mori, K., Nagao, H., and Yoshihara, Y. (1999) The olfactory bulb: coding and processing of odor molecule information. *Science* (Washington), **286**, 711–715.

Morin, P.-P., Dodson, J.J., and Dore, F.Y. (1989) Cardiac responses to a natural odorant as evidence of a sensitive period for olfactory imprinting in young Atlantic salmon, Salmo salar. *Canadian Journal of Fisheries and Aquatic Sciences*, **46**, 122–130.

Müller-Schwarze, D. (2006) *Chemical Ecology of Vertebrates*. Cambridge University Press, Cambridge.

Munday, P.L., Dixson, D.L., Donelson, J.M. *et al.* (2009) Ocean acidification impairs olfactory discrimination and homing of marine fish. *Proceedings of the National Academy of Sciences of the United States of America*, **106**, 1848–1852.

Nordeng, H. (1977) A pheromone hypothesis for homeward migration in anadromous salmonids. *Oikos*, **28**, 155–159.

Olsén, K.H. (1986a) Chemoattraction between juveniles of two sympatric stocks of Arctic charr (*Salvelinus alpinus* (L.)) and their gene frequency of serum esterases. *Journal of Fish Biology*, **28**, 221–231.

Olsén, K.H. (1986b) Emission rate of amino acids and ammonia and their role in olfactory preference behaviour of juvenile Arctic charr, *Salvelinus alpinus* (L.). *Journal of Fish Biology*, **28**, 255–265.

Olsén, K.H. (1986c) Modification of conspecific chemoattraction in Arctic charr (*Salvelinus alpinus* (L.)), by nitrogenous excretory products. *Comparative Biochemistry and Physiology*, **85A**, 77–81.

Olsén, K.H. (1989) Sibling recognition in juvenile Arctic charr (*Salvelinus alpinus* (L.)). *Journal of Fish Biology*, **34**, 571–581.

Olsén, K.H. (1992) Kin recognition in fish mediated by chemical cues, in *Fish Chemoreception* (ed. T.J. Hara), Chapman and Hall, London, pp 229–248.

Olsén, K.H. (1993) The development of the olfactory organ of the Arctic charr *Salvelinus alpinus* (L.) (Teleostei, Salmonidae). *Canadian Journal of Zoology*, **71**, 1973–1984.

Olsén, K.H. (1999) Review - Present knowledge of kin discrimination in salmonids. *Genetics*, **473**, 1–5.

Olsén, K.H. (2011) Effects of pollutants on olfactory mediated behaviors in fish and crustaceans, in *Chemical Communication in Crustaceans* (eds T. Breithaupt and M. Thiel), Springer, New York, Dordrecht, Heidelberg, London, pp. 507–529.

Olsén, K.H. and Höglund, L.B. (1985) Reduction by a surfactant of olfactory mediated attraction between juveniles of Arctic charr (*Salvelinus alpinus* (L.)). *Aquatic Toxicology*, **6**, 57–69.

Olsén, K.H. and Liley, N.R. (1993) The significance of olfaction and social cues in milt availability, sexual hormone status and spawning behaviour of male rainbow trout (*Oncorhynchus mykiss*). *General and Comparative Endocrinology*, **89**, 107–118.

Olsén, K.H. and Winberg, S. (1996) Learning and sibling odour preference in juvenile Arctic char, *Salvelinus alpinus*. *Journal of Chemical Ecology*, **22**, 773–786.

Olsén, K.H., Lohm, J., Grahn, M., and Langefors, Å. (1998) MHC and kin discrimination in juvenile Arctic charr, *Salvelinus alpinus* (L.). *Animal Behaviour*, **56**, 319–327.

Olsén, K.H., Bjerselius, R., Petersson, E. *et al.* (2000) Lack of species specific primer effects of odours from female Atlantic salmon (*Salmo salar* L.) and brown trout (*Salmo trutta*). *Oikos*, **88**, 213–220.

Olsén, K.H., Bjerselius, R., Mayer, I., and Kindahl, H. (2001) Both ovarian fluid and urine have priming effects in mature Atlantic salmon male parr. *Journal of Chemical Ecology*, **27**, 2337–2349.

Olsén, K.H., Johansson, A.-K., Bjerselius, R. *et al.* (2002) Mature Atlantic *salmon (Salmo salar)* male parr are attracted to ovulated female urine but not to ovarian fluid. *Journal of Chemical Ecology*, **28**, 29–40.

Olsén, K.H., Petersson, E., Ragnarsson, B. *et al.* (2004) Downstream migration in *Salmo salar* smolt sibling groups. *Canadian Journal of Fisheries and Aquatic Sciences*, **61**, 328–331.

Olsén, K.H., Sawisky, G.R., and Stacey, N.E. (2006) Endocrine and milt responses of male crucian carp (*Carassius carassius*) to periovulatory females under field conditions. *General and Comparative Endocrinology*, **149**, 294–302.

Persson, E., Larsson, P., and Tjälve, H. (2002) Cellular activation and neuronal transport of intranasally instilled benzo(a)pyrene in the olfactory system of rats. *Toxicological Letters*, **133**, 211–219.

Persson, E., Henriksson, J., and Tjälve, H. (2003) Uptake of cobalt from the nasal mucosa into the brain via olfactory pathways in rats. *Toxicology Letters*, **145**, 19–27.

Quinn, T.P. and Busack, C.A. (1985) Chemosensory recognition of siblings in juvenile coho salmon (*Oncorhynchus kisutch*). *Animal Behaviour*, **33**, 51–56.

Quinn, T.P., Stewart, I.J., and Boatright, C.P. (2006) Experimental evidence of homing to site of incubation by mature sockeye salmon, *Oncorhynchus nerka*. *Animal Behaviour*, **72**, 941–949.

Restrepo, D., Miyamoto, T., Bryant, B.P., and Teeter, J.H. (1990) Odor stimuli trigger influx of calcium into olfactory neurons of the channel catfish. *Science* (Washington), **249**, 1166–1168.

Rouleau, C., Borg-Neczak, K., Gottofrey, J., and Tjälve, H. (1999) Accumulation of waterborne mercury(II) in specific areas of fish Brain. *Environmental Science & Technology*, **33**, 3384–3389.

Saglio, P., Bretaud, S., Rivot, E., and Olsén, K.H. (2003) Chemobehavioral changes induced by short-term exposures to prochloraz, nicosulfuron and carbofuran in goldfish. *Archives of Environmental Contamination and Toxicology*, **45**, 515–524.

Sandhal, J., Baldwin, D., Jenkins, J., and Scholz, N. (2004) Oder-evoked field potentials as indicators of sublethal neurotoxicity in juvenile coho salmon (*Oncorhynchus kisutch*) exposed to copper, chlorpyrifos, or esfenvalerate. *Canadian Journal of Fisheries and Aquatic Sciences*, **61**, 404–413.

Saucier, D., Astic, L., and Rioux, P. (1991) The effects of early chronic exposure to sublethal copper on the olfactory discrimination of rainbow trout, *Oncorhynchus mykiss*. *Environmental Biology of Fishes*, **30**, 345–351.

Scholz, A.T., Horrall, R.M., Cooper, J.C., and Hasler, A.D. (1976) Imprinting to chemical cues: the basis for homestream selection in salmon. *Science* (Washington, DC), **196**, 1247–1249.

Scholz, N.L., Truelove, N.K., French, B.L. *et al.* (2001) Diazinon disrupts antipredator and homing behaviors in chinook salmon (*Oncorhynchus tshawytscha*). *Canadian Journal of Fisheries and Aquatic Sciences*, **57**, 1911–1918.

Scott, A.P., Liley, N.R., and Vermeirssen, E.L.M. (1994) Urine of reproductively mature female rainbow trout, *Oncorhynchus mykiss* (Walbaum), contains a priming pheromone which enhances plasma levels of sex steroid and gonadotrophin II in males. *Journal of Fish Biology*, **44**, 131–147.

Shiraiwa, T., Kamiyama, N., and Kashiwayanagi, M. (2007) Decreases in urinary pheromonal activities male mice after exposure to 3-methylchoranthrene. *Toxicological Letters*, **169**, 137–144.

Siefkes, M.J., Winterstein, S.R., and Li, W. (2005) Evidence that 3-keto petromyzonol sulphate specifically attracts ovulating female sea lamprey, *Petromyzon marinus*. *Animal Behaviour*, **70**, 1037–1045.

Sloman, K.A. and Wilson, R.W. (2006) Antropogenic impacts upon behaviour and physiology, in *Fish Physiology, Vol. 24: Behaviour and Physiology in Fish* (eds A. Sloman, R.W. Wilson, and S. Balshine), Elsevier Academic Press, San Diego, CA and London, UK, pp. 413–468.

Sorensen, P.W. and Caprio, J. (1998) Chemoreception, in *The Physiology of Fishes*, 2nd edn (ed. D.H. Evans), CRC Press, Boca Raton, Florida, pp. 375–405.

Sorensen, P.W., Hara, T.J., and Stacey, N.E. (1991) Sex pheromones selectively stimulate the medial olfactory tracts of male goldfish. *Brain Research*, **558**, 343–347.

Sorensen, P.W., Hoye, T.R., Fine, J.M. *et al.* (2005) *Structural and behavioral characterization of a potent migratory pheromone in the sea lamprey.* Annual Conference on Great Lakes Research, 48, Pages.

Stabell, O.B. (1984) Homing and olfaction in salmonids: a critical review with special reference to the Atlantic salmon. *Biological Reviews of the Cambridge Philosophical Society*, **59**, 333–388.

Stacey, N. (2015) Hormonally-derived pheromones in teleost fishes, in *Fish Pheromones and Related Cues* (eds Peter W. Sorensen and Brian D. Wisenden), John Wiley & Sons, Inc., Hoboken.

Stacey, N. and Sorensen, P.W. (2002) Hormonal pheromones in fish, in *Hormones, Brain and Behavior*, vol. 2 (eds D.W. Pfaff, A.P. Arnold, A.M. Etgen *et al.*), Elsevier, San Diego, CA, pp. 375–434.

Stacey, N. and Sorensen, P.W. (2005) Reproductive pheromones, in *Fish Physiology*, vol. 24: *Behaviour and Physiology in Fish* (eds A. Sloman, R.W. Wilson, and S. Balshine), Elsevier, Academic Press, San Diego, CA and London, UK, pp. 359–412.

Stacey, N. and Sorenen, P.W. (2006) Reproductive pheromones, in *Behaviour and Physiology of Fish* (eds K.A. Sloman, R.W. Wilson, and S. Balshine), Academic Press, London, pp. 359–412.

Stacey, N. and Sorensen, P.W. (2009) Hormonal pheromones in fish, in *Hormones, Brain and Behavior*, vol. 1, 2nd edn (eds D.W. Pfaff, A.P. Arnold, A.M. Etgen *et al.*), Academic Press, San Diego, California, pp. 639–681.

Stacey, N., Chojnacki, A., Narayanan, A. *et al.* (2003) Hormonally derived sex pheromones in fish: exogenous cues and signals from gonads to brain. *Canadian Journal of Physiology and Pharmacology*, **81**, 329–241.

Stacey, N.E., Van Der Kraak, G.J., and Olsén, K.H. (2011) Male primer endocrine responses to preovulatory female cyprinids under natural conditions in Sweden. *Journal of Fish Biology*, **80**, 147–165.

Sutterlin, A.M. (1974) Pollutants and the chemical senses of aquatic animals – perspective and review. *Chemical Senses and Flavor*, **1**, 167–178.

Suzuki, N. (1982) Responses of olfactory receptor cells to electrical and chemical stimulation, in *Chemoreception in Fishes*. Developments in Aquaculture and Fisheries Science, vol. **8** (ed. T.J. Hara), Elsevier, Amsterdam, Oxford, New York, pp. 93–108.

Tallkvist, J., Henriksson, J., d'Argy, R., and Tjälve, H. (1998) Transport and subcellular distribution of nickel in the olfactory system of pikes and rats. *Toxicological Sciences*, **43**, 196–203.

Tallkvist, J., Persson, E., Henriksson, J., and Tjälve, H. (2002) Cadmium-metallothionein interactions in the olfactory pathways of rats and pikes. *Toxicological Sciences*, **67**, 108–113.

Thommesen, G. (1978) The spatial distribution of odour induced potentials in the olfactory bulb of char and trout (Salmonidae). *Acta Physiologica Scandinavica*, **102**, 205–217.

Thommesen, G. (1983) Detection of a blocking effect of low pH in the trout olfactory organ, in *Chemoreception in Studies of Marine Pollution* (ed. K.B. Døving), Reports from a workshop at Oslo, July 13 and 14, 1980. The Norwegian Marine Pollution Research and Monitoring Programme, pp. 44–49.

Tierney, K.B., Taylor, A.L., Ross, P.S., and Kennedy, C.J. (2006a) The alarm reaction of coho salmon parr is impaired by the carbamate fungicide IPBC. *Aquatic Toxicology*, **79**, 149–157.

Tierney, K.B., Ross, P.S., Jarrard, H.E. *et al.* (2006b) Changes in juvenile coho salmon electro-olfactogram during and after short-term exposure to current-use pesticides. *Environmental Toxicology and Chemistry*, **25**, 2809–2817.

Tierney, K.B., Singh, C.R., Ross, P.S., and Kennedy, C.J. (2007) Relating olfactory neurotoxicity to altered olfactory-mediated behaviors in rainbow trout exposed to three currently-used pesticides. *Aquatic Toxicology*, **81**, 55–64.

Tierney, K.B., Baldwin, D.H., Hara, T.J. *et al.* (2010) Olfactory toxicity in fishes. *Aquatic Toxicology*, **96**, 2–26.

Tilton, F.A., Tilton, S.C., Bammler, T.K. *et al.* (2011) Transcriptional impact of organophosphate and metal mixtures on olfaction: copper dominates the chlorpyrifos-induced response in adult zebrafish. *Aquatic Toxicology*, **102**, 205–215.

Tomlinson, A.H. and Esiri, M.M. (1983) Herpes simplex encehpalitis – Immunohistological demonstration of spread of virus via olfactory pathways in mice. *Journal Neurological Sciences*, **60**, 473–484.

Törnquist, M., Kreuger, J., Adielsson, S., and Kylin, H. (2005) Bekämpningsmedel i vatten och sediment från typområden och åar, samt i nederbörd under 2004. Swedish University of Agricultural Sciences. *Ekohydrologi*, **87**, 1–91 (In Swedish).

Torstensson, L., Börjesson, E., and Arvidsson, B. (1999) Treatment of bare toot spruce seedlings with permethrin against pine weevil before lifting. *Scandinavian Journal of Forest Research*, **14**, 408–415.

Waiser, M.J., Humphries, D., Tumber, V., and Holm, J. (2011) Effluent-dominated streams. Part 2: Presence and possible effects of pharmaceuticals and personal care products in Wascana Creek, Saskatchewan, Canada. *Environmental Toxicology and Chemistry*, **30**, 508–519.

Ward, A.J.W., Duff, A.J., Horsfall, J.S., and Currie, S. (2008) Scents and scents-ability: pollution disrupts chemical social recognition and shoaling in fish. *Proceedings of the Royal Society of London, Series B, Biological Sciences*, **274**, 101–105.

Waring, C.P. and Moore, A. (1997) Sublethal effects of a carbomate pesticide on pheromonal mediated endocrine function in mature Atlantic salmon (*Salmo salar*) parr. *Fish Physiology and Biochemistry*, **17**, 203–211.

Waring, C.P., Moore, A., and Scott, A.P. (1996) Milt and endocrine responses of mature Atlantic salmon (*Salmo salar*) male parr to water borne testosterone, 17,20ß-dihydroxy-pregnen-4-en-3-one-20-sulfate and the urine from adult female and male salmon. *General and Comparative Endocrinology*, **103**, 142–149.

Wilson, E.O. (1975) *Sociobiology. The New Synthesis*. Belknap, Harvard University Press, Cambridge, Massachusetts, and London, England.

Winberg, S. and Olsén, K.H. (1992) The influence of rearing conditions on the sibling odour preference of juvenile Arctic charr (*Salvelinus alpinus* (L)). *Animal Behaviour*, **44**, 157–164.

Wisby, W.J. and Hasler, A.D. (1954) The effect of olfactory occlusion on migrating silver salmon (*O. kisutch*). *Journal of the Fisheries Research Board of Canada*, **11**, 472–478.

Wisenden, B.D. (2015) Chemical cues that indicate risk of predation, in *Fish Pheromones and Related Cues* (eds Peter W. Sorensen and Brian D. Wisenden), John Wiley & Sons, Inc., Hoboken.

Yambe, H. and Yamazaki, F. (2000) Urine of ovulated female masu salmon attracts immature male parr treated with methyltestosterone. *Journal of Fish Biology*, **57**, 1058–1064.

Yambe, H. and Yamazaki, F. (2001) Species-specific releaser effect of urine from ovulated female masu salmon and rainbow trout. *Journal of Fish Biology*, **59**, 1455–1464.

Yambe, H., Munakata, A., Kitamura, S. *et al.* (2003) Methyltestosterone induce male sensitivity to both primer and releaser pheromones in the urine of ovulated female masu salmon. *Fish Physiology and Biochemistry*, **28**, 279–280.

Yambe, H., Kitamura, S., Kamio, M. *et al.* (2006) L-Kynurenine, an amino acid identified as a sex pheromone in the urine of ovulated female masu salmon. *Proceedings of the National Academy of Sciences of the United States of America*, **103**, 15370–15374.

Yun, S.S., Scott, A.P., and Li, W. (2003) Pheromones of the male sea lamprey, *Petromyzon marinus* L.: structural studies on a new compound, 3-keto allocholic acid, and 3-keto petromyzonol sulfate. *Steroids*, **68**, 297–304.

Zippel, H.P. (1993) Regeneration in the peripheral and the central olfactory system: a review of morphological, physiological and behavioral aspects. *Journal Hirnsfoschung*, **34**, 207–229.

Chapter 11
Pheromones in Marine Fish with Comments on Their Possible Use in Aquaculture

Peter Hubbard

Centro de Ciências do Mar, Faro, Portugal

11.1 INTRODUCTION TO PHEROMONES IN MARINE FISHES

Knowledge of pheromonal systems in marine fish lags behind that of freshwater fish. The marine environment can be markedly different from freshwater habitats; does this mean that pheromonal systems are different too? First, there is the obvious difference in salinity; marine and freshwater fish must cope with different osmoregulatory challenges. Also, the sheer size of some oceans and the massive volumes of water involved may influence the biology of ocean inhabitants. Some parts of the marine environment may remain highly stable in terms of physicochemical characteristics (e.g., the low light levels and near-constant temperatures of the deep ocean), whereas others may be extremely variable—but, in some respects, predictably so—such as the intertidal zone. The consequent differences in other classes of organisms present, the biological differences (i.e., no insect larvae—an important source of food for many freshwater fish—live in the sea), will also play a role. Clearly, there is a large "gray area" between these two extremes (the North American Great Lakes, for example, are bodies of freshwater as large as many seas). Nevertheless, these differences in habitat and ecological niches have caused the taxa of fish present in the marine environment to be markedly different from their freshwater equivalents. For example, the Gadoid fish (cods)—extremely important in commercial terms—are almost exclusively marine; whereas, the Cyprinids—containing some of the most important experimental fish models, such as the goldfish and zebrafish—are exclusively freshwater. As very little work has been done on alarm pheromones (Manassa and McCormick, 2012) or social cues in marine fish, this chapter will focus on reproductive pheromones.

11.2 WHY PHEROMONES IN MARINE FISH?

Given the biological and physicochemical differences between marine and freshwater environments, is it reasonable to assume that marine fish, such as freshwater fish, use pheromones? And, if so, are similar chemicals used for similar messages? At a conceptual level, the first question is easier to answer: Yes. This is reasonable because (1) much of the ocean is dark and/or turbid, thereby favoring chemical over visual communication; (2) most fish have highly sensitive olfactory systems; (3) most

Fish Pheromones and Related Cues, First Edition. Edited by Peter W. Sorensen and Brian D. Wisenden.
© 2015 John Wiley & Sons, Inc. Published 2015 by John Wiley & Sons, Inc.

species do not show sexual dimorphism, with important implications for reproduction; and (4) in groups with extant marine and freshwater species, there is evidence for use of pheromones in the freshwater relatives—e.g., the gobies. As existing evidence is so patchy, the second question is currently much harder to give a coherent answer to.

11.3   WHAT DO WE KNOW ABOUT PHEROMONES IN MARINE FISH?

11.3.1   Gobies

One of the first fish steroid pheromones—etiocholanolone glucuronide—was discovered in a marine teleost, the black goby *Gobius niger* (or *G. jozo*) (Colombo *et al.*, 1980). In the male's well-developed mesorchial gland, large quantities of 5β-reduced conjugated androgens are produced (Fig. 11.1). Reduction and conjugation are two mechanisms to deactivate circulating steroid hormones, rendering them more water-soluble and thereby easier to excrete in the urine; this is an important consideration in marine fish that produce much less urine than freshwater fish (see below), where steroid glucuronides have previously been shown to constitute an important class of fish pheromone (Stacey and Sorensen, 2006). Etiocholanolone glucuronide attracted females ready to spawn (but not nonovulated females), even inducing some to lay their eggs in a nest in the absence of a male. Similarly, in the freshwater round goby (*Neogobius melanostomus*), males release conjugated 5β-reduced androgens (Katare *et al.*, 2011) to which reproductive females respond at both the olfactory (Laframboise and Zielinski, 2011) and behavioral levels (Corkum *et al.*, 2008). Thus, marine and freshwater gobies may be using similar reproductive pheromonal systems.

11.3.2   The Sea Lamprey

Although the sea lamprey (*Petromyzon marinus*), as the name suggests, is predominantly marine, it spawns in freshwater (anadromous) and then dies (semelparous). However, it has become a problem invasive species in the Great Lakes of North America, where populations have become entirely freshwater, spending their adult parasitic stage feeding on commercially important native species in the lakes—particularly the lake trout (*Salvelinus namaycush*) and burbot (*Lota lota*)—before returning to contributory streams to spawn and die. Therefore, it has attracted attention from the pheromone research community, as pheromone traps may prove to be an effective and specific method for reducing lamprey populations or their ultimate removal from the Great Lakes system (Sorensen and Stacey, 2004; Corkum and Belanger, 2007). Ironically, in much of its native habitat—the coastal systems of the north-west Atlantic Ocean and associated rivers of continental Europe—it is an endangered species.

Olfactory sensitivity to bile acids (a term used to describe steroids produced in the liver), first observed by Døving, Selset, and Thommesen (1980) in salmonids, has since been shown to be widespread in fish, including marine species (Sola and Tosi, 1993; Hubbard, Barata, and Canário, 2003; Velez *et al.*, 2009; Meredith, Caprio, and Kajiura, 2012). However, it is only in the sea lamprey that the underlying reasons for this are understood. Sexually maturing, migratory lamprey are attracted by a mixture of unique bile salts released by conspecific larvae (Sorensen *et al.*, 2005a); this is thought to be an innate attraction (i.e., pheromonal) to high-quality nursery habitat (rather than the "imprinting" of natal stream odors that occur in the Atlantic salmon) as the presence of conspecifics larvae is indicative of suitable spawning sites, and successful spawning in the past. Sexually mature, ovulated females, in their turn, are attracted by bile salts released by the males, including 3-keto petromyzonol sulfate (Fig. 11.1; Li *et al.*, 2002; Johnson *et al.*, 2009; Johnson and Li, 2010). It appears that the males have evolved a specific transport mechanism in the gills to excrete these bile salts to the water (Siefkes *et al.*, 2003); it is clear that the olfactory system shows the necessary high sensitivity and specificity to these compounds (Li, Sorensen, and Gallaher, 1995; Li and Sorensen, 1997; Siefkes and Li, 2004).

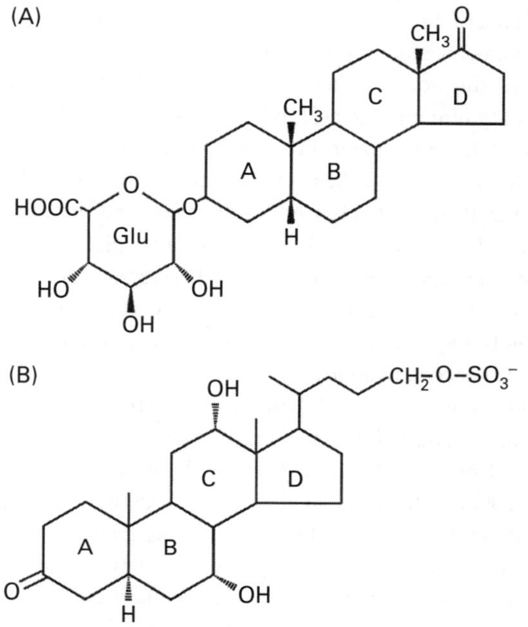

Figure 11.1. Chemical structures of two pheromones from marine fish (*A*) Etiocholan-3α-ol-17-one-glucuronide produced by the black goby (*Gobius niger=jozo*) male, probably in the mesorchial gland, attracts mature females and induces spawning behavior. (*B*) 3-keto petromyzonol sulfate (3-keto 7α,12α,24-trihydroxy-5α-cholan-3-one 24-sulfate), produced by mature male sea lampreys (*Petromyzon marinus*) and released across the gills, attracts mature females to upstream spawning sites. Note how both have the four-ring steroid skeleton (A, B, C, and D) but differ widely according to different groups at different positions, especially the glucuronide group "Glu" at position 3 in (*A*) and the sulfate group at position 24 in (*B*). These conjugates are much more water-soluble than their respective unconjugated forms, which are obviously important for pheromones acting in the aquatic environment. It is likely that neither compound acts alone in nature, but as part of a pheromone mixture.

In contrast to hormonally derived pheromones, bile acids are not considered to have any internal signaling function *per se*, but they are involved in the emulsification of lipids in the gut. However, the fact that they are used as pheromonal signals in the lamprey suggests that their pheromonal function appeared in the vertebrate lineage not long (evolutionarily speaking) after their digestive function. The structural diversity and evolution of bile acids in vertebrates (Haslewood, 1967; Hagey *et al.*, 2010; Hofmann, Hagey, and Krasowski, 2010), combined with their widespread olfactory potency, suggests that the communicative function(s) of this class of compound are not confined to the lamprey. However, much work is yet to be done to clarify what these functions are.

11.3.3 Blennies

Although not commercially important, the blennies (suborder Blennioidei or "true blennies") comprise a group of mainly marine fish that have attracted much attention from the research community because of male parental behavior and alternative reproductive tactics (Gross, 1996;

Henson and Warner, 1997; Taborsky, 1998). Territorial or "bourgeois" nest-holding males attract females to their nest (often a mollusk shell or cleft in the rock) where the females deposit their eggs and the males fertilize, guard, and care for them until hatching. Smaller, parasitic or "sneaker" males will subsequently try fertilizing some of these eggs by mimicking female behavior, and fooling the territorial male into allowing him into the nest, where the sneaker will release its sperm. Along with the differences in behavior, there are clear differences in morphology; territorial males often have secondary sexual characteristics, such as head crests, "cirri' and glands that develop on the fin rays of the unpaired fins (Northcott and Bullock, 1991; Giacomello and Rasotto, 2005). These differences are accompanied by internal differences such as the testicular gland and blind pouches, which are absent or rudimentary in sneaker males (Lahnsteiner, Richtarski, and Patzner, 1990; Lahnsteiner, Nussbaumer, and Patzner, 1993; Neat, Locatello, and Rasotto, 2003). It has long been suspected that these accessory male glands may have some pheromonal function (Laumen, Pern, and Blüm, 1974). Water conditioned by male peacock blennies (*Salaria pavo*) is attractive to females, but less so if the anal glands are removed (Barata *et al.*, 2008; Serrano *et al.*, 2008a). Further, the blind pouches contain the cellular machinery required for the production of steroid glucuronides (Lahnsteiner, Nussbaumer, and Patzner, 1993), and the fluid produced by these glands is a highly potent olfactory stimulus (Serrano *et al.*, 2008b). However, an alternative function proposed for the anal gland is the production of mucous antimicrobials that the male uses in order to protect the developing eggs (Giacomello, Marchini, and Rasotto, 2006; Pizzolon *et al.*, 2010). Nevertheless, females can smell large-molecular-weight secretions (10 kD) from the anal gland (Serrano *et al.*, 2008a). It is possible, therefore, that the female uses the odor of antimicrobials released by the male's anal gland as a basis for her mate-choice. In this respect, it is interesting to note that high molecular-weight mucus-derived odorants are important in the alarm response (Mathuru *et al.*, 2012). Much work remains to be done, but the blennies seem to provide an excellent model for the study of evolution of chemical communication in fish.

11.3.4 Anguilliform Eels

Anguilliforme eels are extremely important fish—both to the aquaculture industry and in scientific terms. Given their importance, the alarming decrease in population sizes (van Ginneken and Maes, 2005) and the difficulties in producing sexual maturity and viable larvae under culture conditions, it is surprising that reproduction and other aspects of their life history are comparatively understudied (Nikolic *et al.*, 2011). Nevertheless, their complex life history (Fig. 11.2) and economic importance make them an ideal species in which to investigate the importance of chemical communication, and olfaction in general. Anguillid eels are catadromous (spawn in seawater; larvae migrate to freshwater) and semelparous (spawn and then die). The larvae migrate from marine spawning grounds (the Sargasso Sea in the case of European and American eels, seamounts on the Marianna Trench in the case of the Japanese eel). Whether olfactory cues are used to guide this journey to freshwater and estuarine habitats is not yet known; however, the final stage of migration into freshwaters appears to be guided by decreasing salinity gradients (Tosi *et al.*, 1990) and olfactory cues released by epiphytic micro-organisms living on the surface of submerged aquatic plants and rocks (Sorensen, 1986; Tosi and Sola, 1993). Movement farther inland (which can be substantial) appears to be enhanced by odors (pheromones) released by immature conspecifics (Miles, 1968). The identities of these odors are unknown. While growing as the "yellow eel" stage—for a variable period—it may be that reproductive pheromones may not be important; this is a period of growth; therefore, the olfactory system is likely to be of more importance in prey location and identification. However, the mechanisms of sex determination and induction of puberty are yet to be fully understood; it is possible that chemical communication is important in these processes. For example, conspecific skin mucus—presumably odorants released by—acts as an attractant (Saglio, 1982), and the odor may depend on both sex and state of sexual maturity (Huertas *et al.*, 2007). However, it is not yet clear what cues—olfactory or other, conspecific-derived or other, and environmental or internal—precipitate "puberty," the development into the silver eel stage and, then, the beginning of the migration to the spawning grounds. Sexual maturation occurs during the migration; it is possible that swimming (Palstra *et al.*, 2008), pressure (Sébert *et al.*, 2007), and chemical communication (Huertas, Canário, and Hubbard, 2008) all play some role in the maturation process. Olfaction may be of prime importance in locating the spawning site (van Ginneken and Maes, 2005) and mature conspecifics with which to spawn (Sorensen and Winn, 1984). Indeed, in laboratory conditions, mature males are highly attracted by

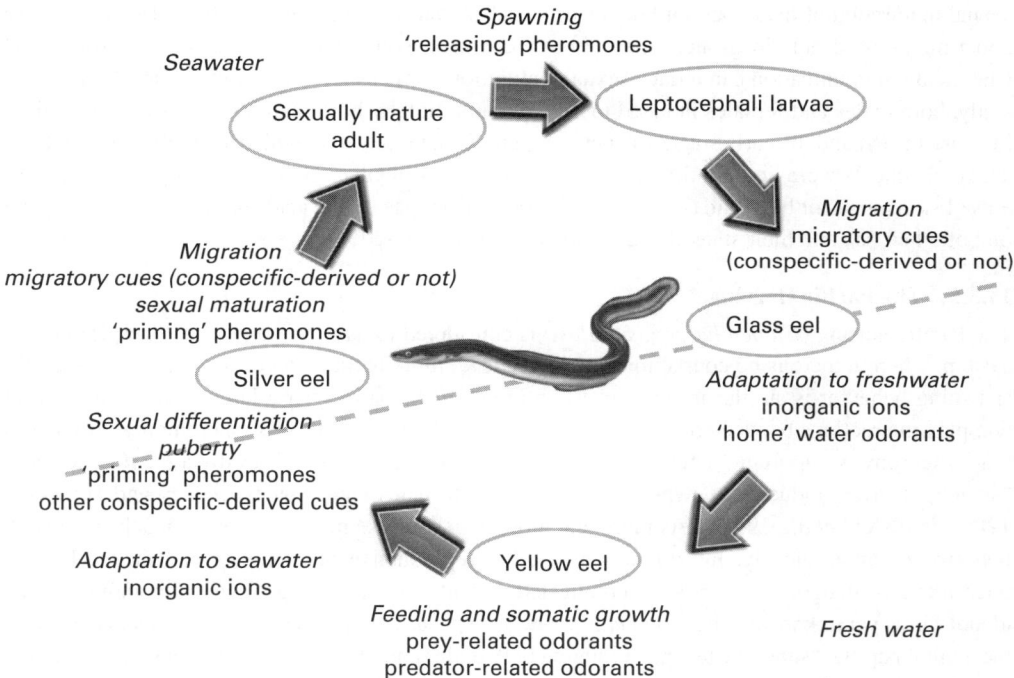

Figure 11.2. The life cycle of the freshwater eel. In addition to the scientific and economic importance of Anguillid eels, their complex life cycle allows "compartmentalization" of many facets of their biology, including olfaction and chemical communication. Somatic growth occurs largely in freshwater ("yellow eel" stage) where olfaction is important for prey detection and predator avoidance; chemical communication may be limited to the alarm response and/or conspecifics recognition. Under the influence of yet-unknown cues, the "silvering" process involves the down-river migration to estuaries; environmental factors, such as changes in salinity and organics, may be important guides during this migration, but do conspecific-derived odorants play a role? During the long oceanic migration, eels mature under the influence of high pressure and the swimming process. But are eels also induced to sexual maturity by conspecific-derived odorants—pheromones—and/or is their migration guided in any part by odorant or "pheromone" trails? Eels do not eat during this migration, but their olfactory system remains functional. What are the identities and roles of reproductive pheromones in spawning (never directly observed under natural conditions)? Do the resultant larvae use the same or different chemical cues to guide their migration to the freshwater habitats of their parent? (*See insert for color representation of the figure.*)

odors (putative pheromones) released by ovulated females, and release sperm while nudging the urogenital opening (Sorensen and Winn, 1984).

Olfaction is thought to be important in "selective tidal stream transport" in eels (i.e., using tidal flows to move in one direction, and then staying immobile on the bottom when the tide changes) (Barbin, 1998). Notably, mucous cell number decreases with heightened sexual maturity in the olfactory epithelium (of females), and advanced sexual maturity is accompanied by "marked degeneration of the sensory epithelium" (Pankhurst and Lythgoe, 1983; Sorensen and Pankhurst, 1988), and this leads the authors to suggest that olfaction is less important at sea. Final sexual maturation is thought to occur at or near the spawning site; olfaction may still be of importance during the migration, or it is only olfactory sensitivity to food-related cues that is affected. Swimming is important for sexual maturation of male, but not female, eels (Palstra and van den Thillart, 2010). However, these experiments used eels in groups; the effect of chemical communication was not controlled for. Chemical communication can induce sexual maturation in males (Huertas *et al.*, 2006). In the latter study, both males and females induced to sexual maturation by hormonal injections caused a slight increase in GSI and the early stages of spermatogenesis in untreated neighboring males, although the active chemicals were not identified. However, candidates involve relatively hydrophobic constituents of skin mucus or bile fluid (Huertas *et al.*, 2007); the types of bile acids produced by eels depend on both their sex and their state of maturation (Huertas *et al.*, 2010).

11.3.5 The Pacific Herring

The Pacific herring (*Clupea harengus pallasi*) is considered to have an "ancestral" (simple) mating system, wherein there is no courtship behavior (indeed, little interaction between individuals at all), spawning behavior is similar in both sexes, and many fish spawn over a large area. In the lack of complex interactions between or among sexes, what could be the role(s) of pheromones? It appears that synchrony is important; a pheromone produced by the testes and released via the milt acts as a "switch," thereby inducing spawning behavior in both males and females (Stacey and Hourston, 1982; Carolsfeld *et al.*, 1997b). As more and more males release milt, positive feedback ensures all fish spawn synchronously. In common with other reproductive pheromones in fish, the herring pheromone is thought to comprise of conjugated steroids and/or prostaglandins, although the exact identity is as yet unknown (Sherwood *et al.*, 1991). Exposure to this pheromone evokes extension of the genital papilla, "substrate testing" (wherein both males and females "test" the substrate with their paired fins) and spawning, either in mid-water or on the substrate (Stacey and Hourston, 1982; Carolsfeld *et al.*, 1997b). These species normally deposit eggs and milt on kelp or similar marine algae. (In captivity, they prefer glass or plastic!) It is possible that these provide the original stimulus for the first few males to release some milt, or even exposure to shallow water (the Pacific herring normally spawns in less than 6 m). Nevertheless, not all fish respond equally to the pheromone: mature fish respond, immature fish do not; females respond later in the season than males; and responsiveness may be related to endocrine status (Stacey and Hourston, 1982; Carolsfeld, Scott, and Sherwood, 1997a). Thus, even in a simple mating system, the pheromonal system remains both context-sensitive and responsive to environmental influences.

11.4 ARE PHEROMONAL SYSTEMS OF MARINE SPECIES LIKELY TO BE DIFFERENT FROM FRESHWATER FISH?

Many aspects of the environment, phylogeny, and physiology of marine fish often differ markedly from freshwater fish (e.g., the complex life history of the eel and migrations of tuna). How might this affect chemical communication systems? A marine environment may influence the routes of release or the nature of the chemical cues themselves (stability in freshwater versus seawater, for example). First, is urine the main vehicle for pheromone release, as in freshwater? Osmoregulatory constraints mean that urine production rates of marine fish are much lower when compared with freshwater fish. Consequently, urine release is more infrequent; spontaneous urination in the plaice (*Pleuronectes platessa*), for example, is less than once every 3 days (Fletcher, 1990). Marine fish, therefore, would need to regulate urine storage and release carefully to maximize any communicative function. Clearly, it is possible that other routes of release are used. The intestinal fluid of the seabream, for example, is a potent odorant; much more so than urine (Hubbard, Barata, and Canário, 2003), and the odorants contained in the mucus and bile fluid of the eel depend on sex and maturity (Huertas *et al.*, 2007). Fish mucus contains antimicrobial agents (Ellis, 2001); it is possible—in blennies at least—that these antimicrobials may act also as chemical cues in mate choice (see above).

Marine fish drink seawater as part of their osmoregulatory strategy (e.g., see Whittamore, 2012) and, as a result, constantly have fluid moving through the digestive system even when not feeding. Although it is probable that intestinal fluid and mucus contain different classes of odorant from the urine, their chemical identities and biological functions must be identified before any firm conclusions can be drawn.

Alternatively, potential pheromones may simply be released across the gills (Vermeirssen and Scott, 1996) whose large surface area and blood flow may provide an efficient release mechanism (Sorensen, Pinillos, and Scott, 2005b). However, in the absence of specific transport mechanisms, such as that of the sea lamprey (Siefkes *et al.*, 2003), this would rely on passive diffusion and would therefore be less effective as an *active* signaling mechanism. Little attention has been paid to the latter issue except for in the sea lamprey.

This begs the question of whether similar classes of compounds are used as pheromones in marine and freshwater fishes The limited evidence available suggests that marine fish use conjugated steroids as pheromones (Colombo *et al.*, 1980), just as many freshwater fish do. Nevertheless, although steroids share a common four-ring core (Fig. 11.1), biological function depends on the *exact* chemical structure; the presence and position of different chemical moieties, such as hydroxyl or keto groups, play significant roles in receptor–ligand interactions and, therefore, the physiological effects of a given steroid. This is likely true for pheromonal steroids as it is for hormonal steroids. Conjugation— the enzymatic binding of a glucuronide or sulfate group to the steroid core (usually at the 3- or 17-position)—inactivates the *hormonal* activity by preventing this ligand–receptor binding. However, the conjugated forms of steroids are those excreted to the water where they may act as *pheromones*; the position and type of conjugating group may confer pheromonal activity. There is, therefore, considerable scope for diversity of structure in this one class of compound; an observation that a given species does not respond to a given steroid cannot rule out a pheromonal role of steroids *per se*. Similarly, there is as yet no evidence for marine fish using prostaglandins as hormonal pheromones, as do some freshwater fish such as cyprinids (Stacey and Sorensen, 2006). However, in this case, lack of evidence is not proof for a lack of function and much research remains to be done.

As mentioned earlier, bile acids and alcohols are well known to act as potent fish odorants in both freshwater and marine species (Hara, 1994; Zhang, Brown, and Hara, 2001; Velez *et al.*, 2009; Zhang and Hara, 2009). However, it is not clear whether this class of compound has a pheromonal role in teleosts—in freshwater *or* seawater—as has been shown for the sea lamprey (Li *et al.*, 2002; Sorensen *et al.*, 2005a). Recently, a "reversal" of bile acid metabolism during artificially induced sexual maturation in the eel, from C_{24} to C_{27} bile salts, suggests that bile acids may play some pheromonal role (see below), perhaps acting as a similar migratory pheromone (Huertas *et al.*, 2010).

Do all released cues have biological relevance? There is no reason *per se* why marine teleosts should be sending different messages via pheromones from freshwater species, even if the chemicals involved and the routes of release differ. In any language, context is important; the precise meaning of chemical messages will depend on the habitat and life history of the species in question. For example, there is a possibility for eels to use olfactory cues—even pheromones—to locate their spawning sites or each other in a similar way to the sea lamprey, despite the difference in migratory behavior. It could be argued that the scale of oceans compared with freshwater bodies may place limitations on chemical communication. However, this can be countered by the large size and huge numbers of some shoals of marine fish, extreme olfactory sensitivity to some important cues, or by the use of microhabitats that would bring them into close proximity to conspecifics.

11.5   WHY ARE PHEROMONES OF MARINE FISH LESS UNDERSTOOD THAN FRESHWATER FISH?

Aquaculture is big business and, therefore, attracts much interest from the research community. So much so that one historian of the industry started his book with "It has been claimed by cynics that the weight of books and papers on various aspects of marine fish and shellfish culture is now well in excess of the actual product" (Kirk, 1987, p. 8). Sadly, this does not hold true for research into pheromones or chemical communication in cultured marine fish. Why should this be? Even though it is well recognized that "[a] greater understanding of pheromones could lead to improvements both in reproductive efficiency and animal welfare" (Wyatt, 2003, p. 252), "[t]he use of fish sex pheromones to control reproduction in fish aquaculture has a promising future but research into commercial applications is in its early stages" (Wyatt, 2003, p. 254). This is probably due to three problems: (1) logistics, (2) diversity (both phylogenetic and in reproductive strategies), and (3) lack of precedent.

Most cultured fish are relatively large, ranging from "plate-sized" bass and seabream to Atlantic cod and tuna, and are consequently more expensive to house, maintain, and manipulate experimentally than popular laboratory fish species such as goldfish, minnows, and zebrafish. Also, marine fish culture requires a source of clean seawater which effectively restricts research institutes working on such species to those on the coast and/or easy access to seawater. Access to healthy broodstocks may be limited as fish-farmers are reluctant for such valuable fish to be man-handled by researchers taking urine, blood samples, etc. except, of course, in those species in which reproduction in culture is not problematic. In that case, obviously, there is consequently less impetus for research. Further, the electro-olfactogram—a favorite method to assess olfactory potency in fish—is less effective in seawater (Hubbard *et al.*, 2011). This is because the high electrical conductivity of seawater reduces the amplitude of the olfactory response (voltage change) compared with that in freshwater fish.

Diversity among teleosts—in both phylogeny and reproductive strategies—means that, in stark contrast to the endocrine system, pheromonal signals are likely to differ widely according to the species and roles that they play. In fact, the goldfish and fathead minnow—the best understood species with regard to reproductive and alarm cues pheromones, respectively—are both cyprinids, an exclusively freshwater taxon. Apart from early work on the black goby (see above; Colombo *et al.*, 1980), very little work has focused on marine species. In many cases, the natural reproductive spawning strategy of important marine aquaculture species is poorly understood (e.g., the Senegalese sole, *Solea senegalensis*; see below); thus, investigation of the potential role of pheromones in reproduction usually comes second to how induction of maturity may be induced with hormone injections (reviewed by Taranger *et al.*, 2010). However, if pheromones play a role in the "ancestral" mating system, such as that found in the herring (Carolsfeld, Scott, and Sherwood, 1997a; Carolsfeld *et al.*, 1997b), then it is likely that pheromones play a fundamental role in many, if not all, teleost mating systems.

Most farmers are conservative by nature, and are unwilling to invest in methods or investigation that may not generate a profit in the short- or medium-term future. Thus, the potential diversity of pheromonal systems and their role(s)—although of enormous interest to the scientific community— is problematic for commercial enterprises; a given pheromone identified and characterized in one species is unlikely to have the same effect(s) in an unrelated species. Consider the history of the modern aquaculture industry; one of the first species to be raised from artificial spawning—the

Atlantic salmon (*Salmo salar*)—essentially arrived at the aquaculturists' facilities primed and ready to spawn. Thus, no precedent was set for investigation into the complex series of events, including chemical communication, that led the fish to arrive at such a condition. The focus, then, was raising the resultant larvae. A great deal of time, effort, and money may have to be invested to fully characterize the identity and roles of pheromones in one species; however, the findings are likely to be applicable to one or a few species at best. There is no "universal" fish pheromone. Therefore, pheromonal research has hitherto largely been ignored by the aquaculture industry; and until significant breakthroughs are made, this is likely to remain the case.

11.6 AQUACULTURE SPECIES AND POSSIBLE APPLICATIONS

11.6.1 Introduction

As stated earlier, understanding of the pheromonal systems of marine fish lags well behind that of freshwater fish, and this is true even for the main aquaculture species, despite much work having been carried out on other aspects of their biology (notably reproduction, feeding, and growth). This section outlines what is known of chemical communication in some of the major marine aquaculture species.

11.6.2 Freshwater (Anguillid) Eels

The two main eel species in aquaculture are the European eel (*Anguilla anguilla*) and the Japanese eel (*A. japonica*). Global eel aquaculture in 2012 was 241 127 tonnes (FAO, 2014). Due to the complex life history and reproductive strategy of eels, artificial reproduction is still not possible on a commercial scale and remains the "Holy Grail" of eel aquaculture. Sexual maturation can be induced by weekly injections of salmon pituitary extract, followed by 17,20β-dihydroxyprogesterone for females, or human chorionic gonadotropin for males (Sorensen and Winn, 1984; Ohta *et al.*, 1996; Ohta *et al.*, 1997; Pedersen, 2003, 2004). However, the subsequent raising of viable larvae has proven highly problematic, especially for the European eel. The aquaculture industry is therefore reliant on the capture of wild glass eels that are then grown on in ponds or recirculation systems. The cost of glass eels is highly variable and constitutes 20–45% of total production costs, thereby becoming a major factor in determining the profitability to the farmer. In fact, the catch of *A. japonica* glass eels is insufficient to supply the industry in Asia, thereby necessitating the costly transport of *A. anguilla* glass eels caught in Europe to, principally, China to cover this shortfall (Tesch, 2003). Chemical communication appears to play various roles in freshwater migration, induction of puberty, sex determination, sexual maturation, and spawning (see above); understanding these systems may aid the artificial maturation of both males and females, thereby improving the quality of resultant larvae and decreasing the commercial reliance on wild-caught glass eels for aquaculture. It may also aid in the management of wild stocks, for example, by guiding descending eels and/or ascending elvers to fish passes around dams.

11.6.3 Gilthead Seabream

The gilthead seabream (*Sparus aurautus*) is a marine Sparid of high economic value in southern Europe and the Mediterranean Sea. It is a protoandrous hermaphrodite: a proportion of males become females in their third year (Buxton and Garratt, 1990). The reproductive system of females is partially under social control—possibly pheromonal control; removal of males results in a decline in circulating LH and sex steroids, and an increased proportion of oocytes undergoing atresia (Meiri *et al.*, 2002). Thus, males are clearly transmitting some sort of signal—possibly chemical—that maintains the female reproductive system in readiness throughout the spawning season. However, males isolated from females continue to spermiate normally, although this does not necessarily mean that females are not transmitting chemical messages to the males. Both males and females release extremely potent odorants; interestingly, it is the intestinal fluid that is more potent than the urine (Hubbard, Barata, and Canário, 2003). This may mean that, because urine production is much less in seawater, urine is a less important route of release for pheromones in marine fish than freshwater fish. The olfactory system of cultured sparids suffers from various developmental abnormalities, such as one large nostril instead of clearly defined inhalant and exhalent nares (Mana and Kawamura, 2002); whether this affects olfactory sensitivity to pheromones or other chemical cues, thereby interfering with the reproductive process, however, is not known.

11.6.4   European Sea Bass

The sea bass (*Dicentrarchus labrax*) is of high economic value in southern Europe and the Mediterranean Sea. The main problem with sea bass aquaculture is the high proportion of males produced under culture conditions; males of this species mature earlier than females (at a marketable size), therefore, somatic growth is diverted to gonadal growth earlier. Thus, the industry would welcome a method for increasing the proportion of females and/or retardation of maturation in males. Sex determination in the sea bass seems to result from a complex interaction of genetic factors (not sex chromosomes), temperature (Pavlidis *et al.*, 2000; Saillant *et al.*, 2002; Saillant *et al.*, 2003; Piferrer *et al.*, 2005), and perhaps pheromones. However, the role of chemical communication in either sex determination or maturation in sea bass has not been addressed. Successful retardation of puberty or increasing the proportion of females by chemical means—simply adding the appropriate chemicals to the water—would have clear commercial applicability.

11.6.5   Senegalese Sole

The Senegalese sole (*S. senegalensis*) and, to a lesser extent, its close relative the common sole (*Solea solea*) are of increasing importance to the aquaculture industry of southern Europe and the Mediterranean Sea. Their slow growth is compensated by their high market value and they can be cocultured with midwater feeders such as the seabream. However, poor reproductive performance of F1 males is one of the main bottlenecks facing the aquaculture industry (Howell *et al.*, 2011). Thus, larvae production is heavily dependent on wild-caught broodstock. No obvious endocrine reason has been found for this, but it seems that the spawning behavior of these captive-bred F1 males is not normal. Sexual development and behavior of the females is less problematic. Whether chemical communication is involved in reproduction is not yet evident. However, males in the presence of females have higher plasma concentrations of the sex steroids 11-ketotestosterone and testosterone, and short-term exposure to females increases sperm viability, motility, and velocity (Cabrita *et al.*, 2011), and various body fluids act as potent odorants (Velez *et al.*, 2007a, 2009). Further, the urine of mature females is more potent an odorant than that from immature females (unpublished observations). Thus, it seems likely that females may be releasing "primer" pheromones that affect the males' brain–pituitary–gonad axis and, consequently, sperm quality. Whether F1 males are deficient in either the detection of, or appropriate response to, chemical cues released by the females is not yet known; however, it remains possible that a problem of chemical communication underlie the poor reproductive performance of these males.

Besides the economic importance of the sole, it is also interesting as it is a flatfish with apparent functional asymmetry in the olfactory system. In sole metamorphosis, unlike the left eye, the left nostril does not migrate to the right side. Given the bottom-living habits of the sole family, this means that the right and left olfactory epithelia are exposed to different water sources and, consequently, odorants. Conspecific-derived odorants are better detected by the upper (right) olfactory epithelium (Velez *et al.*, 2005, 2007a, 2009), whereas food-related odorants are better detected by the lower (left) olfactory epithelium (Velez *et al.*, 2007b, 2011). This raises the intriguing possibility that the upper olfactory epithelium and bulb may be specialized for pheromone detection.

11.7   APPLICATIONS OF PHEROMONES TO THE CULTURE OF MARINE FISH

11.7.1   Introduction

Manipulation of aquaculture finfish stocks usually has one of two main aims: control of maturational status or disease prevention and/or cure. In the case of induction of maturity, this is usually achieved by injection of gonadotropins (GnRH or analogues) or implantation of slow-release devices (Mylonas, Fostier, and Zanuy, 2010; Power *et al.*, 2010). Broodstocks, which in many cases must be caught from the wild, are usually large, and must be maintained under optimal conditions (including diet, temperature and photoperiod), represent a substantial investment. Minimization of the frequency and duration of handling, which often includes anesthesia and invasive procedures, would therefore be advantageous. The use of reproductive pheromones may help in these conditions by (1) artificial control of maturity without recourse to invasive procedures, thereby minimizing stress and the risk of wounding or infection; (2) aiding in the manipulation of sex determination or sorting of the sexes without handling; and (3) noninvasive sampling to assess sexual status and/or stress.

11.7.2   Induction of Maturity and Reproduction

Control of sexual maturation in many cultured marine species currently involves manipulative and expensive injections of hormones (Mylonas, Fostier, and Zanuy, 2010; Power *et al.*, 2010). Could this be replaced or improved by pheromone treatment? More effective stimulation of sexual maturity in eels with pheromones, for example (Huertas *et al.*, 2006), could not only reduce the need for injections but may also improve the quality—and, crucially, survival—of resultant larvae. Pheromone treatment may equally aid in the synchronization or timing of sexual maturity, which is currently achieved through manipulating of temperature and/or photoperiod.

11.7.3   Delaying Precocious Maturation

Early puberty is a problem for the aquaculture industry in many species due to the consequent allocation of energy reserves to the gonads rather than muscular growth. There may also be negative effects on appearance and flesh quality, thus reducing marketability (Taranger *et al.*, 2010). Could precocious puberty in cultured Atlantic cod (*Gadus morhua*), for example, be controlled by pheromone treatment? In aquaculture, cod tend to spawn at an earlier age than in the wild, presumably due to increased food intake and consequent deposition of lipid and protein reserves (Karlsen, Holm, and Kjesbu, 1995; Karlsen *et al.*, 2006). If chemical communication can modulate the onset of puberty, then the high densities of fish in aquaculture are likely to exacerbate this problem. Thus, an understanding of the pheromonal system may prompt the development of remedies; this may be as simple as more effective filtration (to prevent buildup of water pheromone levels) or more sophisticated, such as the development of pheromone antagonists.

11.7.4   Sex Determination

Teleosts show the greatest variety in mechanisms of sex determination of all vertebrates; these may be genetic, or depend on environmental factors, such as temperature and pH, and/or social factors (Devlin and Nagahama, 2002; Guerrero-Estévez and Moreno-Mendoza, 2010). Little work has been directed at exactly what these social factors may be, but pheromones are candidates. In sequential hermaphroditic species, the proportion that changes sex may depend on social conditions. In brightly colored species, such as many reef fish (Munday, White, and Warner, 2006), it may simply be due to visual cues. However, in species such as the seabream (a protandrous hermaphrodite), lack of any external sexual dimorphism would suggest at least a component is chemically mediated (Meiri *et al.*, 2002).

At a more basic level, it has been suggested that sex pheromones could be used to sort males from females, and thereby reduce handling and associated stress. In cases where wild-caught fish are used for broodstock, pheromones could be used to attract those of the sex required (depending, also, on the pheromonal system of the species in question) in a similar way that invasive species may be trapped (Sorensen and Stacey, 2004; Corkum and Belanger, 2007); this has the advantage that only sexually ripe fish would be attracted.

11.7.5   Welfare

Can we use measurements of sex pheromones (e.g., 17,20β-dihyroxyprogesterone) or other released chemicals (e.g., cortisol) to assess reproductive status and/or well-being in cultured fish stocks? Noninvasive sampling of steroids release to the water by fish has been used at the experimental level for several years (Scott and Ellis, 2007; Scott *et al.*, 2008). It would not generate insurmountable technical barriers to apply this to an industrial level. The sea bass (*D. labrax*), for example, is notorious as an easily stressed species; this is reflected in the high rates of release of cortisol to the water (Fanouraki *et al.*, 2008). Given that stress reduces food-conversion efficiency and growth rates in the sea bass (Leal *et al.*, 2011), it is possible that monitoring cortisol in the water may be a noninvasive means of assessing the level of stress in a tank of fish. Thus, although probably not acting as a pheromone, cortisol may form communication with the farmer that all is not well, before disease or increased mortality occur. As yet, however, little work has directly addressed alarm pheromones in marine fish (Manassa and McCormick, 2012).

11.7.6   Potential Problems That Pheromones May Cause

There is indirect evidence that intense aquaculture may affect the behavior or physiology of wild populations via odorants—not necessarily pheromones—released into the natural environment. Norwegian fishermen, for example, claim that migratory routes of cod have changed since the establishment of coastal salmon farms, and laboratory-based behavioral experiments show that cod avoid runoff water from salmon tanks (Sæther, Bjørn, and Dale, 2007). Thus, supplementing cultured stocks with artificial pheromones may exacerbate this problem. An understanding of the pheromones involved, and their persistence and behavior in the natural environment, is clearly necessary before preventative strategies can be developed.

11.8   SUMMARY

It is likely that marine fish use reproductive pheromones to attract, find, and choose mates in similar ways to those used by freshwater fish. Whether the same types of chemicals convey the same types of message is a question that awaits further research, but steroid glucuronides and bile salts are prime candidates. Routes of release could also differ. Pheromones, and olfactory detection of environmental cues, may also prove to be important in some migratory species. However, it is unclear whether, or how, marine fish use alarm pheromones.

Application of pheromones in aquaculture shows enormous potential, and some possible problems. However, most aquaculture research focuses on maximizing growth and reducing cost of feed. Reproduction is the main area in which application of pheromone treatment may prove beneficial. Unfortunately, this is most true in species where reproductive strategies, physiology, and behavior are less understood. However, "the central role of pheromones in the biology of most insects, crustaceans, fish and mammals suggests there is still a vast potential for intervention with pheromones: we have only just begun" (Wyatt, 2003, p. 269). Clearly, much work remains to be done.

REFERENCES

Barata, E.N., Serrano, R.M., Miranda, A. *et al.* (2008) Putative pheromones from the anal glands of male blennies attract females and enhance male reproductive success. *Animal Behaviour*, **75**, 379–389.

Barbin, G.P. (1998) The role of olfaction in homing and estuarine migratory behavior of yellow-phase American eels. *Canadian Journal of Fisheries and Aquatic Sciences*, **55**, 654–575.

Buxton, C.D. and Garratt, P.A. (1990) Alternative reproductive styles in seabreams (Pisces: Sparidae). *Environmental Biology of Fishes*, **28**, 113–124.

Cabrita, E., Soares, F., Beirão, J. *et al.* (2011) Endocrine and milt response of Senegalese sole, *Solea senegalensis*, males maintained in captivity. *Theriogenology*, **75**, 1–9.

Carolsfeld, J., Scott, A.P., and Sherwood, N.M. (1997a) Pheromone-induced spawning of Pacific herring II. Plasma steroids distinctive to fish responsive to spawning pheromone. *Hormones and Behavior*, **31**, 269–276.

Carolsfeld, J., Tester, M., Kreiberg, H., and Sherwood, N.M. (1997b) Pheromone-induced spawning of Pacific herring I. Behavioral characterization. *Hormones and Behavior*, **31**, 256–268.

Colombo, L., Marconato, A., Belvedere, P.C., and Friso, C. (1980) Endocrinology of teleost production: a testicular steroid pheromone in the black goby, *Gobius jozo* L. *Bolletino di Zoologia*, **47**, 355–364.

Corkum, L.D. and Belanger, R.M. (2007) Use of chemical communication in the management of freshwater aquatic species that are vectors of human diseases or are invasive. *General and Comparative Endocrinology*, **153**, 401–417.

Corkum, L., Meunier, B., Moscicki, M. *et al.* (2008) Behavioural responses of female round gobies (*Neogobius melanostromus*) to putative steroidal pheromones. *Behaviour*, **145**, 1347–1365.

Devlin, R. and Nagahama, Y. (2002) Sex determination and sex differentiation in fish: an overview of genetic, physiological, and environmental influences. *Aquaculture*, **208**, 191–364.

Døving, K.B., Selset, R., and Thommesen, G. (1980) Olfactory sensitivity to bile acids in salmonid fishes. *Acta Physiologica Scandinavica*, **108**, 123–131.

Ellis, A.E. (2001) Innate host defence mechanisms of fish against virus and bacteria. *Developmental and Comparative Immunology*, **25**, 827–839.

Fanouraki, E., Papandroulakis, N., Ellis, T. *et al.* (2008) Water cortisol is a reliable indicator of stress in European sea bass, *Dicentrarchus labrax*. *Behaviour*, **145**, 1267–1281.

FAO. (2014) *Species Fact Sheets. Anguilla japonica* (Temminck & Schlegel, 1847). http://www.fao.org/fishery/species/2988/en (accessed August 21, 2014).

Fletcher, C.R. (1990) Urine production and urination in the plaice *Pleuronectes platessa*. *Comparative Biochemistry and Physiology A*, **96A**, 123–129.

Giacomello, E. and Rasotto, M.B. (2005) Sexual dimorphism and male mating success in the tentacled blenny, *Parablennius tentacularis* (Teleostei : Blenniidae). *Marine Biology*, **147**, 1221–1228.

Giacomello, E., Marchini, D., and Rasotto, M.B. (2006) A male sexually dimorphic trait provides antimicrobials to eggs in blenny fish. *Biology Letters*, **2**, 330–333.

Gross, M.R. (1996) Alternative reproductive strategies and tactics: diversity within sexes. *Trends in Ecology & Evolution*, **11**, 92–98.

Guerrero-Estévez, S. and Moreno-Mendoza, N. (2010) Sexual determination and differentiation in teleost fish. *Reviews in Fish Biology and Fisheries*, **20**, 101–121.

Hagey, L.R., Møller, P.R., Hofmann, A.F., and Krasowski M.D. (2010) Diversity of bile salts in fish and amphibians: evolution of a complex biochemical pathway. *Physiological and Biochemical Zoology*, **83**, 308–321.

Hara, T.J. (1994) The diversity of chemical stimulation in fish olfaction and gustation. *Reviews in Fish Biology and Fisheries*, **4**, 1–35.

Haslewood, G.A.D. (1967) Bile salt evolution. *Journal of Lipid Research*, **8**, 535–550.

Henson, S.A. and Warner, R.R. (1997) Male and female alternative reproductive behaviors in fishes: a new approach using intersexual dynamics. *Annual Review of Ecology and Systematics*, **28**, 571–592.

Hofmann, A.F., Hagey, L.R., and Krasowski, M.D. (2010) Bile salts of vertebrates: structural variation and possible evolutionary significance. *Journal of Lipid Research*, **51**, 226–246.

Howell, B., Pricket, R., Cañavate, P. *et al.* (2011) Sole farming: there or thereabouts! *Aquaculture Europe*, **36**, 42–45.

Hubbard, P.C., Barata, E.N., and Canário, A.V.M. (2003) Olfactory sensitivity of the gilthead seabream (*Sparus auratus* L) to conspecific body fluids. *Journal of Chemical Ecology*, **29**, 2481–2498.

Hubbard, P.C., Barata, E.N., Ozório, R.O.A. *et al.* (2011) Olfactory sensitivity to amino acids in the blackspot seabream (*Pagellus bogaraveo*): a comparison between olfactory receptor recording techniques in seawater. *Journal of Comparative Physiology A*, **197**, 839–849.

Huertas, M., Scott, A.P., Hubbard, P.C. *et al.* (2006) Sexually mature European eels (*Anguilla anguilla* L.) stimulate gonadal development of neighbouring males: possible involvement of chemical communication. *General and Comparative Endocrinology*, **147**, 304–313.

Huertas, M., Hubbard, P.C., Canário, A.M., and Cerdà, J. (2007) Olfactory sensitivity to conspecific bile fluid and skin mucus in the European eel *Anguilla anguilla* (L). *Journal of Fish Biology*, **70**, 1907–1920.

Huertas, M., Canário, A.V.M., and Hubbard, P.C. (2008) Chemical communication in the Genus *Anguilla*: a minireview. *Behaviour*, **145**, 1389–1407.

Huertas, M., Hagey, L., Hofmann, A.F. *et al.* (2010) Olfactory sensitivity to bile fluids and bile acids in eel (*Anguilla anguilla*), goldfish (*Carassius auratus*) and Mozambique tilapia (*Oreochromis mossambicus*) suggests a "broad range" of sensitivity not confined to those produced by con-specifics alone. *Journal of Experimental Biology*, **213**, 308–317.

Johnson, N.S. and Li, W. (2010) Understanding behavioral responses of fish to pheromones in natural freshwater environments. *Journal of Comparative Physiology A*, **196**, 701–711.

Johnson, N.S., Yun, S.-S., Thompson, H.T. *et al.* (2009) A synthesized pheromone induces upstream movement in female sea lamprey and summons them into traps. *Proceedings of the National Academy of Sciences of the USA*, **106**, 1021–1026.

Karlsen, Ø., Holm, J.C., and Kjesbu, O.S. (1995) Effects of periodic starvation on reproductive investment in 1st time spawning Atlantic cod (*Gadus morhua* L.). *Aquaculture*, **133**, 159–170.

Karlsen, O., Norberg, B., Kjesbu, O.S., and Taranger, G. (2006) Effects of photoperiod and exercise on growth, liver size, and age at puberty in farmed Atlantic cod (Gadus morhua L.). *Ices Journal of Marine Science*, **63**, 355–364.

Katare, Y.K., Scott, A.P., Laframboise, A.J. *et al.* (2011) Release of free and conjugated forms of the putative pheromonal steroid 11-oxo-etiocholanolone by reproductively mature male round goby (*Neogobius melanostomus* Pallas, 1814). *Biology of Reproduction*, **84**, 288–298.

Kirk, R. (1987) *A History of Marine Fish Culture in Europe and North America*, Fishing News Books Ltd., Farnham.

Laframboise, A.J. and Zielinski, B.S. (2011) Responses of round goby (*Neogobius melanostomus*) olfactory epithelium to steroids released by reproductive males. *Journal of Comparative Physiology A*, **197**, 999–1008.

Lahnsteiner, F., Richtarski, U., and Patzner, R.A. (1990) Functions of the testicular gland in two blenniid fishes, *Salaria (=Blennius) pavo* and *Lipophrys (=Blennius) dalmatinus* (Blenniidae, Teleostei) as revealed by electron microscopy and enzyme histochemistry. *Journal of Fish Biology*, **37**, 85–97.

Lahnsteiner, F., Nussbaumer, B., and Patzner, R.A. (1993) Unusual testicular accessory organs, the testicular blind pouches of blennies (Teleostei, Blenniidae). Fine structure, (enzyme-) histochemistry and possible functions. *Journal of Fish Biology*, **42**, 227–241.

Laumen, J., Pern, U., and Blüm, V. (1974) Investigations on the function and hormonal regulation of the anal appendices in *Blennius pavo* (Risso). *Journal of Experimental Zoology*, **190**, 47–56.

Leal, E., Fernández-Durán, B., Guillot, R. *et al.* (2011) Stress-induced effects on feeding behavior and growth performance of the sea bass (*Dicentrarchus labrax*): a self-feeding approach. *Journal of Comparative Physiology B*, **181**, 1035–1044.

Li, W. and Sorensen, P.W. (1997) Highly independent olfactory receptor sites for naturally occurring bile acids in the sea lamprey, *Petromyzon marinus*. *Journal of Comparative Physiology A*, **180**, 429–438.

Li, W., Sorensen, P.W., and Gallaher, D.D. (1995) The olfactory system of migratory adult sea lamprey (*Petromyzon marinus*) is specifically and acutely sensitive to unique bile acids released by conspecific larvae. *Journal of General Physiology*, **105**, 569–587.

Li, W., Scott, A.P., Siefkes, M.J. *et al.* (2002) Bile acid secreted by male sea lamprey that acts as a sex pheromone. *Science*, **296**, 139–141.

Mana, R.R. and Kawamura, G. (2002) A comparative study on morphological differences in the olfactory system of red sea bream (*Pagrus major*) and black sea bream (*Acanthopagrus schlegeli*) from wild and cultured stocks. *Aquaculture*, **209**, 285–306.

Manassa, R.P. and McCormick, M.I. (2012) Social learning and acquired recognition of a predator by a marine fish. *Animal Cognition*, **15**, 559–565.

Miles, S.G. (1968) Rheotaxis of elvers of the American eel (*Anguilla rostrata*) in the laboratory to water from different streams in Nova Scotia. *Journal of the Fisheries Research Board of Canada*, **25**, 1591–1602.

Mathuru, A.S., Kibat, C., Cheong, W.F. *et al.* (2012) Chondroitin fragments are odorants that trigger fear behavior in fish. *Current Biology*, **22**, 538–544.

Meiri, I., Gothilf, Y., Zohar, Y., and Elizur, A. (2002) Physiological changes in the spawning gilthead seabream, *Sparus aurata*, succeeding the removal of males. *Journal of Experimental Zoology*, **292**, 555–564.

Meredith, T.L., Caprio, J., and Kajiura, S.M. (2012) Sensitivity and specificity of the olfactory epithelia of two elasmobranch species to bile salts. *Journal of Experimental Biology*, **215**, 2660–2667.

Munday, P.L., White, J.W., and Warner, R.R. (2006) A social basis for the development of primary males in a sex-changing fish. *Proceedings of the Royal Society B-Biological Sciences*, **273**, 2845–2851.

Mylonas, C.C., Fostier, A., and Zanuy, S. (2010) Broodstock management and hormonal manipulations of fish reproduction. *General and Comparative Endocrinology*, **165**, 516–534.

Neat, F.C., Locatello, L., and Rasotto, M.B. (2003) Reproductive morphology in relation to alternative male reproductive tactics in *Scartella cristata*. *Journal of Fish Biology*, **62**, 1381–1391.

Nikolic, N., Bagliniere, J.L., Rigaud, C. *et al.* (2011) Bibliometric analysis of diadromous fish research from 1970s to 2010: a case study of seven species. *Scientometrics*, **88**, 929–947.

Northcott, S.J. and Bullock, A.M. (1991) The morphology of the club glands on the dorsal fin of mature male shannies, *Lipophrys pholis* (L) (Blenniidae, Teleostei). *Journal of Fish Biology*, **39**, 795–806.

Ohta, H., Kagawa, H., Tanaka, H. *et al.* (1996) Changes in fertilization and hatching rates with time after ovulation induced by 17,20 beta-dihydroxy-4-pregnen-3-one in the Japanese eel, *Anguilla japonica*. *Aquaculture*, **139**, 291–301.

Ohta, H., Kagawa, H., Tanaka, H. *et al.* (1997) Artificial induction of maturation and fertilization in the Japanese eel, *Anguilla japonica*. *Fish Physiology and Biochemistry* **17**, 163–169.

Palstra, A.P. and van den Thillart, G.E.E.J.M. (2010) Swimming physiology of European silver eels (*Anguilla anguilla* L.): energetic costs and effects on sexual maturation and reproduction. *Fish Physiology and Biochemistry*, **36**, 297–322.

Palstra, A.P., Schnabel, D., Nieveen, M.C. *et al.* (2008) Male silver eels mature by swimming. *BMC Physiology*, **8**, 14.

Pankhurst, N.W. and Lythgoe, J.N. (1983) Changes in vision and olfaction during sexual maturation in the European eel *Anguilla anguilla* (L.). *Journal of Fish Biology*, **23**, 229–240.

Pavlidis, M., Koumoundouros, G., Sterioti, A. *et al.* (2000) Evidence of temperature-dependent sex determination in the European sea bass (*Dicentrarchus labrax* L.). *Journal of Experimental Zoology*, **287**, 225–232.

Pedersen, B.H. (2003) Induced sexual maturation of the European eel *Anguilla anguilla* and fertilisation of the eggs. *Aquaculture*, **224**, 323–338.

Pedersen, B.H. (2004) Fertilisation of eggs, rate of embryonic development and hatching following induced maturation of the European eel *Anguilla anguilla*. *Aquaculture*, **237**, 461–473.

Piferrer, F., Blázquez, M., Navarro, L., and González, A. (2005) Genetic, endocrine, and environmental components of sex determination and differentiation in the European sea bass (*Dicentrarchus labrax* L.). *General and Comparative Endocrinology*, **142**, 102–110.

Pizzolon, M., Giacomello, E., Marri, L. *et al.* (2010) When fathers make the difference: efficacy of male sexually selected antimicrobial glands in enhancing fish hatching success. *Functional Ecology*, **24**, 141–148.

Power, D.M., Guerreiro, P.M., Hubbard, P.C. *et al.* (2010) Endocrinology applied to aquaculture of finfish, in *Recent Advances in Aquaculture Research* (ed. G. Koumoundouros), Transworld Research Network, Kerala, pp. 1–55.

Sæther, B.-S., Bjørn, P.-A., and Dale, T. (2007) Behavioural responses in wild cod (*Gadus morhua* L.) exposed to fish holding water. *Aquaculture*, **262**, 260–267.

Saglio, P. (1982) Piégeage d'anguilles (*Anguilla anguilla* L.) dans le milieu naturel au moyen d'extraits biologiques d'origine intraspécifique. Mise en évidence de l'attractivité phéromonale du mucus épidermique. *Acta Œcologica/Œcologia Applicata*, **3**, 223–231.

Saillant, E., Fostier, A., Haffray, P. *et al.* (2002) Temperature effects and genotype-temperature interactions on sex determination in the European sea bass (*Dicentrarchus labrax* L.). *Journal of Experimental Zoology*, **292**, 494–505.

Saillant, E., Fostier, A., Haffray, P. *et al.* (2003) Effects of rearing density, size grading and parental factors on sex ratios of the sea bass (*Dicentrarchus labrax* L.) in intensive aquaculture. *Aquaculture*, **221**, 183–206.

Scott, A.P. and Ellis, T. (2007) Measurement of fish steroids in water—a review. *General and Comparative Endocrinology*, **153**, 392–400.

Scott, A.P., Hirschenhauser, K., Bender, N. *et al.* (2008) Non-invasive measurement of steroids in fish holding water: important considerations when applying the procedure to behaviour studies. *Behaviour*, **145**, 1307–1328.

Sébert, M.-E., Amérand, A., Vettier, A. *et al.* (2007) Effects of high hydrostatic pressure on the pituitary–gonad axis in the European eel, *Anguilla anguilla* (L.). *General and Comparative Endocrinology*, **153**, 289–298.

Serrano, R.M., Barata, E.N., Birkett, M.A. *et al.* (2008a) Behavioural and olfactory responses of female *Salaria pavo* (Pisces: Blenniidae) to a putative multi-component male pheromone. *Journal of Chemical Ecology*, **34**, 647–658.

Serrano, R.M., Lopes, O., Hubbard, P.C. *et al.* (2008b) 11-ketotestosterone stimulates putative sex pheromone production in the male peacock blenny, *Salaria pavo* (Risso 1810). *Biology of Reproduction*, **79**, 861–868.

Sherwood, N.M., Kyle, A.L., Kreiberg, H. *et al.* (1991) Partial characterization of a spawning pheromone in the herring *Clupea harengus pallasi*. *Canadian Journal of Zoology-Revue Canadienne De Zoologie*, **69**, 91–103.

Siefkes, M.J. and Li, W. (2004) Electrophysiological evidence for detection and discrimination of pheromonal bile acids by the olfactory epithelium of female sea lampreys (*Petromyzon marinus*). *Journal of Comparative Physiology A*, **190**, 193–199.

Siefkes, M.J., Scott, A.P., Zielinski, B. *et al.* (2003) Male sea lampreys, *Petromyzon marinus* L., excrete a sex pheromone from gill epithelia. *Biology of Reproduction*, **69**, 125–132.

Sola, C. and Tosi, L. (1993) Bile salts and taurine as chemical stimuli for glass eels, *Anguilla anguilla*: a behavioural study. *Environmental Biology of Fishes*, **37**, 197–204.

Sorensen, P.W. (1986) Origins of the freshwater attractant(s) of migrating elvers of the American eel, *Anguilla rostrata*. *Environmental Biology of Fishes*, **17**, 185–200.

Sorensen, P.W. and Pankhurst, N.W. (1988) Histological changes in the gonad, skin, intestine and olfactory epithelium of artificially matured male American eels, *Anguilla rostrata* (LeSueur). *Journal of Fish Biology*, **32**, 297–307.

Sorensen, P.W. and Stacey, N.E. (2004) Brief review of fish pheromones and discussion of their possible uses in the control of non-indigenous teleost fishes. *New Zealand Journal of Marine and Freshwater Research*, **38**, 399–417.

Sorensen, P.W. and Winn, H.E. (1984) The induction of maturation and ovulation in American eels, *Anguilla rostrata* (LeSueur), and the relevance of chemical and visual cues to male spawning behaviour. *Journal of Fish Biology*, **25**, 261–268.

Sorensen, P.W., Fine, J.M., Dvornikovs, V. *et al.* (2005a) Mixture of new sulfated steroids functions as a migratory pheromone in the sea lamprey. *Nature Chemical Biology*, **1**, 324–328.

Sorensen, P.W., Pinillos, M., and Scott, A.P. (2005b) Sexually mature male goldfish release large quantities of androstenedione into the water where it functions as a pheromone. *General and Comparative Endocrinology*, **140**, 164–175.

Stacey, N.E. and Hourston, A.S. (1982) Spawning and feeding behavior of captive Pacific herring, *Clupea harengus pallasi*. *Canadian Journal of Fisheries and Aquatic Sciences*, **39**, 489–498.

Stacey, N.E. and Sorensen, P.W. (2006) Reproductive Pheromones, in *Behaviour and Physiology of Fish.*, vol. **24** (eds K.A. Sloman, R.W. Wilson and S. Balshine), Elsevier Inc., San Diego, pp. 359–412.

Taborsky, M. (1998) Sperm competition in fish: "bourgeois" males and parasitic spawning. *Trends in Ecology and Evolution*, **13**, 222–227.

Taranger, G.L., Carrillo, M., Schulz, R.W. *et al.* (2010) Control of puberty in farmed fish. *General and Comparative Endocrinology*, **165**, 483–515.

Tesch, F.-W. (2003) *The Eel*, Blackwell Science Ltd, Oxford.

Tosi, L. and Sola, C. (1993) Role of geosmin, a typical inland water odor, in guiding glass eel *Anguilla anguilla* (L) migration. *Ethology*, **95**, 177–185.

Tosi, L., Spampanato, A., Sola, C., and Tongiorgi, P. (1990) Relation of water odor, salinity and temperature to ascent of glass eels, *Anguilla anguilla* (L.): a laboratory study. *Journal of Fish Biology*, **36**, 327–340.

van Ginneken, V.J.T. and Maes, G.E. (2005) The European eel (*Anguilla anguilla*, Linnaeus), its lifecycle, evolution and reproduction: a literature review. *Reviews in Fish Biology and Fisheries*, **15**, 367–398.

Velez, Z., Hubbard, P.C., Barata, E.N., and Canário, A.V.M. (2005) Evidence for functional asymmetry in the olfactory system of the Senegalese sole (*Solea sensgalensis*). *Physiological and Biochemical Zoology*, **78**, 756–765.

Velez, Z., Hubbard, P.C., Barata, E.N., and Canário, A.V.M. (2007a) Differential detection of conspecific-derived odorants by the two olfactory epithelia of the Senegalese sole (*Solea senegalensis*). *General and Comparative Endocrinology*, **153**, 418–425.

Velez, Z., Hubbard, P.C., Hardege, J.D. *et al.* (2007b) The contribution of amino acids to the odour of a prey species in the Senegalese sole (*Solea senegalensis*). *Aquaculture*, **265**, 336–342.

Velez, Z., Hubbard, P.C., Welham, K. *et al.* (2009) Identification, release and olfactory detection of bile acids in the intestinal fluid of the Senegalese sole (*Solea senegalensis*). *Journal of Comparative Physiology A*, **195**, 691–698.

Velez, Z., Hubbard, P.C., Hardege, J.D. *et al.* (2011) Evidence that 1-methy-L-tryptophan is a food-related odorant for the Senegalese sole (*Solea senegalensis*). *Aquaculture*, **314**, 153–158.

Vermeirssen, E.L. and Scott, A.P. (1996) Excretion of free and conjugated steroids in rainbow trout (*Onchorhynchus mykiss*): evidence for branchial excretion of the maturation-inducing steroid, 17,20b-dihydroxy-4-pregnen-3-one. *General and Comparative Endocrinology*, **101**, 180–194.

Whittamore, J.M. (2012) Osmoregulation and epithelial water transport: lessons from the intestine of marine teleost fish. *Journal of Comparative Physiology B*, **182**, 1–39.

Wyatt, T.D. (2003) *Pheromones and Animal Behaviour*, Cambridge University Press, Cambridge.

Zhang, C. and Hara, T.J. (2009) Lake char (*Salvelinus namaycush*) olfactory neurons are highly sensitive and specific to bile acids. *Journal of Comparative Physiology A*, **195**, 203–215.

Zhang, C., Brown, S.B., and Hara, T.J. (2001) Biochemical and physiological evidence that bile acids produced and released by lake char (*Salvelinus namaycush*) function as chemical signals. *Journal of Comparative Physiology B*, **171**, 161–171.

Chapter 12
Applications of Pheromones in Invasive Fish Control and Fishery Conservation

Peter W. Sorensen

University of Minnesota, St. Paul, USA

12.1 INTRODUCTION

We have an enormous stake in effectively managing the approximately 30 000 species of fish that inhabit our planet. Wild fish are an important source of food, commerce, and recreation, and they play key roles in ecosystem health. However, the condition of many of our fisheries is poor and deteriorating. Native fisheries are collapsing, and both nonnative (also called 'exotic' or 'alien') nuisance and invasive species are becoming overabundant. The causes and consequences of these intertwined problems are complex. Solutions will require accurate information on the abundance and distribution of fish and then gaining control of them in targeted, environmental-friendly ways. At present, very few tools exist to provide information on either distribution or abundance, and the tools available for control are crude, inefficient, and nonspecific. For example, nets, electrofishing, and available fish poisons capture or kill all the fishes they encounter (i.e. they are nontargetable), and generally with low efficiency. New options are desperately needed to manage problematic wild fish.

Pheromones play key roles in the lives of many fishes (Liley, 1982; Sorensen and Caprio, 1998; Sorensen and Baker, 2015; Stacey, 2015; Wisenden, 2015). They are potent, environmentally safe, species (or taxon) specific, and they can be easily and safely added to the water, thereby offering many opportunities for managing native and exotic fishes. Because fish pheromones drive many aspects of behavior and physiology, including both attraction and repulsion, their possible applications are myriad. This is especially true for exotic, invasive fishes for which few tools exist and that have communication systems that are likely to be novel and operate independently of native fishes.

That pheromones may be useful in the management of wild fishes is supported by their successful application in pest insect management. Fish and insects have much in common: both live in huge, three-dimensional habitats that are difficult to manage, both are mobile, and both rely heavily on chemical cues including pheromones to mediate both species-specific attraction and repulsion (Sorensen, Christensen, and Stacey, 1998). Because insects pheromones were identified nearly 25 years before fish pheromones (see below), their use in insect management is far more advanced, although their promise is evident for both. This chapter reviews (Section 12.2) how pheromones have

---

Fish Pheromones and Related Cues, First Edition. Edited by Peter W. Sorensen and Brian D. Wisenden.

come to be used in insect control and what lessons we might apply to fish. Section 12.3 then discusses how pheromones are currently being considered for use in nuisance and invasive fisheries management. Section 12.4 finally reviews how pheromones might be used in the conservation of threatened/endangered fishes. I conclude with thoughts for future studies.

12.2 PHEROMONES IN PEST INSECT CONTROL WITH SOME LESSONS FOR FISH

The first pheromone, bombykol, was identified in the silk worm, *Bombyx mori*, in the late 1950s, and it immediately sparked a search for applications (Hecker and Butenandt, 1984). Thousands of insect pheromones have since been identified and synthesized; and after a difficult start, numerous approaches to using them in monitoring, trapping, and disrupting nuisance insects in agriculture are now in place. This growing success has stimulated parallel interest in fish pheromones, the application of which has nevertheless proceeded at a slower pace. Only a handful of fish pheromones have been conclusively identified, starting with identification of a goldfish (*Carassius auratus*) pheromone in the late 1980s (Sorensen and Hoye, 2010). There are many reasons for these differences between insects and fishes, but most relate to the relative ease of study, funding, and the complexity of the systems involved. Insect pheromones tend to be relatively simple hydrocarbons that are also produced in glands (making their isolation and synthesis relatively easy) and attract great interest (and funding) from the agricultural industry. Still, there are many similarities between how fish and insects use pheromones and thus how these might be applied.

Over the past few decades, pest insect control has shifted from exclusive reliance on pesticides to integrated pest management (IPM). IPM incorporates diverse suites of management tools including pheromones to reduce and control populations to tolerable levels. Eradication is generally not the goal, and scenarios are usually customized to fit local scenarios as well as needs and the biologies of the pests being targeted. Similar goals are now sought for invasive fish control, because development of species-specific piscicides (fish toxins) is no longer viewed as practical, owing to high development and licensing costs in the United States. Generally, IPM aims at achieving targeted (species-specific) control by wedding information on species abundance and life history with targeted control that can include poisons as well as pheromonally mediated trapping and disruption. Control in insects typically addresses different combinations of the pest's life stages including recruitment (introduction of young into a population), adult survival, reproductive success, and immigration. Presumably, IPM in fisheries management should as well because, while some fish have vulnerable juvenile stages (e.g., common carp, *Cyprinus carpio* [Sorensen and Bajer, 2011]), for other species the adults are most vulnerable (e.g., sea lamprey, *Petromyzon marinus*, in the Laurentian Great Lakes [Christie and Goddard, 2003; Sorensen and Bergstedt, 2011]). IPM is rapidly gaining in importance in insect control schemes in third-world countries where insects are developing resistance to insecticides that are also becoming more expensive and tightly regulated. Insect pheromones are presently being used in several ways: (i) as attractive lures for monitoring; (ii) as lures for mass trapping (sometimes paired with killing); and (iii) as disruptants. All of these options are also being considered for fish.

Accurate data on population abundance is a essential component of any IPM scheme, whether for insects or fish and it is a huge challenge. Sex pheromones are proving very useful in pest insect population monitoring for agriculture and are now commonly used as lures in small monitoring traps across fields and orchards worldwide (Witzgall, Kirsch, and Cork, 2010). For example, over 10 million hectares of fields in dozens of countries are now monitored for various Lepidoptera (moths) and Coleoptera (beetles) using pheromone traps (Witzgall, Kirsch, and Cork, 2010). Pheromone traps are also being used to locate outbreaks of invasive insects in forests (e.g., emerald ash borer, *Agrilus planipennis*). An especially useful type of insect pheromone trap has proven to be the "sticky trap," which is inexpensive and easy to deploy and monitor. Similar trapping success has yet to be developed for fish; thus, while traps have been suggested for invasive fish such as the round goby (*Neogobius melanostomus*) in the Great Lakes (Corkum and Belanger, 2007), these have not yet been deployed owing to a lack of both a synthetic pheromone and an effective design. Often monitoring for invasive insects has also been accompanied by either by successful targeted pesticide applications or mass trapping (see below). The success of pheromonally mediated monitoring of insects is grounded in decades of biochemical work that has shown that many insects use mixtures of aromatic hydrocarbons as sex pheromones, and these are also species-specific and relatively simple to synthesize and deploy. This has not proven to be the case for fish pheromones and complex mixtures (see below; Sorensen and Baker, 2015), and only half a dozen fish pheromones have been as yet identified. I will discuss why these differences exist and some ways forward forward for fish after elaborating on insect IPM.

Remediation and control are the second key component of most insect IPM programs. Usually, efforts are highly targeted, and sometimes only pheromones can provide this fine control. Once nuisance insect species have been located; farmers often employ pheromonally mediated mass trapping that often uses larger traps and insecticides in attract-and-kill scenarios [e.g., cotton boll weevil (*Anthonomous grandis*)]. Using pheromones in mass capture requires fully identified (and active) natural pheromones; but because most insect pheromones are blends of relatively simple volatiles (Howse, Stevens, and Jones, 1998), this challenge has often been solvable. For pheromones to be useful in mass trapping, they must also evoke powerful and simple attraction; and while this is common in insects, it has not proven to be the case for fish for responses are not always simple - fish pheromones appear to be complex and cognitive processes appear to be involved. Further, fishes have also not been as well studied (see below).

Finally, some insect IPM schemes also seek to shut down reproduction (recruitment). Because many insects use extremely specific blends of hydrocarbons as sex pheromones, this result can often be accomplished by adding just a few components to the environment. For example, in the case of the boll weevil, it has been discovered that simple blends of pheromone applied to cotton plants will confuse insects and prevent reproduction. The blend need not be perfect. Whether, and how, this approach might be applied to invasive fish control is not yet clear, and it may ultimately depend on how inexpensively fish pheromones can be manufactured and the specific role that mixtures play in pheromonal specificity-both of which are presently not well understood but do appear less straightforward.

In summary, pheromones have proved to be extremely useful in the control of pest insect management largely because they are attractive, potent, species specific, and can target various life stages in many ways. In contrast, the scenario for those fishes studied to date appears much more complex and challenging- although poorly studied.

12.3 PHEROMONES IN NUISANCE AND INVASIVE FISH CONTROL

Many of the applications for pheromones in insects appear in principal to be applicable to fish. Fish, like insects, are highly mobile, are difficult to locate and trap, and rely heavily on species-specific chemical cues to find each other. This is particularly true for the many pest and invasive fishes that evolved in vast, turbid waters that lack light, and from origins that may have favored the evolution of unique cues. However, to date, efforts to develop and test fish pheromones in the field have been limited, probably because of the greater chemical complexity and cost of fish pheromones and regulatory issues (in the United States, aquatic pheromones developed for nuisance fishes are classified as "pesticides"). Nevertheless, because these issues are solvable, it is reasonable that pheromones be considered for use in monitoring, removal and control, and disruption. In addition, pheromonally-mediated disruption of fish movement patterns (without trapping) might be an option for some migratory pest fishes. I review all four uses below after first defining the problems caused by nuisance fishes.

12.3.1 An Introduction to the Problem of Nuisance and Invasive Fish

Unwanted species are generally termed "pests" or "nuisance" species, whereas those that cause economic and or ecological damage are termed "invasive." Fish have been moved all over the world both by accident and on purpose: dozens have become invasive and are now causing severe economic and ecological problems. This issue is acute in many freshwater ecosystems because these systems are typically less diverse and are therefore less resilient than marine systems. In many instances, invasive fish now comprise over half the fish in some freshwater ecosystems, and disrupt food chains and displace native fishes. Species of special concern include the walking catfish (*Clarius batrachus*) in North America, the common carp (*C. carpio*) in most temperate regions of the world, sea lamprey in the Laurentian Great Lakes, bigheaded carps (*Hypophthalmichthys* sp.) in the Mississippi River, brown trout (*Salmo trutta*) in many temperate regions of the world, lionfish (*Pteriois volitans*) in the Caribbean and Atlantic Ocean, tilapia (*Oreochromis* sp.) in many tropical areas of the world, snakehead (*Channa micropeltes*) in North America, rudd (*Scardinius* sp.) in the Southern Hemisphere, mosquitofish (*Gambusia* sp.) in many subtropic regions of the world, and smallmouth bass (*Micropterus dolomieu*) in Asia (http://www.issg.org/database/species/search.asp?st=100ss&fr=1&str=&lang=EN, http://www.environmentalgraffiti.com/news-invasive-fish). With just a few exceptions, pheromones have been at least partially identified in all of these species. In addition, with the possible exception of the mosquitofish, all species are social, lack obvious visual sexual dimorphisms, and thus appear to rely heavily on pheromones.

Reducing and controlling nuisance and invasive fish is a daunting challenge because of the scale of the problem, and no good solutions presently exist; therefore, all options including pheromones must be seriously considered. IPM is especially attractive because its premise of targeted local control seems achievable and is suited to take advantage of many tools, including pheromones. The only invasive fish that is presently controlled is the sea lamprey; this is largely attributable to a relatively novel toxin that was isolated for this ancient species in the 1950s and is now implemented as part of an IPM program that is attempting to use pheromones (Sorensen and Hoye, 2007). No new piscicides are presently being developed because their use in open waters makes testing and registration extremely expensive. The large volumes of water that fishes typically inhabit also represent a challenge, especially because most are public waterways that cannot be easily regulated or treated and contain game fish. Most workers are convinced that the solution(s) to invasive fish control lies in developing suites of tools that can measure the abundance of invasive fishes and remove these fish as part of IPM programs similar to those for insects. These tools and possible roles of pheromones in them are described below.

12.3.2 Using Pheromones for Monitoring Fish Distribution and Abundance

Targeted, sustainable fish management programs require accurate data on the number of fish present, their distribution, and ideally their sex and maturity. These data are difficult, if not impossible, to collect in large and/or flowing waterbodies where trapping, netting, and electrofishing are often ineffective. Even a simple presence–absence measure, which is critical for assessing the distribution limits of invasive species, is very difficult to obtain using conventional methods. As is the case with insects (see above), monitoring is especially important at invasion fronts where invaders are present at very low densities if at all and also hard to measure. One example are the two species of bigheaded (Asian) carps, *Hypophthalmichthys* sp., that are presently invading the upper reaches of the Mississippi and Illinois Rivers, but whose precise abundance and distribution cannot be systematically determined (Jerde *et al.*, 2011). Another example is the sea lamprey, *P. marinus*, whose spawning streams must be monitored by cumbersome electrofishing for re-invasion after toxin (lampricide) treatments (Sorensen and Vrieze, 2003). However, fish species could potentially be monitored by (i) measuring the presence of species-specific pheromones (see Sorensen and Baker, 2015) or (ii) using pheromones to lure fish into traps for counting, or (iii) tracking pheromone-releasing (Judas) fishes as they swim into groups of conspecifics where they might be captured and counted (Table 12.1). All three possibilities are reviewed below. None of these have been implemented, but a few have been tested.

Table 12.1.   Summary of the Different Ways Pheromones Might be Used to Control Nuisance and Invasive Fishes.

Strategy	Advantages	Challenges
1. Monitoring:		
Measuring pheromone	Doable, inexpensive Compliments eDNA Gender information Complete cue not needed	May not be species-specific
Pheromone traps	Doable, complete cue not needed	Need to develop traps
Judas Fish	Doable, complete cue not needed	Need to track fish
2. Remediation and Control:		
Mass trapping	Could compliment other techniques	Need a great deal of high-quality cue
Redirection	Inexpensive Complete cue not needed Could compliment other techniques and approaches	Not 100% effective
3. Disruption:		
	Could compliment other techniques	Need a lot of cue

MEASURING PHEROMONE CONCENTRATION AS AN INDEX OF POPULATION ABUNDANCE

Assessing the presence of invasive fishes using biochemical techniques is appealing because it is relatively easy to collect water samples. The technique is somewhat similar to that used for measuring the DNA (environmental or "eDNA") which is released by fishes and which can now be measured, albeit with difficulty (Jerde *et al.*, 2011). The two techniques could nevertheless complement each other and make up for each other's weaknesses. Like measuring eDNA, measuring pheromone concentrations could provide valuable information on both fish distribution and relative abundance. Pheromone concentration could also inform about reproductive condition and abundance. Further, while DNA can be transported by nonfish vectors (e.g., fish-eating birds can defecate traces of DNA), pheromones cannot; therefore, pheromone measurements may be less prone to "false positives." However, both approaches are susceptible to the variable effects of dilution and degradation (false negatives). Using pheromone concentrations to monitor species abundance is complicated by the fact that fish pheromones now appear to be mixtures of relative common taxon-specific metabolites (Sorensen and Hoye, 2010; Sorensen and Baker, 2015), therefore measurement techniques may have to examine multiple components. Nevertheless, the prospect of measuring a few specific components is not difficult with mass spectrometry (Stewart and Sorensen, 2015) and could have merit. For example, although petromyzonamine disulfate is released by many species of lamprey as a pheromone, sea lampreys are often the only species present; thus, alternative sources of this pheromone are not a concern so it could be used in some instances. Another example is $17\alpha,20\beta$-dihydroxy-4-pregnen-3-one which serves as part of a priming sex pheromone mixture in many carp species (Stacey, 2015). This compound is only released in notable quantities by preovulatory fish in the 12-h period before spawning; therefore, lack of species specificity may not be an issue. Although not yet described, there is a possibility for some invasive fishes to produce novel products such as MHC-derived peptides that could be measured at the same time.

Several techniques are being developed to measure pheromones released by invasive fishes in natural waters, and some proof-of-concept studies have been modestly successful (Fine and Sorensen, 2005). These techniques, reviewed in detail by Stewart and Sorensen (2015), use liquid chromatography to isolate products from their natural matrix and then use mass spectrometry to identity and quantify them (Fine and Sorensen , 2005; Stewart, Baker, and Cooney, 2011; Xi *et al.*, 2011). Tracers are often added to these samples, including synthesized deuterated isomers, to precisely quantify recovery and measurement (Sorensen and Hoye, 2007; Xi *et al.*, 2011). Immunoassay (e.g., ELISA) has also been examined as an alternative, and ultimately has great potential because it can be conducted more quickly and easily (Stewart and Sorensen, 2015). Further, passive sampling devices can be placed into water bodies to quickly extract pheromones (Stewart and Baker, 2012). For example, sea lamprey sex and larval pheromones have already been successfully measured in natural waters (Fine and Sorensen, 2005; Xi *et al.*, 2011), but further information on degradation and dispersal rates is required to make these techniques genuinely useful. Perhaps, the greatest potential use for measuring pheromones may lie in their ability to complement and confirm eDNA that presently suffers from issues of detection limits, marker specificity, and degradation. Ultimately, these issues are solvable.

USING PHEROMONE-LADEN TRAPS TO MONITOR FISH ABUNDANCE

Traps baited with pheromones have proven to be extremely useful in monitoring insect pests at invasion fronts (see above). Might similar strategies be used for fishes? Likely, but it appears that their utility will depend on several factors that are not yet well understood. These include the biology of the species in question (whether it is attracted to pheromonal sources and under what conditions) and whether high quantities of pure pheromone are available. Our recent experiences with the common carp have shed some light on these questions and are described below. It is unfortunate that the common carp experiment—that was designed to be a simple proof of concept study and that did not specifically evaluate monitoring—appears to be the only experiment of this kind.

The common carp is highly invasive across much of the world (Sorensen and Bajer, 2011), and various hormonally derived sex pheromones have now been partially identified for it (Stacey, 2015). A pheromonally mediated trapping scheme for common carp in Midwestern lakes was recently tested and found to have potential (Lim and Sorensen, 2012). This proof of concept study assessed attraction using radiotagged conspecifics as subjects because carps do not readily enter traps while using prostaglandin $F_{2\alpha}$ ($PGF_{2\alpha}$)-implanted carp (a procedure that induces sex pheromone release) placed in traps as donors simply to test attraction. Prostaglandin-implanted fish were used because they are known to naturally and continuously release the complete $PGF_{2\alpha}$-derived sex pheromone complex, which also cannot yet be synthesized (see Lim and Sorensen, 2012; Sorensen and Baker, 2015). This study found that sexually male carp, but not females, were attracted to vicinities of a trap-based source from up to 20 meters (Fig. 12.1), demonstrating sex pheromone activity of $PGF_{2\alpha}$-derived sex pheromones (Stacey, 2015) in the field and suggesting the possibility that such a scheme could, with improvements to trap design, be used to monitor invasive carp abundance and distribution. A version of this monitoring scenario could also be deployed for other fishes such as the bigheaded carps that are presently next to impossible to locate. Cost would not be high. A better understanding of whether and how carps orient to pheromone plumes, and how they respond to traps, would be extremely helpful.

USING PHEROMONE-LADEN JUDAS FISH TO LOCATE OTHERS FOR MONITORING

Our experiences with the common carp (above) have demonstrated that behavioral and environmental context are very important to pheromone function: Pheromones are more active when released in the presence of the donor fish. In addition, many, and perhaps most, fishes including the carps seek each other out while simultaneously attracting others. A mobile pheromone-release mechanism is likely to be much more effective than a stationary one for eliciting attraction-driven aggregation (and counting fish). Previously, we have found that we can make many fish, including invasive carps, sexually

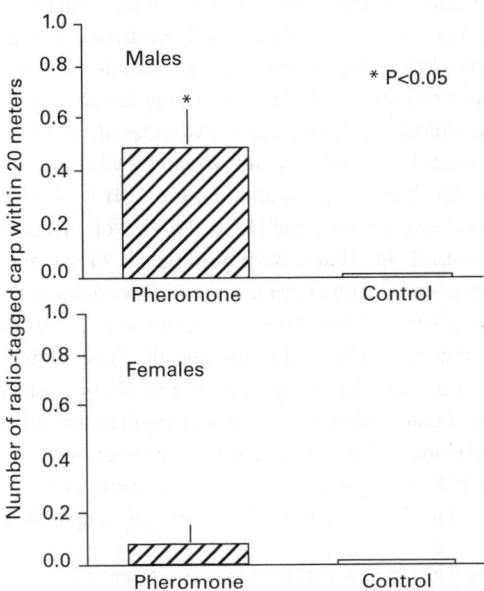

Figure 12.1.   The mean daily number of radiotagged male common carp detected within 20 m of holding cages containing either PGF-implanted female carp or left empty in a carp-infested lake. Lim, H. & Sorensen,P.W. Common carp implanted with prostaglandinF2a release a sex pheromone that attracts conspecific males in both the laboratory and field. Journal of Chemcial Ecology 38, 127–134 (2012).

receptive and attractive by implanting them with $PGF_{2\alpha}$, which functions as a hormone to drive sexual behavior and pheromone release (see Section 12.3.3 and Lim and Sorensen, 2012; Stacey, 2015). If $PGF_{2\alpha}$-implanted fish were tracked using radio or acoustical tags (perhaps after sterilization), then they should rapidly find and induce aggregations of mature conspecifics. Aggregations might then be censused using netting or electrofishing or perhaps even eDNA with pheromones. This is known as the "Judas fish technique"; it has been used (to date without pheromones) for common carp control on several occasions outside of the spawning season (Bajer, Chizinski, and Sorensen, 2011). Although complex, this scheme would take full advantage of fishes' natural behaviors and abilities to find each other across broad areas. It also need not be necessarily expensive.

12.3.3   Using Pheromones for Mass Removal of Nuisance Fishes

Many fishes use shoaling, migratory, and/or sex pheromones to find each other at various points in their lives (Sorensen and Baker, 2015; Stacey, 2015). These types of pheromones are generally species or taxon-specific and are of great importance to both migrating and reproductively active fishes. However, while pheromone-mediated mass trapping of insects has shown promise, results to date for fishes have been mixed. Migratory and sex pheromones (but not shoaling pheromones) have been tested using sea lamprey and brook trout (*Salvelinus fontinalis*). Our limited and to date marginally-successful experiences with pheromonally-mediated mass trapping in fishes are described in the section that follows.

EXPERIENCES WITH THE SEA LAMPREY MIGRATORY PHEROMONE IN TRAPPING

Several hundred species of fish perform spawning migrations guided by odors, which often include pheromones (Sorensen and Baker, 2015). For problematic migratory species (e.g., sea lamprey), pheromones could have great potential in targeted trapping programs. The sea lamprey migratory pheromone is the only well-understood migratory pheromone and is also the only one that has been tested thus far.

The sea lamprey is an ancient, jawless, and parasitic fish that entered the Laurentian Great Lakes during the construction of canal systems and then rapidly destroyed its native fisheries (Sorensen and Bergstedt, 2011). In the Great Lakes this invasive species spends 1–2 years in lakes, growing before maturing and entering streams to spawn and die; its resulting larvae may then spend many years there before metamorphosing and returning to lakes/oceans (Moser *et al.*, 2014). Field and laboratory studies have indicated that stream choice by migratory adult sea lamprey is guided by odor (Vrieze, Bjerselius, and Sorensen, 2010, 2011), Meckley, Wagner, and Gurarie (2012) of which a migratory pheromone released by larval lampreys is key (Sorensen and Hoye, 2007). Notably, anosmic adult lampreys do not even find rivers: this is a powerful cue. Extensive physiological, biochemical, and physiological study has also shown that this pheromone has three primary steroidal components that are extremely active in the laboratory and mimic the activity of larval odor (Sorensen *et al.*, 2005; Sorensen and Baker, 2015). However, when recently synthesized and added to streams, the three components were not as attractive as larval odor (Meckley, Wagner, and Luehring, 2012). The likely explanation is that the migratory pheromone contains additional unidentified component(s) required for full activity, as laboratory experiments had previously hinted (Fine and Sorensen, 2008). The lesson seems to be that fully complete and active pheromonal mixtures will be required if they are to be useful in the field where they will have to "compete" with natural cues. This basic work needs to continue in spite of its challenges.

Experiences Using Brook Trout Sex Pheromone in Trapping

The brook trout is native to eastern regions of North America but invasive in the west. It is difficult to trap; and although primarily a visual animal, it appears to use hormonal sex pheromones (Stacey, 2015). To test this possibility and its possible utility in invasive control, Young, Micek, and Rathbun (2003) conducted an experiment using hoop nets baited with mature male brook trout. These attracted males, which also entered the traps. Studies of pheromone identity combined with synthesis are required to determine full potential.

Experiences With The Sea Lamprey Sex Pheromone

The sea lamprey provides our best example of how sex pheromones might be useful in natural settings for population management. After re-entering streams, adult sea lampreys start to mature, and the males build nests and release a sex pheromone. Initial stream-side maze and laboratory studies and electrophysiological recordings demonstrated that 3-keto-petromyzonal sulfate (3kPZS) is a key component of the pheromone released in urine by spermiating males (Li *et al.*, 2002). However, subsequent tests of this compound in the field found that while 3kPZS is a powerful attractant, neither it alone nor it in combination with a another component (3-ketoallocholic acid [3kACA]) was able to attract migratory females into traps as effectively as the natural odor of spermiating male pheromone (Fig. 12.2) (Johnson *et al.*, 2009; Johnson and Li, 2010; Luehring, Wagner, and Li, 2011). More recent studies also describe an inability of 3kPZS and 3kACA to stimulate nest-tending behaviors (Johnson *et al.*, 2012). As was the case with the sea lamprey migratory pheromone, over a decade of intense research went into identifying and synthesizing these cues; therefore, this result was disappointing (Sorensen and Hoye, 2007). One lesson is that all components, even ones thought to be minor in laboratory studies, may need to be present for synthesized pheromones to be effective lures for mass trapping in the field. Nevertheless, half a century of insect pheromone work argues for patience and persistence because success should ultimately be possible and the great promise it offers.

12.3.4   Using Pheromones to Cause Mating Disruption Among Nuisance Fish

Pheromones have been widely used in large quantities to severely disrupt mating in insects such as the cotton boll weevil that use precise pheromonal blends (Witzgall, Kirsch, and Cork, 2010). Although the precise mechanism by which this strategy works is not clear, the premise seems to be

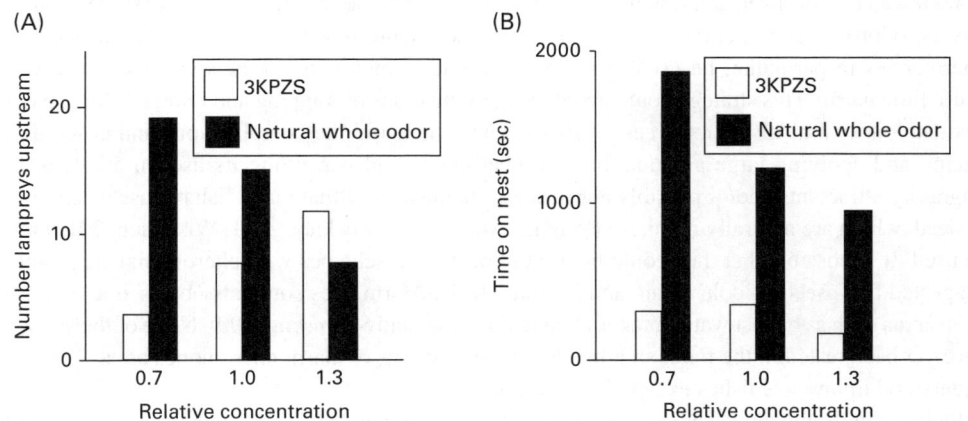

Figure 12.2.   The behavior of mature female sea lamprey to a synthesized component of the male sex pheromone [3ketopetromyzonol sulfate (3kPZS)] and the odor of spermiated males (the entire pheromone). The odors were tested in competition with each other in a natural stream while varying the relative concentration of 3kPZS (a value of 1.0 denotes that each cue was at the same overall concentration with respect to the other). The left-hand panel (A) shows that 3kPZS is almost as active as the natural pheromone at stimulating upstream swimming to a trap when tested at balanced concentrations. By contrast, the right-hand panel (B) shows that while 3ZPZS can stimulate swimming, it is much less able to hold females at nest sites, irrespective of concentration. This limits its utility in control. Data were redrawn from Johnson *et al.* (2009) using additional data described by Johnson *et al.* (2012).

that by adding large quantities of pheromone to the environment, scientists can create odorous noise that both overwhelms and distorts functioning of insect sensory systems. One advantage of this strategy is that it does not necessarily require intimate knowledge of a species' sex pheromone systems as only specific components need be added. Notably, most insects use highly specific blends of volatile hydrocarbons as pheromones; the ratios of these drive specific behaviors; therefore if altered by just a few percent, they are no longer discernible and cause confusion (Sorensen, Christensen, and Stacey, 1998; Sorensen and Stacey, 2004). Although it is not yet clear how fish pheromone mixtures work, they presently do not always appear to be reliant on specific ratios. For example, the ratio of PGFs present in the common carp sex pheromone system seems unimportant while overall composition is key (Lim and Sorensen, 2011). Specific ratios of bile sterols also appear to be relatively unimportant in the sea lamprey (Fine and Sorensen, 2008). Similarly, altered hormonal ratios merely change signal strength and not overarching function in goldfish hormonal sex pheromones (Stacey, 2015). Further, identified fish pheromones are expensive to synthesize (Sorensen and Hoye, 2007), and it is difficult to imagine at the present time that enough product could (or would) be produced for possible use in control schemes that saturate the aquatic environment for disruption although development of synthetic analogs has some promise (Burns, Sorensen, and Hoye, 2011) if regulatory agencies would permit their use. Mating disruption may thus not have great promise for wild fishes.

12.3.5   Using Pheromones to Redirect the Movements of Invasive Fishes

Over a thousand species of fishes migrate (perform highly directed movements) from one place or another at one point in their lives. Other species perform less directed, but still aggressive, movements that are important to range expansion (e.g., bigheaded carps) and invasiveness. Further, as reviewed by Sorensen and Baker (2015), a very large number of these species (e.g., sea lamprey) rely on odorous cues, including pheromones to guide them in such movement. Might odors, and pheromones in particular, be used to redirect these movements to places where these fish would cause little harm? This strategy could avoid the complications of trapping and removal. Also, it could also exploit natural (living) sources of pheromones and avoid the complications and costs of producing and applying large amounts of pure pheromone and permitting its use. In addition, both migratory attractants and/or possibly alarm cues (chemical cues that many fish release when injured or dead, which are naturally repulsive (Wagner, Stroud, and Meckley, 2011; Wisenden, 2015) might be used. It is possible that they could be used in push–pull schemes with pheromonal attractants as suggested for insects (Cook, Khan, and Pickett, 2007). Alarm cues could also be used to divert fish from areas such as ballast water intakes (Maniak, Losing, and Sorensen, 2000). None of these options have yet been tested in the field, and the identity and actions of alarm pheromones are as yet poorly understood in invasive fishes except the sea lamprey.

Pheromonally mediated redirection appears to make particular sense for the sea lamprey in the Laurentian Great Lakes because endangered, native, heterospecific lamprey species are also found in this system. These are sessile (they live in sediments) and release the same migratory pheromone as the sea lamprey (Fine, Vrieze, and Sorensen, 2004). Further, alarm cues have been characterized in larval lampreys; therefore, these could be eventually added to a possible "push–pull" scheme (Wagner, Stroud, and Meckley, 2011; Bals and Wagner, 2012). Hypothetically, larval forms of native species of lamprey could be stocked above natural barriers in Great Lake streams. Their odor should attract migrating adult sea lamprey to areas where they would fail to reproduce or could be managed (Sorensen and Vrieze, 2003; Twohey, Sorensen, and Li, 2003). Conversely, and perhaps in conjunction with native lamprey stocking, lamprey larvae could be removed from regions where adult lampreys are not wanted. Injured or dead lamprey might even be placed in these locations during the brief lamprey spawning runs. These efforts need not be difficult or expensive. Indeed, analyses of adult sea lamprey capture records, and their relationship to larval densities suggests that this scheme is already working (Sorensen and Vrieze, 2003). This scheme would also restore native lampreys.

12.3.6　Summary and Thoughts About Possible uses of Pheromones in Invasive Fish Control

Pheromones appear to have considerable unrealized potential for application in invasive and pest fish control. Options clearly exist for their use in monitoring, trapping, and redirection as part of IPM schemes (Table 12.1). At present, monitoring and redirection appear to be the most promising. Monitoring identified pheromonal components using extent mass spectrometry technology of even a few pheromonal components could provide new, valuable information to supplement eDNA monitoring. Likewise, application of pheromonally enhanced Judas carp could proceed using existing understanding to gain valuable and new information. Diversion schemes that use natural sources of pheromones could proceed and complement nascent IPM schemes at very low cost. Clearly, all would have to be part of a well-coordinated IPM plan that includes other tools. Ultimately, however, as with insects, synthesized complete pheromones will be most useful for nuisance fish control because, although addition of waters may present regulatory complications, only synthetic cues can be produced in quantities adequate for large-scale use on large scales while being quantifiable. Presently, this represents a considerable challenge as fish pheromones appear to be complex multicomponent mixtures (Sorensen and Baker, 2015) whose identification and eventual production will be difficult. However, we have already made excellent progress identifying sea lamprey and carp pheromones, so full elucidation of these pheromones and application should be achievable if adequate time and resources were to be dedicated to these worthwhile goals. A very significant motivator is the unfortunate fact that very few tools exist for invasive fish control at present; the potential is great.

12.4 USING PHEROMONES IN FISHERY CONSERVATION

Pheromones also play significant roles in the migratory and reproductive biologies of many endangered and threatened native fishes. Although possibilities for their use have not yet been formally tested, pheromones could be used for enhancing native fish abundance, as well as for controlling nuisance fishes. Many of the same schemes are applicable and doable. The theory behind some scenarios will be examined next.

12.4.1 Redirection and Restoration of Native Migratory Fishes by Seeding Natural Populations

Many threatened and endangered migratory fishes [e.g., salmon, native lamprey, amphidromous gobies (see Sorensen and Baker, 2015)] locate each other and critical habitat by orienting to the odors of conspecifics. This evolutionary strategy presumably evolved because pheromones are "honest" (i.e., their meaning cannot be obscured or manipulated by either donors or receivers), long-lived indicators of habitat quality. However, if a species is extirpated from its upstream habitat, this strategy will prove problematic because the pheromones that bring fish there will, of course, no longer be present. Conservation and restoration schemes for fish that use these strategies should focus on re-establishing pheromone-releasing conspecifics at strategic locations. One could argue that pheromones need always be considered in native fish restoration simply because of their importance and the fact that their distribution has been impacted both directly and indirectly by many things including pollution, overharvesting, and dams.

The Pacific lamprey (*Lampetra tridenta*) is an excellent example of endangered species that would benefit from an understanding and application of pheromones (Close, Fitzpatrick, and Li, 2002; Yun *et al.*, 2011). Its anadromous life history is similar to that of the sea lamprey: its migratory adults appear to locate spawning and nursery habitat using the odor of upstream resident larvae (i.e., a migratory pheromone). Over the past century, the Pacific lamprey's nursery grounds have suffered from severe habitat deterioration and damming so that the number of stream-resident larvae is now very low. Of course, the situation is self-reinforcing as few larvae result in few adults. This scenario raises the possibility that managers might restore this species by re-establishing pheromone production in the upper reaches of streams either by adding pheromone or ideally by restoring and then protecting the larvae so that they naturally release significant quantities of pheromones (and at low cost!). Similarities with the natural pheromonal redirection scheme for sea lamprey control are evident (see above).

The now-threatened freshwater (Anguillid) eels also might benefit from migratory pheromones in conservation attempts. In particular, the European eel, *Anguilla anguilla*, once an extremely abundant catadromous fish, is now endangered across much of its range in the North Atlantic. It too relies heavily on olfaction to find mates and habitat. In particular, young migratory Anguillid eels orient to conspecific odor when moving upstream from the ocean to find feeding habitat (Sorensen and Baker, 2015). Dams are thus a problem. Recently, Briand, Fatin, and Legault (2002) discovered that migrating young eels can be diverted around dams using conspecific odors released by stocked conspecifics. This seems to solve a problem at very low cost.

12.4.2 Sex Pheromones in Fish Conservation

Many native fishes are now failing to recruit and presumably reproduce. Although many factors may be responsible for this, including pollution (the olfactory sense is destroyed by pollutants at sublethal concentrations [Olsen, 2015]) and poor habitat, it is possible that a lack of conspecifics and pheromones could be a problem too. Thus, adding priming sex pheromones to natural waters might restore reproductive cycling and success (Stacey, 2015) when numbers are low. This would be simple to test.

12.4.3 Summary of Pheromones in Fishery Conservation

Pheromones play significant roles in the lives of many fishes that are now endangered. Simply understanding these roles so that water quality and habitat can be managed and improved could be very beneficial, but this has not yet been explicitly considered. In addition, pheromones could be added either directly (synthesized cues) or indirectly (stocking conspecifics in strategic ways) to alter migratory paths and/or trigger reproduction in fishes that now fail to reproduce.

12.5 SUMMARY AND FUTURE DIRECTIONS

Several proof-of-concept studies clearly demonstrate that as with the insects, pheromones could ultimately prove invaluable in the management of both invasive fishes and endangered native fishes. Advantages include potential ease of application (especially with natural sources), low cost, specificity of action, low-to-no environmental costs, and ability to be used in many types of IPM. This promise is only magnified by the lack of other tools to control nuisance and invasive fishes. Ultimately, synthesized cues will be required for remediation, generating a need for much more detailed studies of pheromonal identity and function in the field. Fishes do appear to use pheromones in different and more complex ways than insects (e.g., mixture recognition and attraction seem less prescribed than insects) but this need not be a "show-stopper". Nevertheless, applied research should continue with fishes using the pheromonal components we have already identified to develop and test pheromones in monitoring (direct measurement and Judas fish) and field attraction. In addition, diversion schemes using natural cues seem well worth testing, especially as native fish restoration would be an added benefit. We need to know more about how pheromones alter fish movement patterns and either attract or repel so that we can create effective pheromone plumes (Vrieze, Bjerselius, and Sorensen, 2010, 2011). The sea lamprey and carp model systems are well developed and could be used for this work, but we need others too. A lesson from the insects is that while this work will be challenging, it ultimately will be extremely beneficial.

REFERENCES

Bajer, P.G., Chizinski, C.J., and Sorensen, P.W. (2011) Using the Judas technique to locate and remove wintertime aggregations of invasive common carp. *Fisheries Management and Ecology*, **18**, 497–505.

Bals, J.D. and Wagner, C.M. (2012) Behavioral responses of sea lamprey (*Petromyzon marinus*) to a putative alarm cue derived from conspecific and heterospecific sources. *Behaviour*, **149**, 901–923.

Briand, C., Fatin, D., and Legault, A. (2002) Role of eel odour on the efficiency of an eel, *Anguilla anguilla*, ladder and trap. *Environmental Biology of Fish*, **65**, 473–477.

Burns, A.C., Sorensen, P.W., and Hoye, T.R. (2011) Synthesis and olfactory activity of unnatural, sulfated 5β-bile acid derivatives in the sea lamprey (*Petromyzon marinus*). *Steroids*, **76**, 291–300.

Christie, G.C. and Goddard, C.I. (2003) Sea Lamprey International Symposium (SLIS II): advances in the integrated management of sea lamprey in the Great Lakes. *Journal of Great Lakes Research*, **29**, 1–14.

Close, D.A., Fitzpatrick, M.S., and Li, H.W. (2002) The ecological and cultural importance of a species at risk of extinction, Pacific lamprey. *Fisheries*, **27**, 19–25.

Cook, S.M., Khan, Z.R., and Pickett, J.A. (2007) The use of push-pull strategies in integrated pest management. *Annual Review of Entomology*, **52**, 375–400.

Corkum, L.D. and Belanger, R.M. (2007) Use of chemical communication in the management of freshwater aquatic species that are vectors of human diseases or are invasive. *General and Comparative Endocrinology*, **153**, 401–417.

Fine, J.M. and Sorensen, P.W. (2005) Biologically-relevant concentrations of petromyzonol sulfate, a component of the sea lamprey migratory pheromone, measured in stream waters. *Journal of Chemical Ecology*, **31**, 2205–2210.

Fine, J.M. and Sorensen, P.W. (2008) Isolation and biological activity of the multi-component sea lamprey migratory pheromone and new information in its potency. *Journal of Chemical Ecology*, **34**, 1259–1267.

Fine, J.M., Vrieze, L.A., and Sorensen, P.W. (2004) Evidence that petromyzontid lampreys employ a common migratory pheromone that is partially comprised of bile acids. *Journal of Chemical Ecology*, **30**, 2091–2110.

Hecker, E. and Butenandt, A. (1984) Bombykol revisted- reflections on a pioneering period and some of its consequences, in *Techniques in Pheromone researched*, (eds H.H. Hummer and T.A. Miller), Springer-Verlag, New York, pp. 1–44.

Howse, P.E., Stevens, I.D.R., and Jones, O.T. (1998) *Insect Pheromones and Their use in Pest Management*, Chapman & Hall, London.

Jerde, C.L., Mahon, A.R., Chadderton, W.L., and Lodge, D.M. (2011) "Sight-unseen" detection of rare aquatic species using environmental DNA. *Conservation Letters*, **4**, 150–157.

Johnson, N.S. and Li, W. (2010) Understanding behavioral responses of fish to pheromones in natural freshwater environments. *Journal of Comparative Physiology A*, **96**, 701–711.

Johnson, N.S., Yun, S.-S., Thompson, H.T. *et al.* (2009) A synthesized pheromone induces upstream movement in female sea lampreys and summons them into traps. *Proceedings National Academy of Science USA*, **106**, 1021–1026.

Johnson, N.S., Yun, S.-S., Thompson, H.T., and Li, W. (2012) Multiple functions of a multi-component mating pheromone in sea lamprey *Petromyzon marinus*. *Journal of Fish Biology*, **80**, 538–554.

Li, W., Scott, A.P., Seifkes, M. *et al.* (2002) Bile acid secreted by male sea lamprey that acts as a sex pheromone. *Science*, **296**, 138–141.

Liley, N.R. (1982) Chemical communication in fish. *Canadian Journal of Fisheries and Aquatic Sciences*, **39**, 22–35.

Lim, H.K. and Sorensen, P.W. (2011) Polar metabolites synergize the activity of prostaglandin $F_{2\alpha}$ in a species-specific hormonal sex pheromone released by ovulated common carp. *Journal of Chemical Ecology*, **37**, 695–704.

Lim, H.K and Sorensen, P.W. (2012) Common carp implanted with prostaglandin $F_{2\alpha}$ release a sex pheromone complex that attracts conspecific males in both the laboratory and field. *Journal of Chemical Ecology*, **38**, 127–134.

Luehring, M.A., Wagner, C.M., and Li, W. (2011) The efficacy of two synthesized sea lamprey sex pheromone components as a trap lure when placed in direct competition with natural male odors. *Biological Invasions*, **13**, 1589–1597.

Maniak, P.J., Lossing, R., and Sorensen, P.W. (2000) Injured Eurasian ruffe, *Gymnocephalus cernuus*, release an alarm pheromone which may prove useful in their control. *Journal of Great Lakes Research*, **26**, 183–195.

Meckley, T.D., Wagner, C.M., and Luehring, M.A. (2012) Field evaluation of larval odor and mixtures of synthetic pheromone components for attracting migrating sea lampreys in rivers. *Journal Chemical Ecology*, **38** (8), 1062–1069.

Meckley, T.D., Wagner, C.M., and Gurarie, E. (2015) Coastal movements of migrating sea lamprey (Petromyzon marinus) in response to a partial pheromone added to river water: implications for management of invasive populations. *Canadian Journal of Fisheries and Aquatic Sciences*, **71**, 533–544.

Moser, M.L., Almeida, P.R., Kemp, P.S., and Sorensen, P.W. (2015) Lamprey spawning migration, in *Lampreys: Biology, Conservation and Control, Fish and Fisheries Series* (ed. M.F. Docker), Springer, in press.

Olsen, H. (2015) Effects of pollutants on olfactory detection and responses to odors including pheromones in fishes, in *Fish Pheromones and Related Cues* (eds Peter W. Sorensen and Brian D. Wisenden), John Wiley & Sons, Inc., Hoboken.

Sorensen, P.W. and Bajer, P.G. (2011) The common carp, in *Encyclopedia of Invasive Introduced Species*, (eds. D. Simberloff and M. Rejmanek), University of California Press, Berkeley, pp. 100–104.

Sorensen, P.W. and Baker, C. (2015) Species-specific pheromones and their roles in shoaling, migration and reproduction: a critical review and synthesis, in *Fish Pheromones and Related Cues* (eds Peter W. Sorensen and Brian D. Wisenden), John Wiley & Sons, Inc., Hoboken.

Sorensen, P.W. and Bergstedt, R. (2011) The sea lamprey, in *Encyclopedia of Invasive Introduced Species*, (eds. D. Simberloff and M. Rejmanek), University of California Press, Berkeley, pp. 619–622

Sorensen, P.W. and Caprio, J. (1997) Chemoreception in fish. Chapter 15, in *The Physiology of Fishes*, 2nd edn (ed. R.E. Evans), CRC Press, Boca Raton, pp. 375–406.

Sorensen, P.W. and Hoye, T.E. (2007) A critical review of the discovery and application of a migratory pheromone in an invasive fish, the sea lamprey, *Petromyzon marinus* L. *Journal of Fish Biology*, **71** (Supplement D), 100–114.

Sorensen, P.W. and Hoye, T.E. (2010) Pheromones in Vertebrates, in *Comprehensive Natural Products Chemistry II* (eds L. Mander and H.-W. Lui), Elsevier, Oxford, pp. 225–262.

Sorensen, P.W. and Stacey, N.E. (2004) Brief review of fish pheromones and discussion of their possible uses in the control of non-indigenous teleost fishes. *New Zealand Journal of Marine and Freshwater Research*, **38**, 399–417.

Sorensen, P.W. and Vrieze, L.A (2003) Chemical ecology and application of the sea lamprey migratory pheromone. *Journal of Great Lakes Research*, **29** (Supplement 1), 66–84.

Sorensen, P.W., Christensen, T.A., and Stacey, N.E. (1998) Discrimination of pheromonal cues in fish: emerging parallels with insects. *Current Opinion in Neurobiology*, **8**, 458–467.

Sorensen, P.W., Fine, J.M., Dvornikovs, V. *et al.* (2005). Mixture of new sulfated steroids functions as a migratory pheromone in the sea lamprey. *Nature Chemical Biology*, **1**, 324–328.

Stacey, N.E. (2015) Hormonal-derived pheromones in teleost fishes, in *Fish Pheromones and Related Cues* (eds Peter W. Sorensen and Brian D. Wisenden), John Wiley & Sons, Inc., Hoboken.

Stewart, M. and Baker, C.F. (2012) A sensitive analytical method for quantifying petromyzonol sulfate in water as a potential tool for population monitoring of the Southern Pouched Lamprey, Geotria Australis, in New Zealand Streams. *Journal of Chemical Ecology*, **38**, 135–144.

Stewart, M. and Sorensen, P.W. (2015) Measuring and identifying fish pheromones, in *Fish Pheromones and Related Cues* (eds Peter W. Sorensen and Brian D. Wisenden), John Wiley & Sons, Inc., Hoboken.

Stewart, M., Baker, C.F., and Cooney, T. (2011) A rapid, sensitive, and selective method for quantitation of lamprey migratory pheromones in river water. *Journal of Chemical Ecology*, **37**, 1203–1207.

Twohey, M.B., Sorensen, P.W., and Li, W. (2003) Possible applications of pheromones in an integrated sea lamprey control program. *Journal of Great Lakes Research*, **29** (Supplement 1), 794–800.

Vrieze, L.A., Bjerselius, R.K., and Sorensen, P.W. (2010) The importance of the olfactory sense to migratory sea lampreys seeking riverine spawning habitat. *Journal of Fish Biology*, **76**, 949–964.

Vrieze, L.A., Bergstedt, R.A., and Sorensen, P.W. (2011) Olfactory-mediated stream finding behavior of migratory adult sea lamprey (*Petromyzon marinus*). *Canadian Journal of Fisheries and Aquatic Science*, **68**, 523–533.

Wagner, C.M., Stroud, E.M., and Meckley, T.D. (2011) A deathly odor suggests a new sustainable tool for controlling a costly invasive species. *Canadian Journal of Fisheries and Aquatic Sciences*, **68**, 1157–1160.

Wisenden, B.D. (2015) Chemical cues that indicate risk of predation, in *Fish Pheromones and Related Cues* (eds Peter W. Sorensen and Brian D. Wisenden), John Wiley & Sons, Inc., Hoboken.

Witzgall, P., Kirsch, P., and Cork, A. (2010) Sex pheromones and impact on pest management. *Journal of Chemical Ecology*, **36**, 80–100.

Young, M.K., Micek, B.K., and Rathbun, M. (2003) Probable pheromonal attraction of sexually mature brook trout. *North American Journal Fisheries Management*, **23,** 2376–2382.

Yun, S.-S., Wildbill, A.J., Siefkes, M.J. *et al.* (2011) Identification of putative migratory pheromones from Pacific lamprey (*Lampetra tridentata*). *Canadian Journal of Fisheries and Aquatic Sciences*, **68**, 2194–2203.

Xi, X., Johnson, N.S., Brant, C.O. *et al.* (2011) Quantification of a male sea lamprey pheromone in tributaries of Laurentian Great Lakes by liquid chromatography-tandem mass spectrometry. *Environmental Science and Technology*, **45**, 6437–6443.

Afterword

The process of editing and writing this book highlighted both how little we understand about fish pheromones and related chemical cues, and how interesting and important they are. There appear to be hundreds of examples of conspecific odors which exert powerful effects on conspecific behavior and physiology in the fishes. In at least a few cases, specific chemical structures as well as key components of the neural systems that detect them have been identified: there is no doubt that pheromones exist within this, the largest and most diverse group of vertebrates. Further, it is now clear that many of these conspecific odors play key roles in the lives of these fishes and that they warrant further study because of their key roles in ecology, evolution, toxicology, and aquaculture. Some pheromones even appear to have potential for use in fisheries management. Nevertheless, despite all that we know, the bottom line is that we really understand very little. Only for a few dozen species have pheromonal chemical identity, release, detection, and response all been documented. Such complete suites of data are key because they are needed to answer fundamental questions about ecological, evolutionary, and neurobiological specialization. Only for a handful of species is there presently clear data on the chemistry of active chemical structures (e.g., 3-keto-petromyzonal sulfate in sea lamprey, $17\alpha,20\beta$-dihydroxy-4-pregnen-3-one in goldfish) and none of these structures are species-specific (although some appear to genus-specific) or appear to compromise the entire pheromone. Thus, neither the full complexity nor the range of pheromonal odors is known in the fishes so I argue that their complete range of biological function is also not known. Further, not a single study seems to have directly addressed whether and how a fish's body chemistry might be innately discerned (or perhaps predisposed to being learned), although this appears to be a common assumption of pheromone function. Finally, only a few dozen families of the nearly 500 that comprise the fishes have been examined for pheromonal function. Thus, we really have only scratched the surface in this field.

Given how little we still know, and these uncertainties, is it any wonder then that the top experts in the field including the authors of this book have chosen to use various definitions of the term pheromone in their chapters spite of my suggestions in the Introduction? I wonder if we are really looking at a spectrum of scenarios and chemical communication strategies when we discuss the topic of pheromones. It seems likely that different groups of fishes might very well have evolved to employ different types of pheromones in different manners for different situations. For example, some types of sex pheromones may comprise simple hormonal compounds that may or may not be complimented by

Fish Pheromones and Related Cues, First Edition. Edited by Peter W. Sorensen and Brian D. Wisenden.
© 2015 John Wiley & Sons, Inc. Published 2015 by John Wiley & Sons, Inc.

species-specific arrays of body odors or complexes, while some alarm odors may comprise rather simple compounds that are predisposed to be recognized and even learned. Why not? After all, the fishes are the most diverse group of vertebrates in terms of taxonomy, behavior, ecology, and neuro-anatomy. Answers will come only when we can definitively identify complete sets of pheromones (and related cues) and elucidate the precise neural mechanisms responsible for their discrimination amongst multiple taxa and then prove their function in the natural world in which they evolved. This exercise will no doubt prove fascinating and important to both the fishes and our species. I hope this book will stimulate this endeavor.

Peter W. Sorensen
August 9, 2014

Index

Page numbers in *italics* denote figures, those in **bold** denote tables.

Fish Pheromones and Related Cues, First Edition. Edited by Peter W. Sorensen and Brian D. Wisenden.
© 2015 John Wiley & Sons, Inc. Published 2015 by John Wiley & Sons, Inc.